AF307397

PROGRESS IN COLLOID & POLYMER SCIENCE

Editors: F. Kremer (Leipzig) and G. Lagaly (Kiel)

Volume 99 (1995)

Analytical Ultracentrifugation

Guest Editor:

J. Behlke (Berlin)

ISBN 978-3-662-15602-5
DOI 10.1007/978-3-7985-1666-3

ISBN 978-3-7985-1666-3 (eBook)

ISSN 0340-255 X

Die Deutsche Bibliothek –
CIP-Einheitsaufnahme

Analytical ultracentrifugation / [9th
Symposium on Analytical Ultracentrifugation (AUC), held at the Max
Delbrück Center for Molecular Medicine
in Berlin-Buch, March, 1995]. Guest
ed.: J. Behlke. – Darmstadt : Steinkopff ;
New York : Springer, 1995
    (Progress in colloid & polymer science ;
    Vol. 99)

NE: Behlke, Joachim [Hrsg.]; Symposium
    on Analytical Ultracentrifugation
    <9, 1995, Berlin>; Max-Delbrück-
    Centrum für Molekulare Medizin
    <Berlin>; GT

This work is subject to copyright. All
rights are reserved, whether the whole or
part of the material is concerned,
specifically those rights of translation,
reprinting, reuse of illustrations,
recitation, broadcasting, reproduction on
microfilms or in other ways, and storage
in data banks. Duplication of this
publication or parts thereof is only
permitted under the provisions of the
German Copyright Law of September 9,
1965, in its version of June 24, 1985,
and a copyright fee must always be paid.
Violations fall under the prosecution act
of the German Copyright Law.

The use of registered names, trademarks,
etc. in this publication does not imply,
even in the absence of specific statement,
that such names are exempt from the
relevant protective laws and regulations
and therefore free for general use.

© Springer-Verlag Berlin Heidelberg 1995
Originally published by Dr. Dietrich
Steinkopff-Verlag GmbH & Co. KG.,
Darmstadt in 1995
Softcover reprint of the hardcover 1st edition 1995

Chemistry editor: Dr. Maria Magdalene
Nabbe; English editor: James C. Willis;
Production: Holger Frey, Bärbel Flauaus.

Type-Setting: Macmillan Ltd.,
Bangalore, India

Progr Colloid Polym Sci (1995) V
© Steinkopff Verlag 1995

PREFACE

The series of symposia on analytical ultracentrifugation (AUC) which take place every 2 years at different universities and institutions in Germany, continued with the 9th meeting held at the Max Delbrück Center for Molecular Medicine in Berlin-Buch, held March 2 & 3, 1995.

The increased interest in ultracentrifugation methods, additionally fueled by the recent development of the new Optima XL-A machine, brought together about 100 scientists from leading groups around the world. More than 50 contributions representing a survey of recent developments and results was given in the frame of the following topics:

– reversible association reactions
– analysis of supramolecular structures including modeling
– size distribution analysis and gel formation
– hardware and software developments.

In several contributions it was demonstrated that methods of analytical ultracentrifugation are based on thermodynamic theory and provide a quantitative description of macromolecular interactions in solution. This has been used for analysis of self-associating systems or heterologous interactions between macromolecules. Furthermore, by combined application of spectroscopic, scattering electron microscopic as well as hydrodynamic methods, interaction and solvation processes, distribution events, and the recognition of supramolecular structures were described. Considerable effort has been devoted to developing new computer-assisted methods for data analysis and interpretation. In these areas a great step forward was made in recent years and one can expect new interesting results, especially by application of these methods to biotechnologically important substances in the near future.

About half of the contributions presented at the symposium were selected for publication in this special volume of Progress in Colloid and Polymer Science. It is hoped that they may be of help for the groups working in this field.

The 9th Symposium on Analytical Ultracentrifugation was generously sponsored by BASF AG (Ludwigshafen), Bayer AG (Leverkusen), Beckman Instruments (München), Biometra GmbH (Göttingen), Max Delbrück Center for Molecular Medicine (Berlin), Röhm GmbH (Darmstadt), and Shimadzu Europa GmbH (Duisburg). Support of this volume by the editor of the journal Colloid and Polymer Science, F. Kremer, is also gratefully acknowledged.

J. Behlke (Berlin)

CONTENTS

Progr Colloid Polym Sci (1995) 99:1–6
© Steinkopff Verlag 1995

M.M. Ladjimi
N. Benaroudj
G. Batelier
F. Triniolles

Self-association of the molecular chaperone HSC 70 as assessed by analytical ultracentrifugation

Received: 5 April 1995
Accepted: 23 June 1995

Dr. M.M. Ladjimi (✉) · N. Benaroudj
G. Batelier · F. Triniolles
Laboratoire d'Enzymologie, C.N.R.S.
91198 Gif-sur-Yvette Cedex, France

Abstract The self-association properties of the molecular chaperone HSC 70 have been assessed by analytical ultracentrifugation. Sedimentation velocity analysis indicates the presence of three species, whose proportions were dependent on protein concentration, but whose sedimentation coefficients, $s_{20,w}$, of 4.3 S, 6.6 S and 8.5 S did not vary with concentration, which is indicative of a slowly equilibrating system. Sedimentation equilibrium studies indicate a dissociation into monomers at low HSC 70 concentrations and an association into dimers and trimers at high concentrations. Multiple sets of sedimentation equilibrium data, obtained at various initial loading concentrations and rotor speeds, were adequately fitted to a single set of equilibrium constants by a monomer–dimer–trimer association model in which the association constants for the monomer–dimer and dimer–trimer equilibrium,
$K_{1-2} = 1.1 \cdot 10^5 \text{ M}^{-1}$ and
$K_{2-3} = 0.9 \cdot 10^5 \text{ M}^{-1}$ respectively,
were nearly identical. Interestingly, an isodesmic, indefinite type of association describes the data almost equally well with a single constant of $1.2 \cdot 10^5 \text{ M}^{-1}$. These results might have important implications for the chaperone function of HSC 70.

Key words Heat shock proteins – HSC 70 – molecular chaperone – self-association – analytical ultracentrifugation

Introduction

The 70 kDa Heat Shock Cognate protein (HSC 70), a constitutively expressed member of the highly conserved 70 kDa Heat Shock Protein family (HSP70), plays an essential role in several cellular processes such as protein folding, assembly and transport. The protein is thought to act as a molecular chaperone, by transiently binding to presumably hydrophobic regions of polypeptide chains, thereby preventing incorrect intra- and inter-molecular interactions, and dissociating from the bound polypeptide upon ATP binding (see reviews by Hendrick and Hartl, 1993; McKay, 1993; McKay et al., 1994; Hightower et al., 1994).

HSC 70 consists of two domains, an amino-terminal ATPase domain which binds and hydrolyses ATP and a carboxy-terminal domain involved in the binding of the protein substrates (Chappell et al., 1987; Wang et al., 1993). The three-dimensional structure of the entire protein is unknown. However, the structure of the isolated amino-terminal fragment of 44 kDa has been solved to a resolution of 2.2 Å (Flaherty et al., 1990), and a hypothetical structure of the carboxy-terminal domain has been modeled based on the structure of the human leucocyte antigen A2 of the class I major histocompatibility complex (Flajnik et al., 1991; Rippmann et al., 1991).

HSC 70 binds tightly to ATP and ADP (Schmid et al., 1985; Palleros et al., 1991; Gao et al., 1993) and has a weak intrinsic ATPase activity (Sadis and Hightower, 1992) that is stimulated two- to five-fold upon binding to unfolded proteins (Palleros et al., 1991; Sadis and Hightwoer, 1992; Benaroudj et al., 1994).

Previous studies using non-denaturing gel electrophoresis (Kim et al., 1992), size-exclusion chromatography and chemical crosslinking (Schlossmann et al., 1984; Palleros et al., 1991; Benaroudj et al., 1994) have shown that HSC70 self-associates in solution to form dimers and trimers. Oligomers of various sizes have also been observed by electron microscopy (Heuser and Steer, 1989), and most members of HSP70 family studied to date such as the bacterial DnaK, HSC70 from plants, the bovine endoplasmic reticulum resident BiP or the human heat shock inducible HSP70, self-associate to form multiple species (Palleros et al., 1993; Carlino et al., 1992; Blond-Elguindi et al., 1993; Brown et al., 1993; Anderson et al., 1994; Schoenfeld et al., 1995), suggesting that self-association is a general, conserved structural feature of the HSP70 family that must be important for function.

In this work, analytical ultracentrifugation have been used to assess the thermodynamic properties of HSC70 self-association and to define the mechanisms and equilibrium constants involved in this process.

Methods

Protein expression and purification

Recombinant HSC70 was expressed and purified as described previously (Benaroudj et al., 1994), except that following the ATP-agarose affinity column step, fractions containing HSC70 were pooled, concentrated by ultrafiltration using YM10 membrane on an ultrafiltration cell (Amicon), and applied onto a PD10 desalting column, equilibrated with 20 mM Tris-HCl pH 7.5, 20 mM KCl, 10 mM $(NH_4)_2SO_4$, 3 mM $MgCl_2$ and 1 mM β-mercaptoethanol, to remove free nucleotide. The protein was snap frozen and stored at $-80\,^{\circ}$C. Protein concentration was determined by the method of Lowry using bovine serum albumin as a standard or by using an extinction coefficient at 280 nm of 0.62 (Greene and Eisenberg, 1990) for a 1 mg/ml protein solution.

Sedimentation velocity

Sedimentation velocity experiments were performed at 20 °C on a Beckman Optima XL-A analytical ultracentrifuge equipped with a An Ti 60 titanium four-hole rotor with two-channel, 12 mm path-length centerpieces. Sample volumes of 400 μl were centrifuged at 60 000 rpm and radial scans of absorbance were taken at 10 min intervals.

Data analysis was performed using the computer program SVEDBERG (Philo, 1994) provided by John Philo. Frictional coefficients and Stokes radii were calculated by the program AXIAL provided by Les Holladay.

Sedimentation equilibrium

Sedimentation equilibrium experiments were carried out at 4 °C using three loading concentrations (0.3 mg/ml, 0.6 mg/ml and 1.2 mg/ml) and three rotor speeds (8000 rpm, 12 000 rpm and 16 000 rpm). Radial scans of absorbance at 280 nm were taken at 2 h intervals, and samples were judged to be at equilibrium by the absence of systematic deviations in overlayed successive scans and when a constant average molecular weight (Mw) was obtained in plots of Mw versus centrifugation time.

Data analysis according to discrete self-association models was performed using the appropriate functions by nonlinear least-squares procedures provided in the Beckman Optima XL-A software package (McRorie and Voelker, 1993).

Data analysis according to an unlimited isodesmic association model (Adams and Lewis, 1968), in which the equilibrium constants for the adition of monomer to any aggregate are equal, as performed using SEDPROG software package provided by Greg Ralston (Ralston and Morris, 1993, and refs. therein).

Results and discussion

Sedimentation velocity

Sedimentation velocity data were directly fitted by nonlinear least squares procedures as described by Philo (1994). The data fitted poorly to a one- or two-component system. However, as shown in Fig. 1, the data fit relatively well to a three-component model system involving a 4.2 S species, a 6.4 S species, and an 8.5 S species with proportions of about 45%, 35%, and 15% respectively at a protein concentration of 1.6 mg/ml. Nevertheless, some deviation of the fit relative to the experimental data is observed due probably to the fact that faster sedimenting species are present that are not taken into account in the fit, as indicated by the positive slope in the upper plateau of the experimental data relative to the fit, and/or a minor change of the equilibrium. Using the relation $(s_1/s_2)^3 = (M_1/M_2)^2$ and bovine serum albumin as a reference (Lin et al., 1991), one can obtain apparent molecular weights of about 67 kDa, 125 kDa and 182 kDa for the three species, which is compatible with HSC70 monomer, dimer and trimer. Moreover, the ratio of the sedimentation coefficient

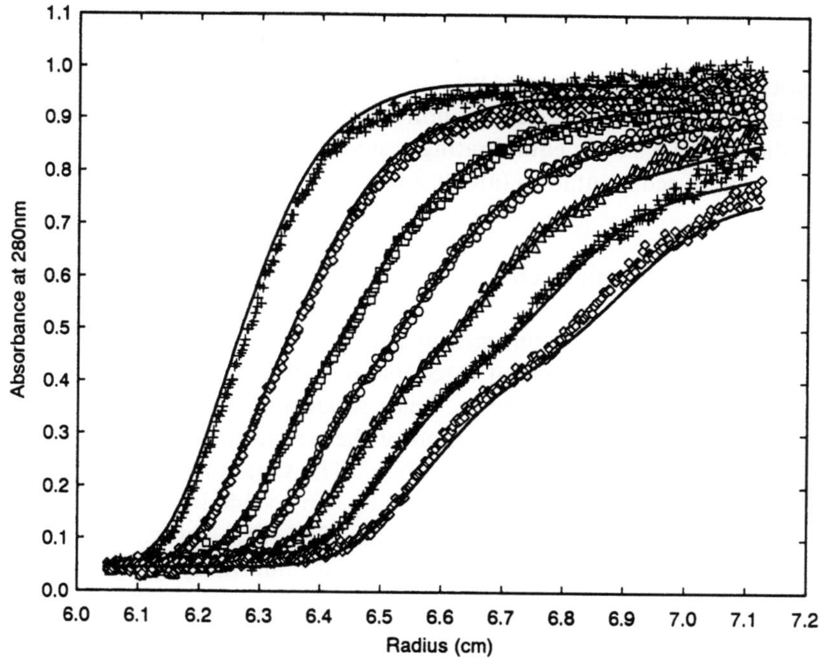

Fig. 1 Analysis of HSC 70 by sedimentation velocity. HSC 70 (1.6 mg/ml) was sedimented, and the data, recorded at increasing sedimentation time (symbols), were analyzed as described in "Materials and Methods". The fitted data curves are represented by a solid line. 36 min of sedimentation time (+), 46 min (◇), 56 min (□), 66 min (○), 76 min (△), 86 min (+) and 96 min (◇)

of the HSC 70 dimer or trimer over that of the monomer is 1.53 and 1.97 respectively, values close to the expected theoretical values of 1.5 and 2 for a dimer and a triangular trimer (Van Holde, 1975).

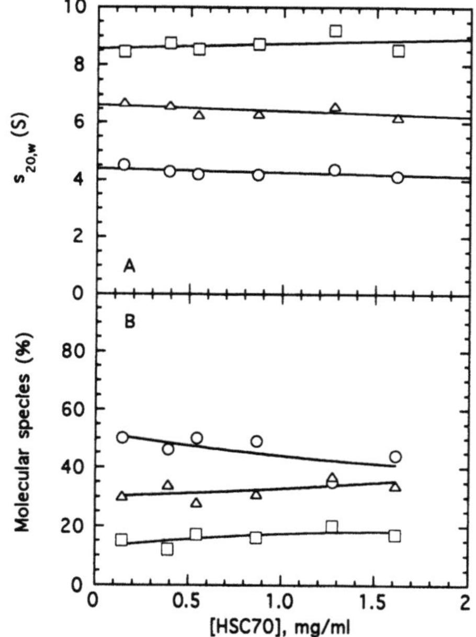

Fig. 2 Dependence of sedimentation coefficient and relative proportions of the various HSC 70 species on protein concentration. $s_{20,w}$ (A), and relative amounts (B), for HSC 70 monomer (○-○), dimer (△-△) and trimer (□-□) were determined as described in "Materials and Methods" using various initial concentrations as indicated on the figure

The plot of the corrected sedimentation coefficients, $s_{20,w}$, values as a function of protein concentration are shown in Fig. 2A. The $s_{20,w}$ value for each species is independent of HSC 70 concentration up to 1.6 mg/ml. Extrapolation to infinite dilution gives $s_{20,w}^0$ values of 4.3 S, 6.6 S and 8.5 S for HSC 70 monomer, dimer and trimer respectively. The frictional ratio values (f/f_0) suggest that all three species are asymmetrical (Table 1).

The variation of the mass fraction of each sedimenting species as a function of HSC 70 concentration is shown in Fig. 2B. Whereas the proportion of the 4.3 S species decreased, that of the 6.6 S and 8.5 S increased with increasing HSC 70 concentration, indicating that HSC 70 dimers and trimers are formed at the expense of monomers as expected for a mass-action law equilibrium. In addition,

Table 1 Hydrodynamic parameters of the HSC 70 Monomer, Dimer and Trimer

Parameter	Monomer	Dimer	Trimer
Molecular weight, Da	70 870	141 740	212 610
v, cm³·g⁻¹	0.735	0.735	0.735
$s_{20,w}$, S	4.3 ± 0.1	6.6 ± 0.1	8.5 ± 0.2
f/f_0	1.22	1.28	1.30
R_S, Å	30.9	38.8	40.4

The conformational parameters were calculated as described in "Material and Methods", using the molecular weight and the partial specific volume values determined from the amino acid composition. $s_{20,w}$, is the sedimentation coefficient; f and f_0 are the frictional coefficients and R_S, the Stokes radius.

the fact that each species of the equilibrium could be characterized by a distinct sedimentation coefficient that is not affected by increasing HSC 70 concentration, as if it existed in a mixture of several non interacting species, is indicative of a slowly equilibrating system as compared to the time of sedimentation (about 2 h).

Sedimentation equilibrium

Sedimentation equilibrium measurements of HSC 70 self-assembly were performed at three initial loading concentrations and three rotor speeds. Least squares analysis of each dataset for the determination of the weight average molecular weight is reported in Table 2. The weight average molecular weight increased slightly, up to 140 kDa, with increasing HSC 70 concentration for each rotor speed, indicative of self-association. This is confirmed by plotting the variation of the weight average molecular weight as a function of HSC 70 concentration, for a single initial loading concentration (Fig. 3), which shows a dissociation into monomers at low concentrations and an association into at least dimers at high concentrations. However, dimers are clearly not the end products of the associative reaction since the weight average molecular weight does not seem to reach a plateau at high concentrations.

Nine datasets, obtained with three initial loading concentrations and three rotor speeds, were simultaneously fitted to a single set of association constants common to all cells. Two classes of models were considered, discrete self-association models with defined stoichiometry and an

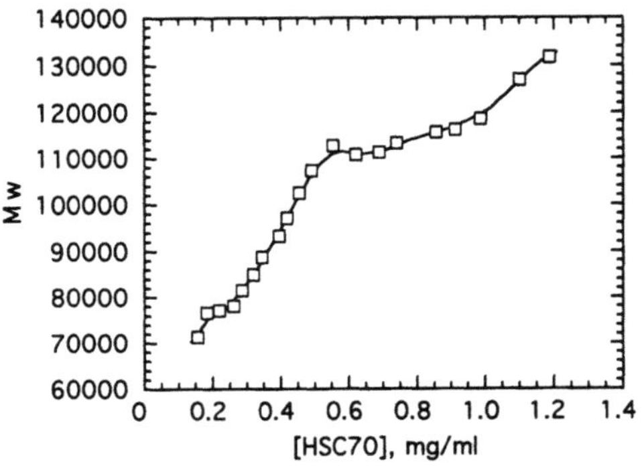

Fig. 3 Variation of the weight-average molecular weight of HSC 70 as a function of protein concentration. HSC 70 (1 mg/ml) was centrifuged at 4 °C at 12 000 rpm and analyzed as described in "Materials and Methods". The weight-average molecular weight (Mw) is calculated from $d \ln c/dr^2$ data, on a point by point basis, using a window of 20 points that moves through the entire data point range. The Mw are given in daltons and only selected points ($\square$-$\square$) are shown

unlimited isodesmic model. As shown in Table 3, a monomer–trimer model offers a better fit relative to a monomer–dimer model based on the square root of the variance and the distribution of the residuals. However, since sedimentation velocity (Figs. 1 and 2), and sedimentation equilibrium (Table 2 and Fig. 3) all indicated not only the presence of dimers but also that the association goes beyond the dimer, the data were fitted to a monomer–dimer–trimer model. Figure 4 shows the results of such an analysis which gave the best fit, as shown by the randomness and the small variation of the residuals, and the square root of the variance of the fit (Table 3). However, the higher degree of oligomerisation of the protein is not adequately taken into account by this model, as indicated by the large residuals at the higher end of the concentration gradient. Interestingly, it appeared that the free energies for adding a monomer to an existing monomer or to a dimer are almost identical (Table 3), suggesting that the data might equally well be fitted by an unlimited isodesmic model, in which the free energies for adding successive monomers would be equal. Fitting of the data by an isodesmic model gives association constant and free energy values nearly identical to those obtained for the monomer–dimer–trimer model (Table 3).

Table 2 Concentration dependence of HSC 70 weight average molecular weight (Mw)

Speed rpm	HSC 70 mg/ml	Mw, (Da)	rms
8000	0.30	129 700 ± 1900	0.006
	0.60	141 000 ± 1700	0.007
	1.20	140 800 ± 1900	0.023
12 000	0.30	111 900 ± 1600	0.006
	0.60	127 100 ± 2000	0.013
	1.20	127 800 ± 2300	0.030
16 000	0.30	114 300 ± 1800	0.014
	0.60	117 600 ± 2100	0.020
	1.20	123 200 ± 3200	0.024

The weight average molecular weight is obtained assuming no self-association reaction, by fitting the equilibrium sedimentation data to a single ideal species as described under "Material and Methods". The rms is defined as the square root of the variance of the fit and expressed in optical density units.

Conclusion

Altogether, these results indicate that HSC 70 self-associates, in a slow and reversible manner, to form dimers, trimers and larger oligomers with the monomer as the basic assembly unit. In addition, they suggest that HSC 70 might assemble in an isodesmic fashion. Either the association is unlimited, or alternatively, it is possible that the

Progr Colloid Polym Sci (1995) 99:1–6
© Steinkopff Verlag 1995

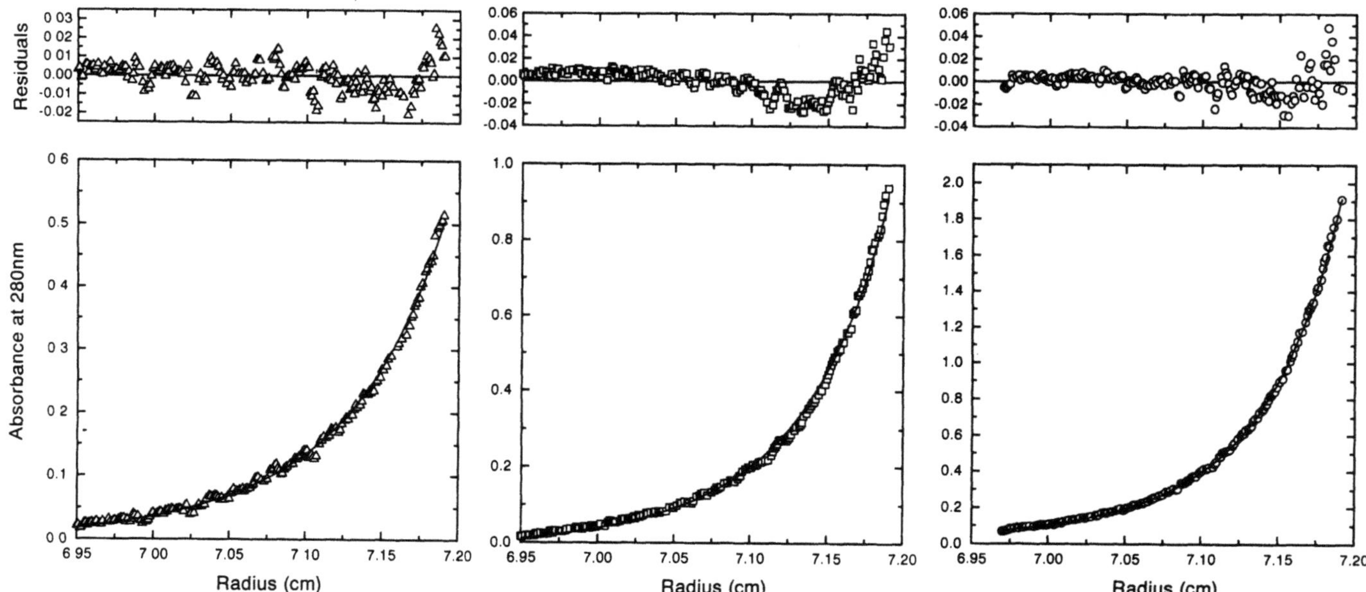

Fig. 4 Monomer–dimer–trimer fit of the HSC 70 equilibrium sedimentation data. Equilibrium sedimentation data, obtained at 4 °C and at 8000, 12 000 and 16 000 rpm using 0.3 mg/ml ($\triangle$-$\triangle$), 0.6 mg/ml ($\square$-$\square$) and 1.2 mg/ml (o-o), were simultaneously fitted to a monomer-dimer-trimer model as described in "Materials and Methods". The symbols represent the experimental data from the 12 000 rpm run whereas the solid line is the result of the simultaneous fit of nine data sets. The residuals representing the variation between the experimental data and those generated by the fit are also shown (see also Table 3)

Table 3 Association constants and free energies of HSC 70 self assembly

Model	Association constant	ΔG^0 (kcalmol^{-1})		rms
monomer–dimer	$K_{1-2} = 25.9 \ 10^5 \ \mathrm{M}^{-1}$	-8.1		0.014
monomer–trimer	$K_{1-3} = 3.3 \ 10^9 \ \mathrm{M}^{-2}$		-12.0^*	0.013
monomer–dimer–trimer	$K_{1-2} = 1.1 \ 10^5 \ \mathrm{M}^{-1}$	-6.3		0.011
	$K_{2-3} = 0.9 \ 10^5 \ \mathrm{M}^{-1}$	-6.2		
	$K_{1-3} = 9.5 \ 10^9 \ \mathrm{M}^{-2}$		-12.6^*	
Isodesmic	$K = 1.2 \ 10^5 \ \mathrm{M}^{-1}$	-6.4		0.013

Equilibrium sedimentation data were analyzed according to "Material and Methods". Molecular weight was fixed to that of the monomer, calculated from the amino acid composition (70870 Da). Association constants were calculated by a simultaneous fit of nine data sets to a single set of constants. All the ΔG^0 values are given on a "per monomer" basis, i.e., the free energy of ading successive monomers, except for "*" that are given for the trimer. The rms is defined as the square root of the variance of the fit and expressed in optical density units.

protein polymerizes until a stable structure, such as a ring for example, is reached. Additional work is needed in order to have a more precise idea about HSC 70 assembly.

Acknowledgements This work was supported by the the Association pour la recherche sur le cancer (ARC) and the Ligue Nationale contre le cancer (FNCLCC). N.B. is supported by an ARC predoctoral fellowship. We are very grateful to Les Holladay, John Philo, Greg Ralston and Walter Stafford for making their computer programs available and Borries Demeler, John Philo, Greg Ralston and Paul Voelker for helpful discussions. We would also like to thank Marie-France Carlier, Joel Janin, Marc Lemaire, Dominique Pantaloni and Jean-Pierre Waller for their comments, suggestions and continuous support.

References

1. Adams ET, Lewis MS (1968) Sedimentation equilibrium in reaction systems. VI. Some applications to indefinite self-associations. Studies with b-lactoglobulin A* Biochemistry 7:1044–1052
2. Benaroudj N, Fang B, Triniolles F, Ghelis C, Ladjimi MM (1994) Overexpression in Escherichia coli, purification and characterization of the molecular chaperone HSC 70, Eur J Biochem 221:121–128
3. Blond-Elguindi S, Fourie AM, Sambrook JF, Gething M-JH (1993) Peptide-dependent stimulation of the ATPase activity of the molecular chaperone BiP is the result of conversion of oligomers to active monomers. J Biol Chem 268:12730–12735
4. Brown CR, Martin RL, Hansen WJ, Beckmann RP, Welch WJ (1993) The constitutive and stress inducible forms of hsp 70 exhibit functional similarities and interact with one another in an ATP dependent fashion. J Cell Biol 120: 1101–1112
5. Carlino A, Toledo H, Skaleris D, DeLisio R, Weissbach H, Brot N (1992) Interactions of liver Grp 78 and Escherichia coli recombinant Grp 78 with ATP: Multiple species and disaggregation. Proc Natl Acad Sci USA 89: 2081–2085
6. Chappell TG, Konforti BB, Schmid SL, Rothman JE (1987). The ATPase core of a clathrin uncoating protein. J Biol Chem 262:746–751
7. Flaherty KM, De-Luca-Flaherty C, McKay DB (1990) Three-dimensional structure of the ATPase fragment of a 70 K heat-shock cognate protein. Nature 346:623–628
8. Flajnik MF, Canel C, Kramer J, Kasahara M (1991) Which came first, MHC class I or class II? Immunogenetics 33:295–300
9. Gao B, Emoto Y, Greene LE, Eisenberg E (1993) Nucleotide binding properties of bovin brain uncoating ATPase. J Biol Chem 268:8507–8513
10. Greene LE, Eisenberg E (1990) Dissociation of clathrin from coated vesicles by the uncoating ATPase. J Biol Chem 265:6682–6687
11. Hendricks JP, Hartl FU (1993) Molecular chaperoning functions of heat shock proteins. Annu Rev Biochem 62:349–384
12. Heuser J, Steer CJ (1989) Trimeric binding of the 70-kD uncoating ATPase to the vertices of clathrin triskelia: A candidate intermediate in the vesicle uncoating reaction. J Cell Biol 109:1457–1466
13. Hightower LE, Sadis SE, Takenaka IM (1994) Interactions of vertebrate hsc 70 and hsp 70 with unfolded proteins and peptides. In: The Biology of Heat Shock Proteins and Molecular Chaperones (Morimoto, RI, Tissières A, Georgeopoulos C (eds), pp. 179–187, Cold Spring Harbor Laboratory Press, New York
14. Kim D, Lee YJ, Corry PM (1992) Constitutive HSP 70: Oligomerization and its dependence on ATP binding. J Cell Physiol 153:353–361
15. Lin T-H, Quinn T, Walsh M, Grandgenett D, Lee JC (1991) Avian myeloblastosis virus reverse trascriptase. J Biol Chem 266:1635–1640
16. McKay DB (1993) Structure and mechanism of 70-kDa heat-shock-related proteins. Advances in Protein Chemistry 44:67–98
17. McKay DB, Wilbanks SM, Flaherty KM, Ha J-H, O'Brien MC, Shirvanee LL (1994) Stres-70 proteins and the interaction with nucleotides. In: The Biology of Heat Shock Proteins and Molecular Chaperones. (Morimoto RL, Tissières A, Georgeopoulos C (eds) Cold Spring Harbor Laboratory Press, New York, pp 153–177
18. McRorie DK, Voelker PJ (1993) Self associating systems in the analytical ultracentrifuge. Beckman Instruments Inc, California
19. Morris M, Ralston GB (1985) Determination of the parameters of self-association by direct fitting of the omega function. Biophys Chem 23:49–61
20. Palleros DR, Welch WJ, Fink AL (1991) Interaction of hsp 70 with unfolded proteins: Effects of temperature and nucleotides on the kinetics of binding. Proc Natl Acad Sci USA 88:5719–5723
21. Palleros DR, Reid KL, Shi L, Fink AL (1993) DnaK ATPase activity revisited. FEBS 336:124–128
22. Philo JS (1994) Measuring sedimentation, diffusion and molecular weights of small molecules by direct fitting of sedimentation velocity profiles. In Modern analytical ultracentrifugation: acquisition and interpretation of data for biological and synthetic polymer systems Schuster TM, Laue TM, (eds) pp 156–170. Birkhäuser Boston
23. Ralston GB, Moris MB (1992) The use of the omega function for sedimentation equilibrium analysis. In Analytical ultracentrifugation in Biochemistry and polymer Science. Harding SE, Rowe AJ, Horton JC (eds), The Royal Society of Chemistry, Cambridge, pp 253–274
24. Rippmann F, Taylor WR, Rothbard JB, Green NM (1991) A hypothetical model for the peptide binding domain of hsp70 based on the peptide binding domain of HLA. EMBO J 10:1053–1059
25. Sadis SE, Hightower LE (1992) Unfolded proteins stimulate molecular chaperone Hsc70 ATPase by accelerating ADP/ATP exchange. Biochemistry 31:9406–9412
26. Schlossman DM, Schmid SL, Braell WA, Rothman JE (1984) An enzyme that removes clathrin coats: Purification of an uncoating ATPase. J Cell Biol 99: 723–733
27. Schmid SL, Braell WA, Rothman JE (1985) ATP catalyzes the sequestration of clathrin during enzymatic uncoating. J Biol Chem 260:10057–10062
28. Schoenfeld HJ, Schmid D, Schröder H, Bukau B (1995) The Dnak chaperone system of Escherichia coli: quaternary structures and interactions of the Dnak and GrpE components. J Biol Chem 270:2183–2189
29. Van Holde KE (1975) Sedimentation analysis of proteins. In The Proteins, 3rd ed, vol I. Neurath H, Hill R (eds) Academic Press, New York, pp 225–291
30. Wang T-F, Chang J-H, Wang C (1993) Identification of the peptide binding domain of hsc 70. J Biol Chem 268: 26049–26051

Progr Colloid Polym Sci (1995) 99:7–10
© Steinkopff Verlag 1995

H.-J. Schönfeld
D. Schmidt
M. Zulauf

Investigation of the molecular chaperone DnaJ by analytical ultracentrifugation

Received: 13 March 1995
Accepted: 23 May 1995

Dr. H.-J. Schönfeld (✉) · D. Schmidt
M. Zulauf
Hoffmann-La Roche Limited
Pharmaceutical Research –
New Technologies
4002 Basel, Switzerland

Abstract The *E. coli* heat shock proteins DnaK (Hsp70), DnaJ and GrpE constitute a cellular chaperone system for protein folding. In the context of a rigorous investigation of the structure-function relationships within this complex system we investigated the quaternary structure of DnaJ by analytical ultracentrifugation under conditions of sedimentation equilibrium. DnaJ appeared heterogeneous under all tested conditions. The observed heterogeneity depended mainly on the pH of the preparation. At pH 5.5 the sedimentation profiles were well fitted by two exponential functions, indicating the presence of a low and a high mole mass component. The distribution of the two components was independent of the protein concentration.

Key words Chaperone – DnaJ – Hsp70 – quaternary structure – analytical ultracentrifugation – sedimentation equilibrium – protein folding

Introduction

Molecular chaperones are proteins that can prevent the misfolding of other proteins without forming part of the folded substrate protein [1]. They often act as homo- and hetero-complexes [2]. The three chaperones DnaJ, DnaK (Hsp70) and GrpE work together as a "chaperone machine" [3, 4]. For example, they can prevent irreversible denaturation of fire fly luciferase after heat treatment *in vivo* and *in vitro* [5] and support its efficient refolding after denaturation with guanidine hydrochloride *in vitro* (H.-J. Schönfeld, unpublished data). To understand the mechanisms of chaperone mediated protein folding the quaternary structures of all components of the folding reaction should be known, although interactions between components may affect these structures. Recently we showed that oligomeric DnaK is monomerized by dimeric GrpE forming well defined DnaK–GrpE complexes of a molar ratio of 1:2, respectively [6].

DnaJ by itself binds to nascent polypeptides [7] and modulates the ATPase activity of DnaK [8]. Only limited information exists on the quaternary structure of DnaJ. The monomer mole mass of DnaJ expected from its amino acid sequence is 41 kDa. It was found to form dimers in glycerol gradient sedimentation and in gel filtration [9]. However, both methods rely on calibration with reference proteins and on assumptions concerning molecular shape [10].

In the present study we investigated DnaJ with analytical ultracentrifugation by sedimentation equilibrium. This technique enables the determination of mole masses or mole mass distributions without calibration and is independent of shape assumptions.

Experimental

Preparation of DnaJ

An efficient purification method for DnaJ was recently established (H.-J. Schönfeld et al., unpublished). Briefly, DnaJ was overexpressed in *E. coli* and purified by cation exchange chromatography and subsequent hydroxy-apatite chromatography in the presence of the detergent Brij58 (Pierce; Rockford, IL) and the reducing agent dithiothreitol (DTT). The purified protein was more than 95% homogenous as judged by SDS-PAGE. DnaJ has hydrophobic properties and because of its 10 cysteine residues that may form improper disulfide bridges, self-aggregation may occur. In our study, the purified DnaJ was soluble up to 5 mg/ml after dialysis against buffers free of detergents or reducing agents as described below. The dialyzed preparation had chaperone activity as assayed by refolding of luciferase (data not shown). In order to investigate the dependence of self-aggregation on pH, we prepared DnaJ at pH 7.7, 5.5 and 9.5 as described below.

DnaJ[pH 7.7]

Purified DnaJ was extensively dialyzed and diluted to desired concentrations using 50 mM Tris-HCl, pH 7.7, 100 mM NaCl.

DnaJ[pH 5.5]

DnaJ[pH 7.7] was successively dialyzed against 50 mM BisTris-HCl, pH 6.5 and 50 mM BisTris-HAc, pH 5.5, both supplemented with 100 mM NaCl. During the first dialysis step we observed protein precipitation. At pH 5.5 the protein was completely resolubilized as judged visually. The DnaJ was further fractionated by gel filtration chromatography (Superose 12, 1×30 cm; Pharmacia LKB) running at 0.5 ml/min in 50 mM BisTris-HAc, pH5.5, 100 mM NaCl. The region around the maximum of the DnaJ peak (recorded at 280 nm) was pooled. The pool, referred to as "DnaJ[pH 5.5]", was finally diluted with the buffer used in gel filtration to the desired concentrations.

DnaJ[pH 9.5], DnaJ[ME] and DnaJ[UF]

DnaJ[pH 7.7] was dialyzed against 50 mM ethanol amine/HCl, pH 9.5, 100 mM NaCl (ethanol amine buffer) and then further fractionated by gel filtration in ethanol amine buffer as described above. The DnaJ peak of the Superose 12 column was pooled and termed "DnaJ[pH 9.5]". One aliquot of DnaJ[pH 9.5] was supplemented with 10 mM 2-mercaptoethanol and referred to as "DnaJ[ME]". Another aliquot of DnaJ[pH 9.5] was ultra filtrated using a membrane with a molecular weight cut off of 100 kDa (YM100, Amicon) and referred to as "DnaJ [UF]".

Analytical ultracentrifugation

All experiments were carried out as sedimentation equilibrium runs with a Beckman analytical ultracentrifuge XL-A using the rotor with three sample cells and sample volumes of about 100 μl.

The experiments with DnaJ[pH 7.7] were run at 7000 rpm at 10 °C for 22.5 h. The other experiments were run at 9000 rpm at 20 °C for 15 h for DnaJ[pH 5.5] and for 8 h for DnaJ[pH 9.5], DnaJ[ME] and DnaJ[UF]. The obtained concentration profiles were analyzed using the program package DISCREEQ (discreet fitting of equilibrium) provided by P. Schuck (NIH, Bethesda; [11]). This program offers the possibility to fit data to a maximum of two exponentials plus a background term. In addition, modeling is possible by imposing any number of mole masses (e.g. monomers, dimers, trimers, etc.) and fitting the relative concentrations thereof. The partial specific volume of DnaJ at 20 °C was computed as 0.725 ml/g from the amino acid composition and the density of the buffer was computed as 1.0038 g/ml according to Laue *et al.* [12].

Results

In all experiments the measured concentration profiles could not be described by a single exponential. We therefore used a model approach for data analysis. Several exponentials corresponding to multiples of the monomer mass were considered and their respective concentrations were determined by linear regression. Results of the prevalent oligomers present at every investigated pH are given below.

DnaJ at pH 7.7

DnaJ[pH 7.7] was analyzed in one run at three different concentrations: 1.2, 0.46 and 0.23 mg/ml. The obtained concentration profiles were independent of the respective loading concentrations. Good fits necessitated the presence of at least three species: monomers, 10-mers and 20-mers, suggesting broad heterogeneity of the analyzed samples.

DnaJ at pH 5.5

DnaJ[pH 5.5] was also analyzed in one run at three different concentrations: 0.6, 0.3 and 0.15 mg/ml. The measured concentration profiles were well described by two exponentials corresponding to 55% (w/w) dimers and 45%

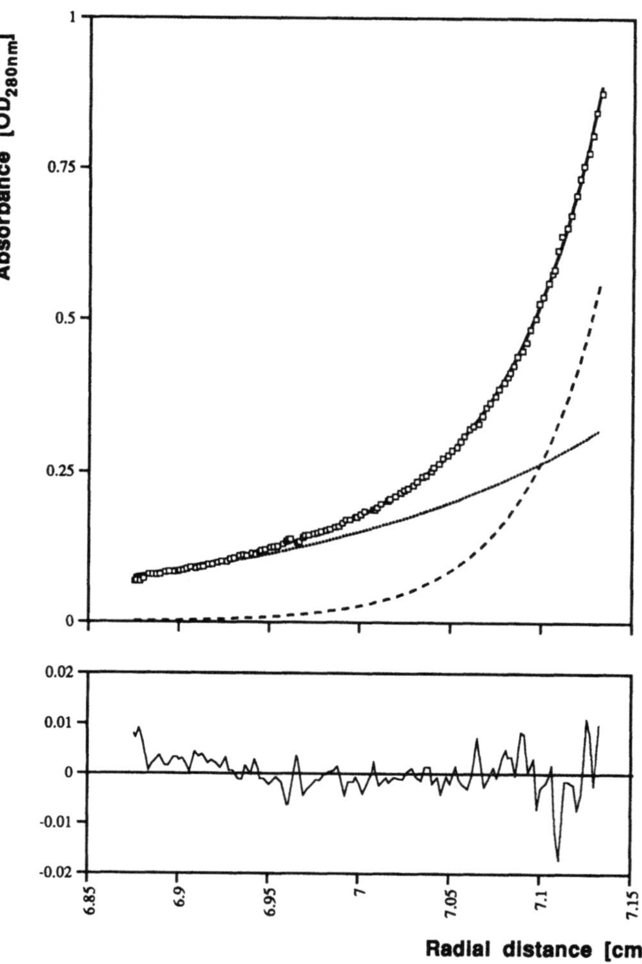

Fig. 1 Upper diagram: radial dependent absorbance of DnaJ [pH 5.5] at 0.6 mg/ml (□) and best fit to a model assuming 55% dimers and 45% octamers (solid line). The computed exponential functions for the dimer (dotted line) and the octamer (dashed line) are plotted. Lower diagram: radial dependent residuals representing the deviations of the experimental data from those generated by the model

octamers for all concentrations. Figure 1 shows how the sum of the theoretical dimer and octamer profiles matched the experimental profile.

DnaJ at pH 9.5

DnaJ[pH 9.5] and DnaJ[ME] at 0.8 mg/ml and DnaJ [UF] at 0.3 mg/ml were analyzed in one run. DnaJ [pH 9.5] and DnaJ[ME] were partially precipitated after centrifugation, whereas DnaJ[UF] remained completely soluble, presumably because of its lower concentration and the removal of large aggregates before the centrifugation. For all three samples, as with DnaJ[pH 7.7], we observed a broad distribution of oligomeric species and

Table 1 The pH dependent heterogeneity of DnaJ

pH	apparent quaternary structures
7.7	broad distribution of oligomers (1–20mers)
5.5	about equal distribution of dimers and octamers
9.5	broad distributions of oligomers, aggregation leading to precipitation

three exponentials were necessary to obtain a reasonable fit to a theoretical sedimentation profile.

Discussion and conclusion

We determined the quaternary structure of DnaJ at various pH values using a technique, analytical ultracentrifugation, that unambiguously yields the molar masses of the protein complexes in solution. The main results of this investigation are summarized in Table 1.

At pH 7.7, close to the isoelectric point of pH 7.5 (determined by isoelectric focusing; H.-J. Schönfeld, unpublished data), and under alkaline conditions (pH 9.5), we found a broad distribution of oligomers of the protein, which could not be resolved into single components. Variation of the concentration, addition of a reducing agent (2-mercaptoethanol) or removal of large oligomers by ultra filtration before the centrifugation experiment did not result in monomodal distributions.

Aggregation was found to be less complex at pH 5.5 where data analysis suggested the presence of only dimers and octamers.

The ratio of oligomers did not significantly change with concentration, indicating the absence of specific interactions. Presumably, at least a certain fraction of DnaJ oligomers was formed at a fixed ratio during protein preparation by intermolecular disulfide formation via cysteine residues. Variations dependent on pH may be explained by unspecific hydrophobic interactions. We conclude that at the present stage of purification, the "natural" quaternary structure of DnaJ cannot be unambiguously determined. Our results raise the question as to which oligomeric DnaJ species are active in supporting protein folding. The preparative separation of the oligomers, if practicable, will help to clarify this question and also facilitate the analysis of interactions between DnaJ and other components of the Hsp70 chaperone machine.

Acknowledgements We thank E.A. Kusznir and B. Pöschl for excellent technical assistance and H. Etlinger for reading the manuscript.

References

1. Ellis RJ (1990) Semin Cell Biol 1:1–9
2. Martin J, Hartl FU (1993) Structure 1:161–164
3. Georgopoulos C (1992) Trends Biochem Sci 17:295-299
4. Morimoto RI, Tissieres A, Georgopoulos C (1994) In: Morimoto RI, Tissieres A, Georgopoulos C (eds) The Biology of Heat Shock Proteins and Molecular Chaperones. Cold Spring Harbor Laboratory, Cold Spring Harbor, NY, pp 1–30
5. Schroder H, Langer T, Hartl FU, Bukau B (1993) EMBO J 12: 4137–4144
6. Schönfeld H-J, Schmidt D, Schröder H, Bukau B (1995) J Biol Chem 270: 2183–2189
7. Hendrick JP, Langer T, Davis TA, Hartl FU, Wiedmann M (1993) Proc Natl Acad Sci USA 90:10216-10220
8. Jordan R, McMacken R (1995) J Biol Chem 270:4563–4569
9. Zylicz M, Yamamoto T, McKittrick N, Sell S, Georgopoulos C (1985) J Biol Chem 260:7591–7598
10. Schönfeld HJ, Pöschl B, Frey JR, Lötscher H, Hunziker W, Lustig A, Zulauf M (1991) J Biol Chem 266: 3863–3869
11. Schuck P (1994) Progr Colloid Polym Sci 94:1–13
12. Laue TM, Shah BD, Ridgeway TM, Pelletier SL (1992) In: Harding SE, Rowe AJ, Horton JC (eds) Analytical Ultracentrifugation in Biochemistry and Polymer Science. Royal Society of Chemistry, Cambridge, pp 90–125

Progr Colloid Polym Sci (1995) 99:11–16
© Steinkopff Verlag 1995

K. Jumel
S.E. Wilson
M.C.M. Smith
S.E. Harding

Investigations of the oligometric state of the 42 kDa repressor isoform from the *streptomyces* temperate bacteriophage φC31

Received: 29 March 1995
Accepted: 18 May 1995

Dr.K. Jumel (✉) · S.E. Harding
National Centre for Macromolecular
Hydrodynamics
The University of Nottingham
Department of Applied Biochemistry
and Food Science
Sutton Bonington LE12 5RD
United Kingdom

S.E. Wilson · M.C.M. Smith
Department of Genetics
Queens Medical Centre
University Park, Nottingham NG7 UH
United Kingdom

Abstract The repressor gene from the *Streptomyces* phage φC31 expresses three N-terminally different, in-frame protein isoforms of polypeptide molecular masses 74, 54 and 42 kDa. Their precise role in the lysis versus lysogeny decision is currently being investigated and a study of their structural and interactive properties was considered an important aid in this investigation. The preliminary data presented here shows that the native 42 kDa isoform and a Histidine (His-)-tagged form exist as dimers or tetramers, depending on the conditions as determined by sedimentation equilibrium in the analytical ultracentrifuge.

Key words Repressor protein – Molecular mass – Sedimentation equilibrium

Introduction

The repressor gene, 'c' of the *Streptomyces* phage, φC31, expresses three in-frame, N-terminally different protein isoforms of polypeptide molecular mass (predicted from the DNA sequence) 74, 54 and 42 kDa [1] (see Fig. 1). *S. lividans* expressing only the 42 kDa protein can confer immunity to some strains of superinfecting φC31 whereas those expressing both the 54 and 42 kDa proteins are fully immune [1]. Our model for the roles of the three isoforms states that the 54 and 42 kDa proteins are the true repressors of the lytic developmental cycle and the 74 kDa protein has some other role, perhaps as an anti-repressor. Both the 54 and the 42 kDa proteins have recently been shown to bind to a conserved inverted repeat (CIR) sequence repeated at least 18 times throughout the phage genome [2, 3, 4]. In order to understand the nature of this novel genetic switch the properties of the repressors are being studied in vitro. Separation of the isoforms from each other is greatly facilitated by the use of translational fusions of the 42, 54 or 74 kDa open reading frames to a run of 6 histidines at their N-terminal ends. His-tagged repressors can then be separated from contaminating untagged isoforms by a single affinity chromatography procedure using Ni-NTA resin (see Experimental Section). We intend to use the purified His-tagged isoforms to study the DNA binding and interactive properties of the three proteins. In order to discover the oligomeric state of the smallest, 42 kDa isoforms and whether the His-tag has had any effect on its oligomeric structure, the molecular masses of the native 42 kDa and the His-tagged 42 kDa proteins were determined by sedimentation equilibrium. We have found that, depending on the conditions, these molecules exist as dimers or tetramers.

Experimental

The native or "wild type" 42 kDa repressor protein was prepared as described in ref. [1]. The native form was

Fig. 1 Amino acid sequence of the SpHI-G fragment of Streptomyces ϕC31. * indicates N-terminus of 54 kDa sequence, * indicates N-terminus of 42 kDa sequence. In the His tagged 42 kDa form, the following amino acids precede the indicated terminus: MRGSHHHHHHGS

```
  1 MKRVTLGGGK AVHYSTTPDG FMASPACGGN RASERYVPTD ADVTCKRCAK

 51 ILAAEAEREE RLNRDPRGDE WMGRTIGDAV TVTLHGRTFD TELTGADHIT

101 PGWTVAYVDE DGQRNGTFVV VTDADIQDGD KVSDPRKDAF DKARALGMDW

151 AEALDYANAK TAEMAQPTHV SSVESATHDN DDNKGTGTM*A TKKLKLKDVR

201 GDVRIGAVPG ADAIHALRNA VDENGRNLPM CRTRTKNPIQ YWGPAAEQKP

251 ELELCAGCSK VVPTGEVSVS EESVEVPGLS MTVSQKSYTP VEGDDKGEN#M

301 AAKNDTQDVD AQISAVHGHV DNIKTAETVE AVKEAAEAAE GIITTLPTKH

351 RNTLRSTVKE ARTARETELT PVTPEAEAAK AEVESRRSAD VAEDFNDIEG

401 VPDLIKDGVK LFSQGVDLGL KLTNAGEKLA HVMLTMRQKI VNPATGLPDL

451 TAERKTTKNA AAEVYAQAKK RIADDDVERQ GAHNSLVRAT QNKASDVLVD

501 WLRAFDGPDR KESLAVASEL FGDKLDGLKD DASISEAIYR LYAGQGIELP

551 RYGRTELARY DRRVKAIEGA TKELETLTDG DKDANPKDVE ALEEKIKELK

601 AEVPEEILTE KLEPKAEKSD AEKTADALKV IRAQVDKAGK RFAKVKTANE

651 KRKAKAELYS IIRAAADAFD LDLSALVTAD EDE
```

purified on a heparin/agarose column using a linear buffer/NaCl gradient (buffer composition: 20 mM Tris-HCl pH 8.0, 5 mM MgCl$_2$, 1 mM EDTA, 0.1 mM DTT, 1 mM benzamidine) and eluted at approx. 400 mM NaCl. It was found to be >95% homogenous for 42 kDa protein and contained ~5% contamination by low molecular weight nucleic acids. Samples were kept frozen at −20 °C until immediately prior to use.

The His-tagged 42 kDa repressor, His-42 was prepared using a "Qiaexpress" kit supplied from Qiagen as follows: DNA from the c gene was cloned into pQE30 to make pDBTF42, such that the repressor protein expessed from this plasmid contained the amino acid sequence MRGSHHHHHHGS fused to the second amino acid residue of the 42 kDa protein open reading frame [5]. Soluble protein (in 50 mM Na-phosphate pH 7.8, 300 mM NaCl) from sonicated, induced cultures of *E. coli* M15pRep4 (pDBTF42) was then loaded onto a Ni-NTA column, washed with wash buffer (50 mM Na-phosphate pH 7.8, 50 mM NaCl, 10 mM imidazole, 10% glycerol) and eluted with wash buffer containing 100 mM imidazole. The His-42 fusion protein was 95% pure and contained no contaminating nucleic acids.

Low speed sedimentation equillibrium experiments were carried out on a Beckman Optima XL-A analytical ultracentrifuge (Beckman, Palo Alto, USA) at 4 °C and a rotor speed of 7000 rpm (His-tagged samples were also run at 10000 rpm). Sample cells (12 mm path length) were filled with 100 μl solution containing 3, 2, and 1 mg/ml of the native 42 kDa isomer and 2, 1 and 0.6 mg/ml of the His-tagged 42 kDa isomer in the sample channel and 120 μl of solvent in the respective solvent channel.

Results and discussion

Apparent weight average molecular masses of the native 42 kDa protein

Figures 2a and 2b show plots of ln$A$ and M^* versus the normalized radial displacement squared parameter (ξ) respectively where A is the absorbance at a wavelength of 280 nm and M^* is an operational point average molecular mass with some useful properties [6]. These plots are typical for the concentration range measured. As Figure 2a indicates, the sample behaves pseudo-ideal. The M^* function has the property that at the cell base $M^* = M_{w,app}$, where $M_{w,app}$ is the apparent weight average molecular mass over the whole distribution in the ultracentrifuge cell and 'app' means at a finite concentration. The extrapolation of the plot shown in Fig. 2b gives an $M_{w,app}$ of 85 000 kDa. The values obtained for the different initial loading concentrations of the native isoform are shown in Table 1. The plot of point average molecular weight versus absorbance (see Fig. 2c), although rather noisy (an unavoidable feature of the absorption optical system), indicates a mean weight average molecular mass of approx. 80 000 Da which would suggest that this protein isoform may be dimerizing. The molecular mass values obtained under the experimental conditions indicate that self-association - reversible or otherwise - may have taken place and with the apparent weight average molecular mass ranging from 70 000 and 85 000 Da dimerization of the protein appears to be the most likely explanation.

Apparent weight average molecular mass of the His-tagged 42 kDa protein

Plots of ln A versus the normalized radial displacement squared (ξ) and $M_{w,app}$ (ξ) versus absorbance for the His-tagged protein at the lowest initial loading concentration used (0.6 mg/ml) are shown in Figs. 3a and 3b respectively. These plots are typical for the concentration range investigated. The very slight downward curvature apparent in Fig. 3a indicates that there may be significant contribution from thermodynamic non-ideality which is confirmed by the decrease in the apparent point weight average molecular mass with concentration as shown in Fig. 3b.

The plot of apparent weight average molecular mass versus initial loading concentration, c^0, at the two speeds is shown in Fig. 4. The initial increase in $M_{w,app}$ with concentration is a clear demonstration of the existence of species of molecular mass > 84 kDa, i.e., the dimer value. Although the apparent maximum M_w (ξ) in Fig. 4 of ~ 130 kDa is between the dimer and tetramer forms, after allowance for non-ideality it could be reasonably inferred that tetrameric species (M ~ 168 kDa) are present. The decrease in molecular mass as $c^0 \rightarrow 0$ is a clear indication of a self-association, i.e., that the species of different molecular mass are in chemical equilibrium.

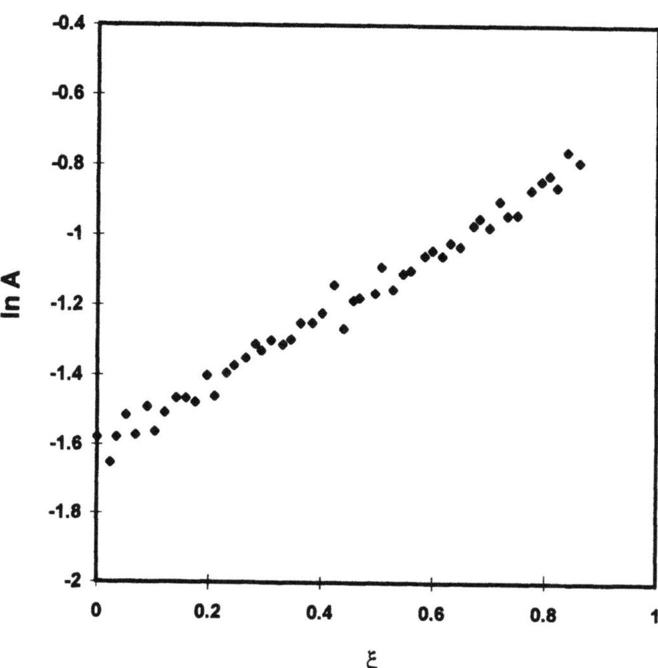

Fig. 2A Plot of ln absorbance versus ξ for native 42 kDa repessor protein ($c^0 = 3$ mg/ml, $\lambda = 280$ nm)

Fig. 2B Plot of M^* versus ξ for native 42 kDa repessor ($c^0 = 3$ mg/ml, $\lambda = 280$ nm)

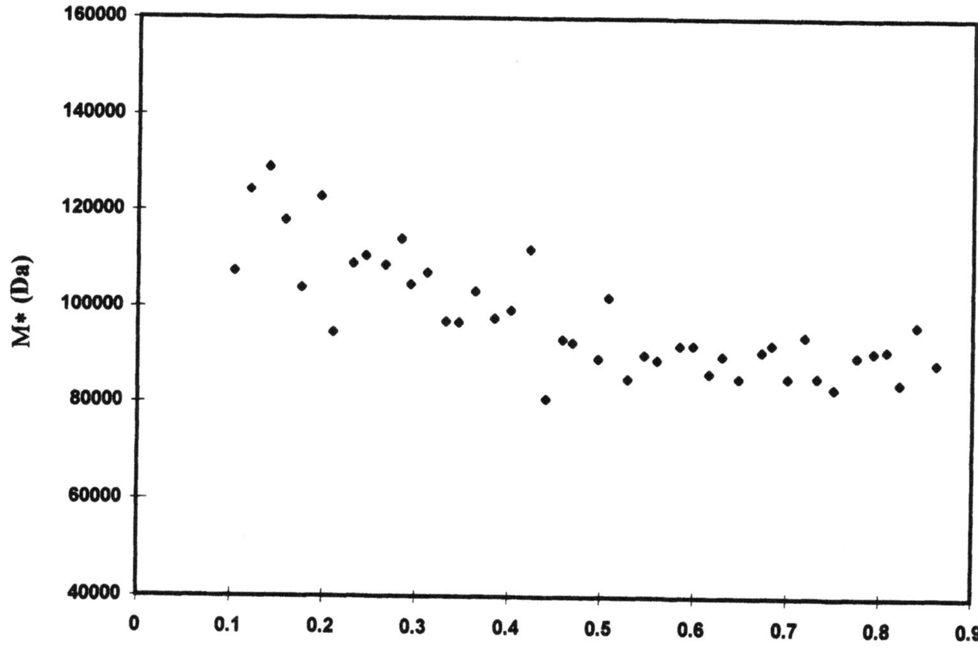

Fig. 2C Plot of point average molecular weight (ξ) versus absorbance for native 42 kDa repressor protein

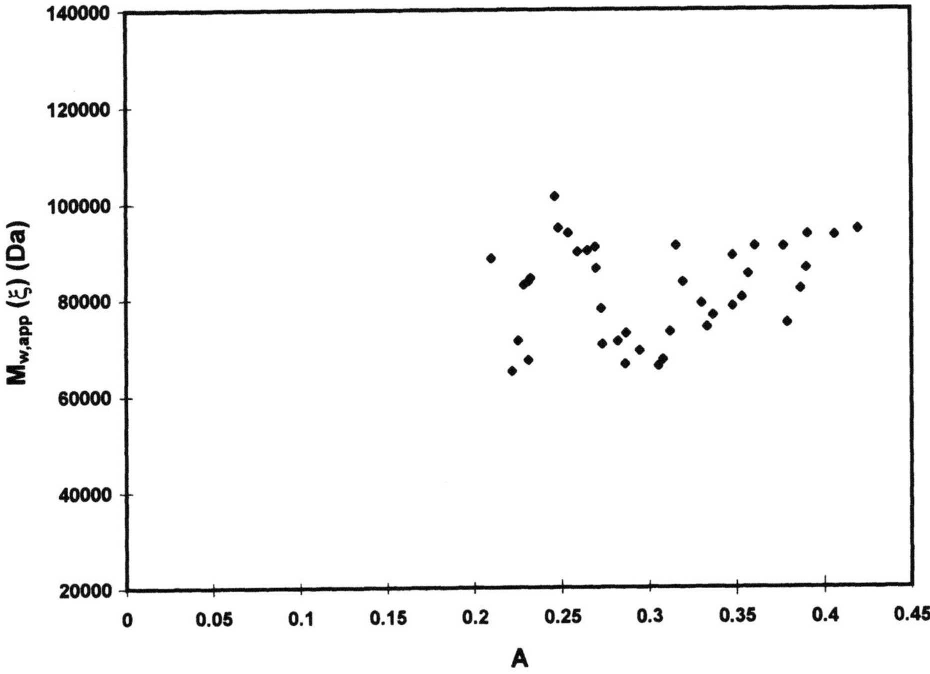

Table 1 Apparent weight average molecular masses ($M_{w,app}$) of the wild-type (WT) and His-tagged (His) 42 kDa repressor isoform (experimental conditions as in text)

Protein concentration, c^0, (mg/ml)	$M_{w,app}$ (Da)		
	WT (7000 rpm)	His (7000 rpm)	10 000 rpm
0.6	ni	99 000 ± 7500	77 000 ± 6000
1.0	70 000 ± 7500	145 000 ± 13 500	127 000 ± 7500
2.0	82 000 ± 5000	134 000 ± 10 000	130 000 ± 10 000
3.0	85 000 ± 5000	ni	ni

ni = not investigated.

Fig. 3A Plot of ln absorbance versus ξ for His-tagged 42 kDa repressor protein (c^0 = 0.6 mg/ml, λ = 280 nm)

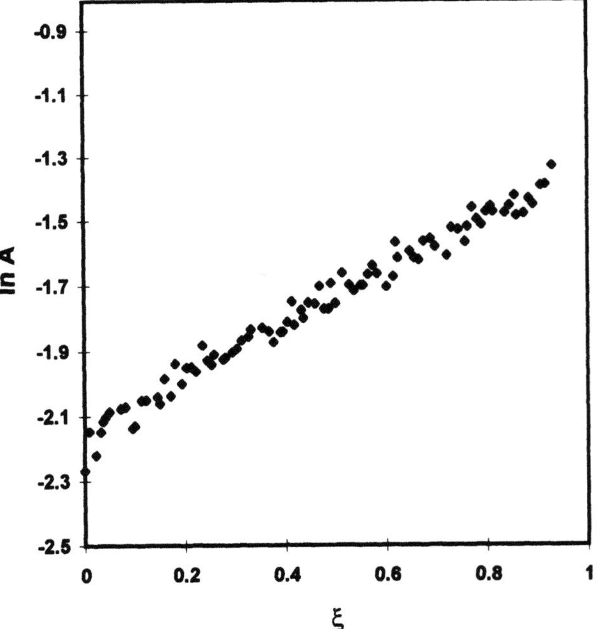

Conclusions

Sedimentation equillibrium evaluation of molecular masses of the 42 kDa isomer of the repressor protein from the temperate *Streptomyces* phage, ϕC31, both, in its native form and in the Histidine-tagged form has indicated that this protein exists in an oligomeric rather than monomeric form. At lower concentrations both isomers are present as dimers whereas at high concentrations (approx. 20 μM) the His-tagged isoform appeared to form higher oligomers (most likely tetramers) whereas the wild-type form appears to remain as dimers. This discrepancy may be explained by the fact that the wild-type form is contaminated by approx. 5% of small molecular mass nucleic

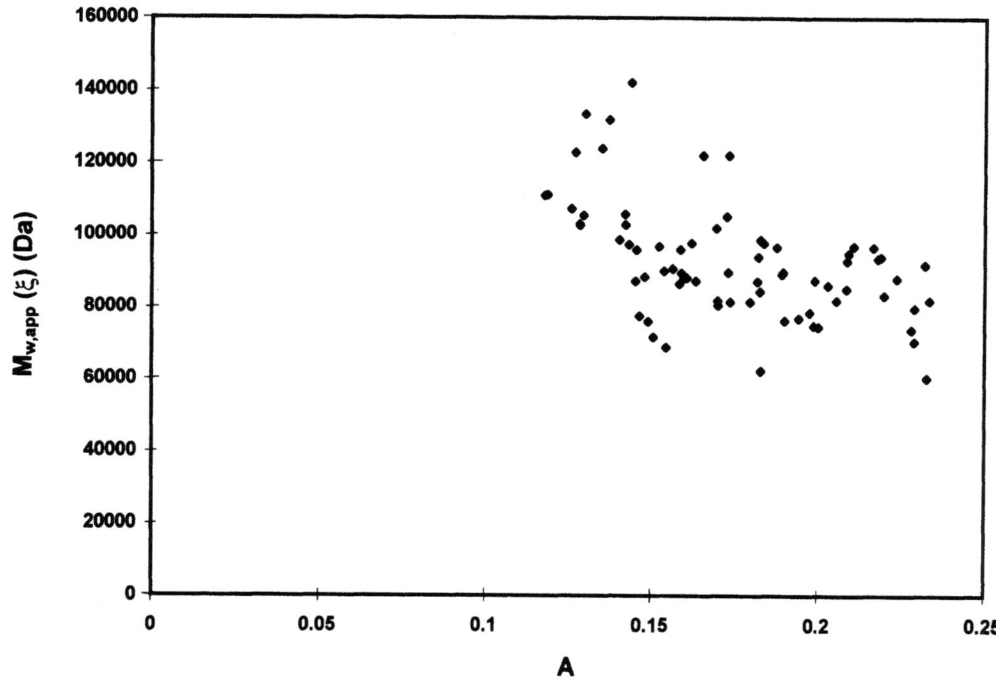

Fig. 3B Plot of point average molecular weight (ξ) versus absorbance for His-tagged 42 kDa repressor protein

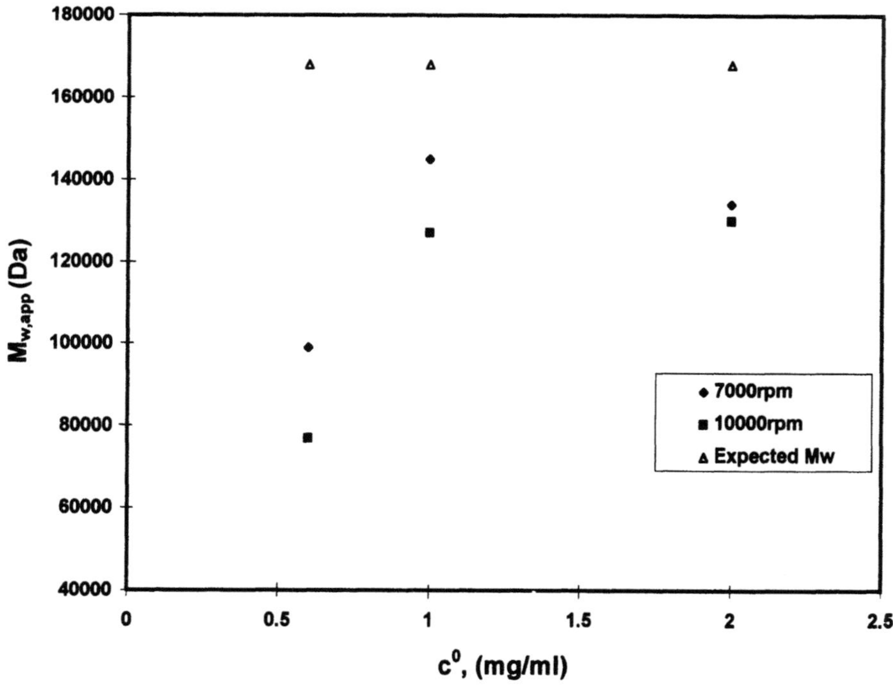

Fig. 4 Dependence of apparent weight average molecular weight of His-tagged 42 kDa isoform on initial loading concentration and rotor speed

acids and these (i) would reduce the weight average molecular mass, and (ii) could well compete with the sites for tetramerization. To address these features and establish the strength of the association process (via dimerization/tetramerization constants etc.) and the contribution from thermodynamic non-ideality will require considerably more measurements on more material than is currently available. This will be a subject for future study.

Acknowledgement The authors wish to thank the Wellcome Trust for supporting this work through their grant (No. 034787/Z/91). This paper was presented at the IX. Symposium on Analytical Ultracentrifugation, Berlin-Buch, March 2/3, 1995.

References

1. Smith, MCM, Owen CE (1991) Mol Microbiol 5:2833–2844
2. Ingham CJ, Owen CE, Wilson SE, Hunter IS, Smith MCM (1994) Nucleic Acids Res 22:821–827
3. Wilson SE, Ingham CJ, Hunter IS, Smith MCM (1995) Mol Microbiol 16:131–143
4. Wilson SE, Smith MCM unpublished results
5. Broughton D, Smith MCM, unpublished results
6. Creeth JM, Harding SE (1982) J Biochem Biophys Methods 7:25–34

Progr Colloid Polym Sci (1995) 99:17–23
© Steinkopff Verlag 1995

C. Ebel

Characterisation of the solution structure of halophilic proteins. Analytical centrifugation among complementary techniques (light, neutron and X-ray scattering, density measurements)

Received: 3 March 1995
Accepted: 16 May 1995

Dr. C. Ebel (✉)
Institut de Biologie Structurale
41 avenue des Martyrs
38 027 Grenoble Cedex 1, France

Abstract This mini-review shows how complementary techniques (light, neutron and x-ray scattering, analytical centrifugation, density measurements) have been used in the study of the solvation of halophilic proteins, proteins which are only stable at high salt. The halophilic proteins studied so far are all solvated by large amounts of salt, compared to non-halophilic proteins.

Key words Halophilic proteins – analytical centrifugation – scattering – densimetry – solvation

Extreme halophilic bacteria are part of the domain of the *Archaea*, one of the three domains, with *Bacteria* and *Eukaria*, into which living organisms have been divided [1]. In the *Archaea* are found, apart from extreme halophiles, methanogens, thermophilic sulfate reducers, extreme thermophiles. The characteristics of the *Archaea* have similarities with both *Bacteria* and *Eukaria*. Proteins from extreme halophiles are studied not only because of their interest from the evolutionary point of view, but also because they provide good models for the study of effects of the environment on protein folding, stability, dynamics and interactions with nucleic acids. Halophilic bacteria grow in media nearly saturated in salt, mainly sodium chloride. Their cytoplasm is nearly saturated in potassium chloride and all their biochemical reactions occur in this extreme medium. Halophilic proteins are inactivated and unfolded when the salt concentration is lowered below one or two molar salt [2, 3]. The examination of their amino-acid content reveals that in the majority of cases, halophilic proteins are composed of more acidic residues (D and E), less basic residues (K, especially), more small hydrophobic residues (A, V, G) and less large hydrophobic ones (I, L) [4, 5]. The recent resolution of the three-dimensional structure of halophilic malate dehydrogenase shows high densities of acidic residues at the surface of the protein [6]. The characterisation of the solution structure of halophilic proteins in their native state began when protocols for their purification (at high salt) were developed [7]. By solution structure, is meant the molar mass of the protein in solution and its solvent interactions. The way to determine the solvent interactions will be described in the second part of this paper. The three proteins studied in sodium or potassium chloride are characterized by usual or extensive hydration, and above all, by unusually extensive solvation of salt [8–10]. The solvation shell is close to saturation or supersaturation in salt.

The effect of salts on protein stability has been reviewed by Von Hippel and Schleich [11]. Salts such as potassium phosphate or ammonium sulfate that stabilise protein structure are also those which favour protein precipitation; they are called "salting-out" salts. Salts such as magnesium chloride or guanidinium hydrochloride destabilise protein structure and also increase protein solubility; this process is called "salting-in". In this classification, sodium chloride and potassium chloride are neutral or moderately salting-out reagents. Arakawa and Timasheff [12], Arakawa et al. [13, 14] (and references therein) have demonstrated that, in general, salting-out salts favour preferential protein interaction with water, while salting-in salts favour preferential interaction with salt. Preferentially hydrated proteins adopt a folded structure and precipitate in order to minimise their surface accessible to the solvent, while proteins interacting preferentially with salt unfold in order to maximise their interactions with the solvent. Halophilic proteins are exceptional since their unusually large salt interactions are correlated with the stability and solubility of the folded state.

The inactivation of halophilic malate dehydrogenase has been studied as a function of the concentration of various salts (KCl, NaCl, potassium phosphate, ammonium sulfate, magnesium chloride), in H_2O and, in some cases, in D_2O and as a function of temperature [15, 16]. It was proposed that different mechanisms dominate in the stabilisation of the folded structure depending on the nature of the solvent. In KCl or NaCl, the stabilisation would involve mainly a network of hydrated ions associated cooperatively with acidic residues localised in patches at the surface of the protein. In potassium phosphate of ammonium sulphate, or when H_2O is exchanged with D_2O, more classical hydrophobic interactions are responsible for the stabilisation of the tertiary and quaternary structure. Results of a first site-directed mutagenesis study involving surface acidic residues are in accordance with this model [4]. The relative importance of the mechanisms of stabilisation would depend also on the proteins. Thermostability measurements on halophilic aldolase in KCl for example suggest predominant hydrophobic interactions [17]. A theoretical modelling study of halophilic dihydrofolate reductase suggested a charge distribution forming a molecular dipole that played a role in substrate interaction [18].

An important result for the study of halophilic protein stabilisation was the experimental determination of the amount of water and salt binding to the proteins. Eisenberg [19, 20] has developed a thermodynamic approach which allows to apprehend the solution structure of macromolecules by various techniques, in the same formalism. The fluctuations in concentration of macromolecules (to which are associated fluctuations of solvent components) produce fluctuations in refractive index, mass density, electron density or neutron scattering length density. This gives rise either to the scattering phenomena, or to the sedimentation of a macromolecule in a centrifugal field. Concentration fluctuations of the macromolecule are related to the osmotic work done against the concentration gradient, which can be expressed as a function of the molar mass of the particle in solution. Density fluctuations of the solution (either in mass, or in electron number, or in neutron scattering length or in refractive index) are expressed by a term, the density or refractive index increment. This term is what is called the buoyancy term in a sedimentation experiment or the contrast in a scattering experiment. Table 1 shows that, for an ideal homogeneous solute, the experimentally accessible parameters obtained from very different techniques are expressed in a very simple way using the concentration of the macromolecule: c_2, the molar mass of the macromolecule: M_2 and the term of contrast which is, depending on the property studied: the refractive index increment $(\partial n/\partial c_2)_\mu$, the neutron scattering length density increment $(\partial \rho_{el}/\partial c_2)_\mu$, or the density: $(\partial \rho/\partial c_2)_\mu$.

The contrast is related not only to the characteristics (composition and volume) of the macromolecule and of the solvent, but also to all the perturbations of the solvent induced by the presence of the macromolecule. This term is of particular interest in the case of a complex system, a system not only containing water (by convention component 1) and the macromolecule (component 2), but also a small solute (component 3). This solute can be a salt, a small organic compound, a detergent. The contrast term can be expressed, from either of the techniques, as a function of other thermodynamic parameters: the partial specific volumes $(\bar{v}_2, \bar{v}_1, \bar{v}_3)$, a preferential interaction parameter with water or with the small solute, ξ_1 or $\xi_3 \cdot \xi_1$ or ξ_3 are thermodynamical parameters which do not have a structural meaning *per se*. They can be interpreted in terms of positive or negative binding of solvent components to the macromolecule, as will be presented below in the study of halophilic proteins. They could also involve steric exclusion of the additive [21, 22], or effects of electrostatic charge [19, 23]. The mechanisms of solvent interaction are in the general case dependent on the composition of the solvent (nature and concentration of the small solute). In the following, we will focus on a particular situation, in which the particle can be considered as invariant, in volume and composition, when the composition of the solvent is changed.

Tardieu et al. [24] have shown that when a linear variation of the density increment is observed as a function of solvent density, it is possible to define an equivalent macromolecule which is invariant with respect to solvent composition. This invariant particle is characterised by five parameters, B_1 (gram of water *per* gram of protein), B_3 (gram of small solute *per* gram of protein), and $\bar{v}_2, \bar{v}_1$ and $\bar{v}_3$, the partial specific volumes for the three components, respectively. The composition of the particle can be considered as constant, whatever the solvent composition, and composed of 1 gram of protein, B_1 gram of water and B_3 gram of small solute (*per* gram of protein) in a volume

Table 1 Calculation of the increments of refractive index $(\partial n/\partial c_2)_\mu$, of neutron scattering length density $(\partial \rho_N/\partial c_2)_\mu$, of electron density $(\partial \rho_{el}/\partial c_2)_\mu$, of density: $(\partial \rho/\partial c_2)_\mu$ from the experimentally accessible parameters for an ideal homogeneous solute.

The experimental parameters are the forward scattered intensity in excess of solvent scattering in light scattering: $R_0(0)$, small angle neutron scattering: $l_N(0)$ and small-angle x-ray scattering: $l_{el}(0)$; the gradient of concentration of the macromolecule c_2 in a centrifuge at the equilibrium: $d(\ln c_2)/dr$; the ratio between sedimentation and diffusion coefficient: s/D; the density of the solutions containing or not containing protein: ρ or ρ^0. c_2 is the concentration of the macromolecule. M_2 is the mass of the macromolecule. The subscript μ expresses the fact that the experiments are performed at constant chemical potential of all solvent components (this can be realised by extensive dialysis of the sample against the solvent). N_A is Avogadro's constant, ω the angular velocity, R the gas constant, T the absolute temperature

Scattering:							
-light scattering:	$R_0(0) =$	$(1/N_A)$	·	c_2	·	$(\partial n/\partial c_2)_\mu^2$	·	M_2
-small angle neutron scattering:	$l_N(0) =$	$(1/N_A)$	·	c_2	·	$(\partial \rho_N/\partial c_2)_\mu^2$	·	M_2
-small angle x-ray scattering	$l_{el}(0) =$	$(1/N_A)$	·	c_2	·	$(\partial \rho_{el}/\partial c_2)_\mu^2$	·	M_2
Equilibrium sedimentation	$d(\ln c_2)/dr =$	$(\omega^2/2RT)$			·	$(\partial \rho/\partial c_2)_\mu$	·	M_2
Hydrodynamic measurements:	$s/D =$	$(1/RT)$			·	$(\partial \rho/\partial c_2)_\mu$	·	M_2
sedimentation velocity +							
quasi-elastic light scattering							
densimetry:	$\rho - \rho^0 =$			c_2	·	$(\partial \rho/\partial c_2)_\mu$		

$\bar{v}_2, + B_1 \bar{v}_1 + B_3 \bar{v}_3$. This particle approach allows to treat the experimental physical parameters without the introduction of the interaction parameters ξ_1 or ξ_3. It allows the treatment of the various techniques in a complementary fashion. Consider a solvated protein in a solvent composed of water and salt (Fig. 1). The increment of density at constant chemical potential for all solutes except the protein, $(\partial \rho/\partial c_2)_\mu$, is expressed as a function of the solvent density and of the five parameters characterising the particle:

$$(\partial \rho/\partial c_2)_\mu = 1 + B_1 + B_3 - \rho^0 . v_T = 1 + B_1 + B_3$$
$$- \rho^0(\bar{v} + B_1 \bar{v}_1 + B_3 \bar{v}_3).$$

As a consequence (Fig. 2), if the density of the solvent is varied, by variation of the solvent composition, the plot of the density increments as a function of the solvent density gives a straight line, where the intercept is the mass of the solvated protein (per 1 gram of protein): $1 + B_1 + B_3$, and where the slope is the volume of the solvated protein (per 1 gram of protein): $1 + B_1 + B_3$, and where the slope is the volume of the solvated protein (per 1 gram of protein): v_T. This is an experiment in contrast variation. The corresponding relations for the increment of electron density: $(\partial \rho_{el}/\partial c_2)_\mu$ and the increment of neutron scattering length density: $(\partial \rho_N/\partial c_2)_\mu$ are in Fig. 2. Depending on the physical parameter studied, these relations contain the electron densities l_1, l_2 and l_3 (in electrons per gram), or the neutron scattering length densities b_1, b_2 and b_3 (in cm per gram).

The complementarity of the techniques is apparent. If we use neutrons or x-rays, we will "weigh" the particle in terms of neutron scattering lengths or electrons instead of mass. Thus the relative "weights" of the particle constituents will be different depending on the technique used. On the contrary, the total volume will be the same, and the slope of the plot of the increments of density as a function of the densities will be the same, whatever the technique used.

The extensive study of solvent interactions for halophilic malate dehydrogenase (hMDH) illustrates the complementarity of the techniques. It has demonstrated that 1 gram of protein is associated with 0.4 gram of water and 0.08 gram of NaCl [8]. To determine this solvation, a contrast variation approach was used between 1 and 5 M NaCl with various techniques (see Fig. 3).

If we examine the contrast variation obtained in terms of mass density, we find that the points are very well aligned. As a consequence, the value of the total volume of the solvated protein is obtained in a precise way. But the points are a long distance away from the y axis, and only a rough estimate of the mass of the particle can be obtained from this technique.

The contrast variation obtained by x-ray scattering offers similar features and thus the two techniques do not give independent information.

Neutron scattering shows a very different feature: the points are very close to the y-axis, and thus provide very precise information on the "weight" of the solvated protein (in terms of neutron scattering length density), but because of scattering in the measurements, the slope which determines the total volume of the article cannot be defined precisely.

The complementarity, therefore, is between mass and x-ray, on the one hand, to define v_T and neutrons, on the other, to define particle composition.

Neutron scattering experiments on protein solutions are often performed in heavy water because of increased contrast between the protein and the solvent and of

Fig. 1 Relation between the density increment $(\partial\rho/\partial c_2)_\mu$, the solvent density ρ^0, and the five parameters characterising a solvated protein, B_1, B_3, $\bar{v}_1$, $\bar{v}_2$ and $\bar{v}_3$, in a solvent composed of water and salt. B_1 and B_3 are the amounts of water and salt associated to the protein (in gram of water and salt *per* gram of protein), $\bar{v}_1$, $\bar{v}_2$ and $\bar{v}_3$ the partial specific volumes of water, protein and salt

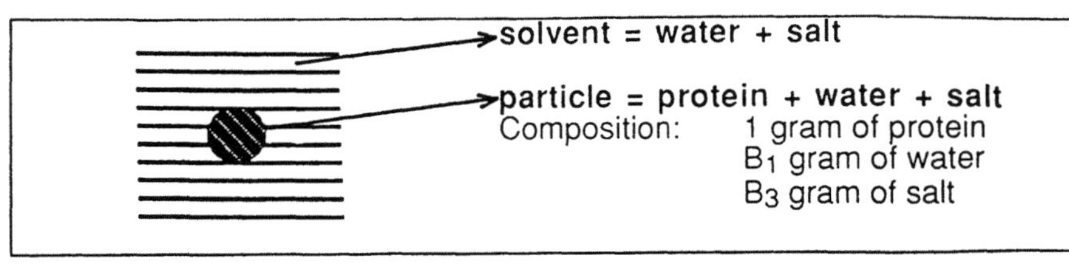

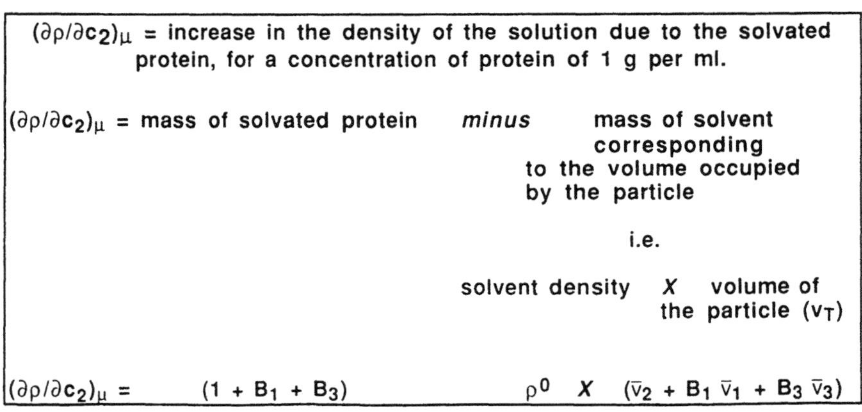

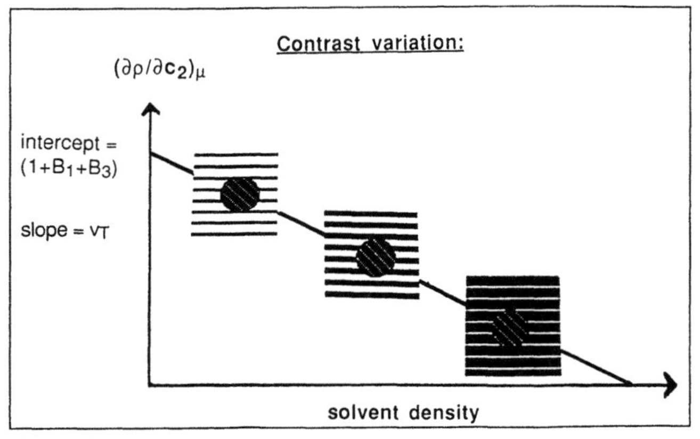

mass:	$(\partial\rho/\partial c_2)_\mu = (1+B_1+B_3) - \rho^0 . v_T$
X-rays:	$(\partial\rho_{el}/\partial c_2)_\mu = (l_2+l_1B_1+l_3B_3) - \rho_{el}^0 . v_T$
neutrons:	$(\partial\rho_N/\partial c_2)_\mu = (b_2+b_1B_1+b_3B_3) - \rho_N^0 . v_T$

Fig. 2 Case of an invariant particle: linear variation of the density increments plotted as a function of the solvent density. The linear variation is obtained when mass density increments $(\partial\rho/\partial c_2)_\mu$, electron density increments $(\partial\rho_{el}/\partial c_2)_\mu$, or neutron scattering length density increments $(\partial\rho_N/\partial c_2)_\mu$, are plotted against the solvent density ρ^0, the electron density ρ_{el}^0 or the neutron scattering length density ρ_N^0. l_1, l_2 and l_3 are the electron densities, b_1, b_2 and b_3 the neutron scattering length densities of respectively water, protein and salt

increased signal over noise ratio due to the fact that heavy water does not have the large incoherent scattering of light water. Neutron scattering experiments on hMDH in solutions containing NaCl in heavy water were performed in order to determine if the solvation of the protein was similar in H_2O and D_2O [16]. It was shown (using also density measurements) that the solvation of hMDH was essentially the same in H_2O and D_2O. But what is interesting from the experimental point of view is that in D_2O, the neutron scattering length densities and density increments are very close whatever the salt concentration (there is negligible contrast variation in D_2O as a function of salt concentrations). However, since now the solvation in H_2O and D_2O of the protein is known to be the same, solvent interactions at one salt concentration could be studied by performing contrast variation by H_2O and D_2O exchange.

We have seen how to use the complementary of the techniques for the interpretation of the density increments. But, depending on the techniques, the density increments are not obtained with the same accuracy. The relationships of density increments to experimental parameters are given in Table 1.

The first point is the precision in the measured parameter. For neutron scattering and x-ray scattering, the value of the forward scattered intensity is obtained from an extrapolation to zero angle of the measurements and has to be put on absolute scale. The gradient of concentration measured in an equilibrium sedimentation equilibrium will be affected by the residual absorption of the solvent. Densimeters allow a very good precision in density measurements, but extreme caution has to be observed during the manipulation of the sample: for example, if the dialysis bag is extracted from the bath, evaporation of water will greatly perturb the measurement.

The second point is that scattering and density measurements require for their interpretation the knowledge of the absolute concentration of the protein. This parameter is the source of systematic errors.

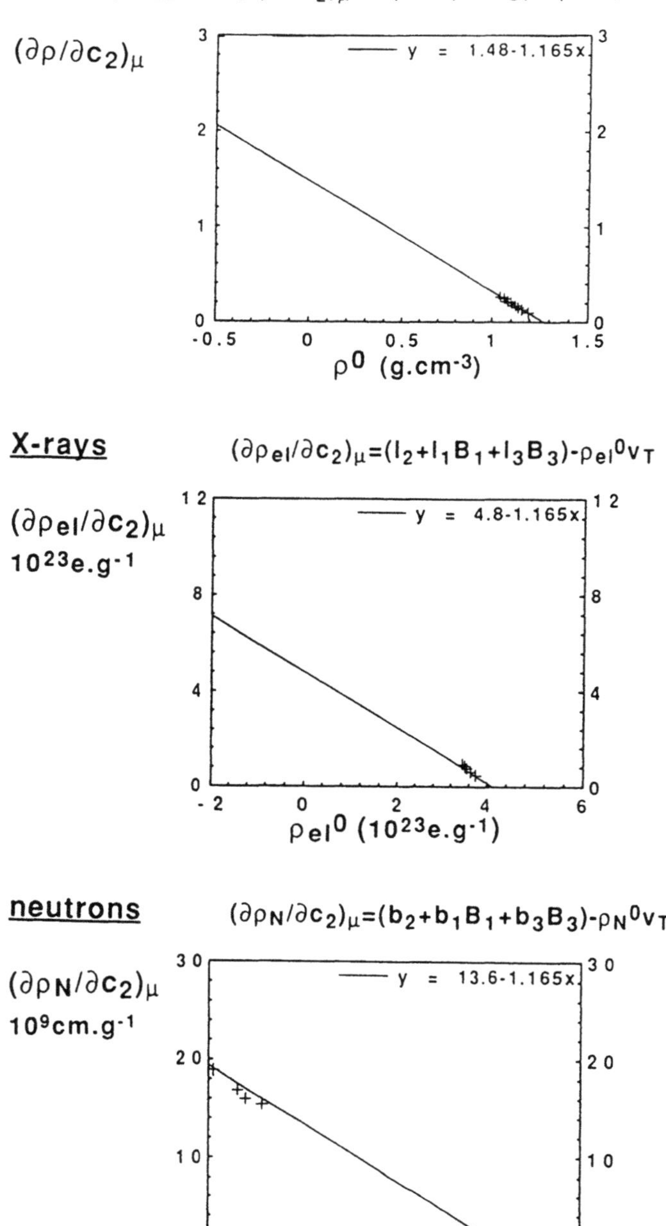

Fig. 3 Contrast variation on halophilic malate dehydrogenase performed in NaCl solutions containing 1 to 4–5 M salt from different techniques. From top to bottom: data are obtained from Pundak and Eisenberg [25], Reich et al. [26] and Zaccai et al. [27]. The lines plotted correspond to a solvation of 0.41 gram of water and 0.08 gram of salt *per* gram of protein corresponding to a total volume for the particle of 1.165 cm$^3 \cdot$ g^{-1} [8]

The third point is that, apart from the densimeter measurements, all the measured physical quantities are related to both the molar mass and the density increment, while these techniques are usually used to determine molar mass alone. In principle, the use of different techniques can lead to the determination of both. For example, density measurements can be used together with either hydrodynamic measurements (sedimentation and diffusion coefficients) or equilibrium sedimentation to determine the mass of the protein. The molar mass in solution of halophilic glyceraldehyde-3-phosphate dehydrogenase was determined in this way (see below). On the other hand, since in hydrodynamic or equilibrium sedimentation measurements, the experimental quantity obtained is $(\partial\rho/\partial c_2)_\mu M_2$, while in scattering experiments it is $(\partial\rho_N/\partial c_2)_\mu \sqrt{M_2}$ or $(\partial\rho_{el}/\partial c_2)_\mu \sqrt{M_2}$, the square of the molar mass can be obtained from the ratio of the slopes ($v_T M_2$ and $v_T \sqrt{M_2}$) of the plots of these values against the solvent densities (see Table 1 and Fig. 2). This was done for halophilic malate dehydrogenase by Zaccai et al. [27], but the accuracy of the value obtained was very poor.

Usually, after the molar mass of the protein in solution is determined, its value is used for the determination of the density increments which are in turn interpreted as solvation. The use of different techniques allows to detect artefacts due to the introduction of an erroneous molar mass. Mass density increments have to be the same, whatever the technique used: density measurements, hydrodynamic measurements (sedimentation and diffusion coefficients) or equilibrium sedimentation. On the same hand, for an invariant particle, the plot of $(\partial\rho/\partial c_2)_\mu$, or of $(\partial\rho_{el}/\partial c_2)_\mu$, or of $(\partial\rho_N/\partial c_2)_\mu$, as a function of ρ^0, or of ρ_{el}^0, or of ρ_N^0, are straight lines, whatever the technique used (see Fig. 2), which have to provide the same slope, the total volume of the solvated protein, and lead to the same values of B_1 and B_3, the amount of water and salt associated to the protein. Each of the techniques except densimetry detects the presence of heterogeneities in a more or less powerful way.

Three examples in the study of solvent interactions of halophilic proteins illustrate the complementarity of the techniques.

The subunit molar mass of halophilic glyceraldehyde-3-phosphate dehydrogenase was determined by mass spectrometry to be 35990 ± 80 g mol^{-1}. The molar mass in solution of the native protein was determined at three salt concentrations by complementary densimetry and hydrodynamic measurements to be 140000 ± 17000 g mol^{-1} corresponding to a tetramer. The values of the density increments obtained from densimetry were not used for the determination of the solvent interactions of the protein, since it was difficult to obtain accurate values with limited amounts of protein. From the linear plot of the density increments obtained from sedimentation and diffusion coefficients as a function of the solvent density, the native tetramer could be assumed to be an invariant particle between 1 and 3.5 M KCl, with the protein binding of 0.18 ± 0.07 gram of water and 0.07 ± 0.02 gram of salt *per* gram [10].

The study of the solvation of the native halophilic elongation factor Tu was undertaken in high KCl solutions [9]. This protein was described as a monomer. Small-angle neutron scattering and hydrodynamic (sedimentation velocity and quasi elastic light scattering) experiments performed after a standard purification procedure were compatible with an invariant particle between 2.2 and 3.5 M KCl. From the experimental data, the samples were suspected to be slightly heterogeneous. The solvation results obtained from the plots of the density increments as a function of the densities for the two types of measurements did not agree with a monodisperse solution. They were interpreted by considering the presence of 15% dimers in the solution. Similar experiments, plus densimetry measurements, were then performed with samples subjected to gel filtration just before the measurements. We obtained self-consistent results from the different techniques, assuming mono-disperse solutions (100% of monomers solvated by 0.4 gram of water and 0.2 gram of salt by gram of protein).

In a third example, the mass density increments of tetrameric halophilic malate dehydrogenase were measured for its native form between 1 M and 5 M NaCl by densimetry and hydrodynamic measurements [8, 25].

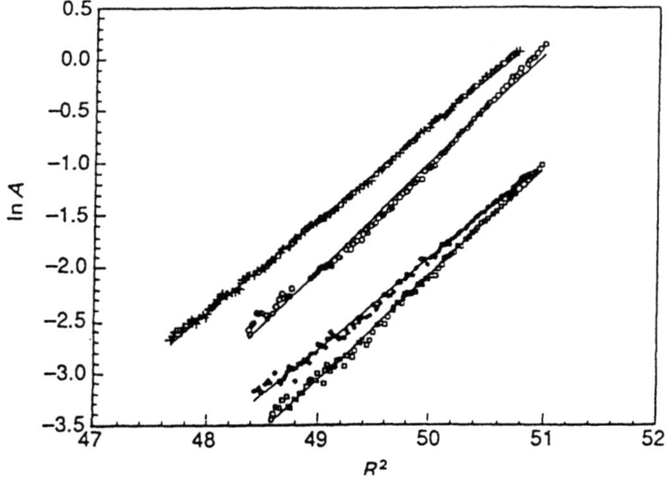

Fig. 4 Sedimentation equilibrium of halophilic malate dehydrogenase in NaCl solutions. (reprinted from F. Bonneté, C. Ebel, G. Zaccai and H. Eisenberg, J. Chem. Soc. Faraday Trans., (1993)). Plot of the logarithm of the absorbance measured in the ultracentrifuge as a function of the square of the distance from the axis of rotation. The four experiments described here were performed at 15 000 r.p.m. at 20 °C, and correspond to: O, 2.5 mol l^{-1} NaCl; □, 3 mol l^{-1} NaCl; + and ■, 4 mol l^{-1} NaCl, with initial concentrations of hMDH of 0.50, 0.16, 0.16, and 0.55 mg ml^{-1}, respectively. All the solutions were buffered with 20 mmol l^{-1} Na phosphate to ph 7. The plot for 3 mol l^{-1} NaCl has been shifted downwards by ln A = 0.25 for clarity. The two lower plots were performed with the protein just eluted from a gel filtration column. Lines are linear fits to the data

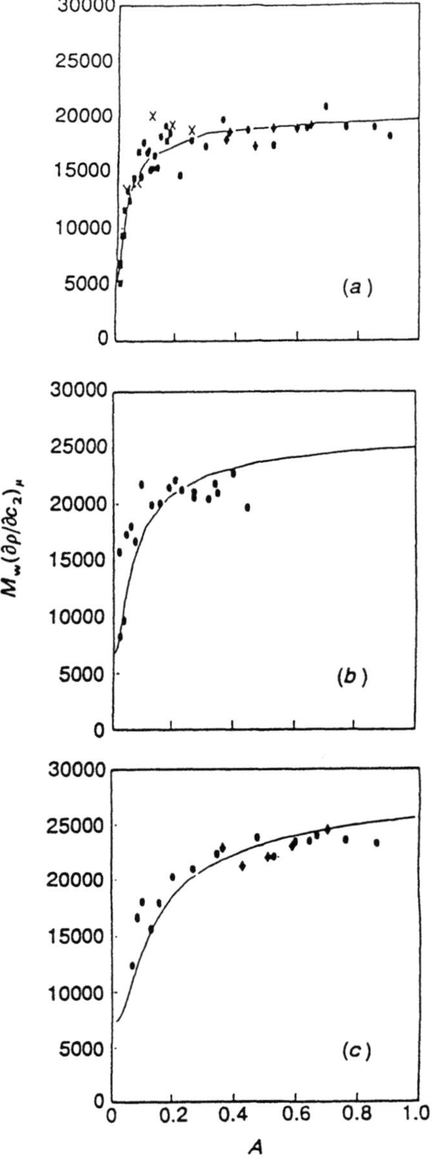

Fig. 5. $M_w \cdot (\partial \rho / \partial c_2)_\mu$ distribution as a function of absorbance of the material in the ultracentrifuge observed for halophilic malate dehydrogenase. (reprinted from F. Bonneté, C. Ebel, G. Zaccai and H. Eisenberg, J. Chem. Soc. Faraday Trans., (1993)) M_w is the weight-average molar mass. ♦, 8000; ● and X, 15000; ■, 30000 r.p.m. in (a) 4 mol l^{-1} NaCl: ●, concentrated sample; X, diluted sample; (b) 3 mol l^{-1} NaCl; (c) 2.5 mol l^{-1} NaCl. The lines are fits to the data for an equilibrium between monomers and tetramers using measured values of $(\partial \rho / \partial c_2)_\mu$ and $M_2 = 130552$ g mol^{-1} for the tetramer calculated from the gene sequence [28]

Both plots of density increments as a function of density were straight lines, indicating an invariant particle, of same slope (leading to a unique value for the total partial specific volume) but slightly shifted. These two lines yield the same solution for the solvation of the protein (since the slopes are the same and the intercepts are not significantly

Progr Colloid Polym Sci (1995) 99:17–23
© Steinkopff Verlag 1995

different) of 0.4 gram of water and 0.1 gram of salt by gram of protein. In order to determine the reason for the discrepancy between these sets of results, we decided to perform complementary equilibrium sedimentation experiments as a function of salt concentration. We found (Fig. 4) that the system was not perfectly homogeneous. The whole set of data was successfully fitted by considering an equilibrium between monomer and tetramer (Fig. 5), thus providing an explanation for the small discrepancy between the different density increments measurements. As a consequence, while very sensitive, equilibrium sedimentation experiments were unable to provide accurate values for the density increments because they produce gradients of protein concentration which depend also on the dissociation of the tetramer. It is impossible in this case to determine precisely both the density increment and the association constant of the protein.

From these examples it appears that the study of the interaction parameters of a protein is not easy. The complementary use of various techniques provides a good control for the interpretation of the results, which is especially important if the protein is not very well characterised with respect to its absolute concentration, or the effective molar mass in solution. The systems described above as examples are composed of water, protein and salt. The same treatment applies for ternary complexes containing nucleic acids or other small solutes such as detergents or other additives.

References

1. Woese CR (1994) Microbiol Rev 58:1–9
2. Lanyi JL (1974) Bacteriol Rev 38:272–290
3. Eisenberg H, Mevarech M, Zaccai G (1992) Adv Prot Chem 43:1–61
4. Madern D, Pfister C, Zaccai G submitted
5. Böhm G, Jaenicke R (1994) Int J Pept Protein Res 43:97–106
6. Dym O, Meravech M, Sussmann J (1995) Science, 267:1344–1346
7. Mevarech M, Eisenberg H, Neumann E (1977) Biochemistry 16:3781–3785
8. Bonneté F, Ebel C, Zaccai G, Eisenberg H (1993) J Chem Soc Faraday Trans 89:2659–2666
9. Ebel C, Guinet F, Langowski J, Urbanke C, Gagnon J, Zaccai G (1992) J Mol Biol 223:361–371
10. Ebel C, Altekar W, Langowski J, Urbanke C, Forest E, Zaccai G (1995) Biophys Chem 54:219–227
11. Von Hippel PH, Schleich T (1969) In: Timasheff SN, Fasman GD (eds) Structure and Stability of Biological Macromolecules. Marcel Dekker, New York, Vol. 2, pp 417–574
12. Arakawa T, Timasheff SN (1984) Biochemistry 23:5912–5923
13. Arakawa T, Bhat R, Timasheff SN (1990) Biochemistry 29:1914–1923
14. Arakawa T, Bhat R, Timasheff SN (1990) Biochemistry 29:1924–1931
15. Zaccai G, Cendrin F, Haik Y, Borochov N, Eisenberg H (1989) J Mol Biol 208:491–500
16. Bonneté F, Madern D, Zaccai G (1994) J Mol Biol 244:436–447
17. Krishnan G, Altekar W (1993) Biochemistry 32:791–798
18. Böhm G, Jaenicke R (1994) Protein Engineering 7:213–220
19. Eisenberg H (1976) In: Biological Macromolecules and Polyelectrolytes in Solution. Clarendon Press, Oxford
20. Eisenberg H (1981) Quart Rev Biophys 14:141–172
21. Greulich KO, Ausio J, Eisenberg H (1985) J Mol Biol 186:167–173
22. Arakawa T, Timasheff SN (1985) Biochemistry 24:6756–6762
23. Zaccai G, Xian SY (1988) Biochemistry 27:1316–1320
24. Tardieu A, Vachette P, Gulik A, Lamaire M (1981) Biochemistry 20:4399–4406
25. Pundak S, Eisenberg H (1981) Eur J Biochem 118:463–470
26. Reich MH, Kam Z, Eisenberg H (1982) Biochemistry 21:5189–5195
27. Zaccai G, Wachtel E, Eisenberg H (1986) J Mol Biol 190:97–106
28. Cendrin F, Chroboczek J, Zaccai G, Eisenberg H, Meravech M (1993) Biochemistry 32:4308–4313

Progr Colloid Polym Sci (1995) 99:24–30
© Steinkopff Verlag 1995

α-helical coiled coils: simple models for self-associating peptide and protein systems

R.M. Thomas
H. Wendt
A. Zampieri
H.R. Bosshard

Received: 16 March 1995
Accepted: 6 June 1995

Dr. R.M. Thomas (✉) · A. Zampieri
Institut für Polymere
ETH-Zentrum
8092 Zürich, Switzerland

H. Wendt · H.R. Bosshard
Biochemisches Institut
Universität Zürich-Irchel
8057 Zürich, Switzerland

Abstract Protein–protein recognition and protein oligomerization functions are among the important roles played by the α-helical coiled coil motif. As part of a study into the equilibrium and kinetic aspects of the folding of model coiled coils, an investigation into the stoichiometry and magnitude of the associative constants by sedimentation equilibrium in the ultracentrifuge has been initiated. A series of interrelated model peptides, with sequence lengths of about 30 residues, has been designed, synthesized and characterized. Each of the peptides has been subjected to conventional sedimentation equilibrium analysis under a variety of conditions. The stoichiometry of self-association has been established and it has been shown that, contrary to the expected result, most of the peptides formed trimeric systems of varying stability. The effect of substituting large hydrophobic constituents of the interacting interfaces of the systems with other similar residues has been shown to have minimal effect on trimer formation. However, substitution of such residues in the central sequence positions with either alanine or asparagine leads to a marked decrease in the stability of the trimeric state.

Key words Sedimentation equilibrium – circular dichroism– oligomerization state

Introduction

The α-helical coiled-coil motif, in which two or more polypeptide chains in an approximately right-handed α-helical conformation are wound around one another to produce a left-handed superhelix, is widespread in the protein world. It is found in long, filamentous structural proteins and in short α-helical bundles in both soluble and membrane proteins. It is, however, becoming increasingly apparent that one of its most important rôles is that of protein-protein recognition. Perhaps the best-understood example of this function is at the dimeric interface of some DNA transcriptional regulators, in which the coiled-coil is formed by the mutual interaction of the helical regions of two separate monomer subunits.

Coiled coils are composed of amphipathic α-helices that have sequences based on a repetitive heptad of amino acids, usually denoted 'abcdefg' [1], in which hydrophobic residues are frequent at positions a and d and positions e and g are often occupied by oppositely charged residues. The high frequency with which leucine appears in d positions in the coiled coil regions of DNA-binding proteins has led to the use of the descriptive term "leucine zipper" [2, 3]. The underlying features of the coiled coil were first described by Crick [4] who deduced that α-helices could pack together efficiently to form a superhelix if their amino acid sequences consisted of recurrent heptads. He noted that the proportion of apolar residues was lower in fibrous than in globular proteins and proposed that the side chains of the apolar residues were packed in the interactive interface in a "knobs-into-holes" manner, a prediction that has been borne out by subsequent experimental findings. Reviews on the coiled coil are available [5, 6] and the current state of research in the field has been summarized recently [7]. Schematic representations of the structures are shown in Fig. 1.

Coiled coils occur in different oligomeric forms. While dimers predominate in the leucine zipper regions of the transcription factors, many of the model peptides studied, as well as some naturally occurring motifs, preferentially assemble to trimeric as well as tetrameric structures. High-resolution structures from x-ray crystallographic analysis are now available for the leucine zipper region of the

GCN4 transcriptional activator GCN4-p1 [8], which is dimeric, as well as for the trimeric "coil-ser" peptide of DeGrado and co-workers [9] and the tetrameric p-LI analogue of the GCN4-p1 sequence [10].

Although it has become possible to predict the occurrence of a coiled coil from sequence data with some confidence [11], little is known about the factors that govern the final oligomerization state. Experimental validation of the number of strands in coiled coils is problematic and the stoichiometry cannot be deduced from spectroscopic methods like CD or NMR alone. The x-ray crystallographic data available is sparse and, in any case, reflects the properties of the molecules at infinite concentration. A method that circumvents these problems and enables the determination of the state of oligomerization directly at biologically relevant concentrations is the measurement of sedimentation equilibrium in the analytical ultracentrifuge. Previously reported studies using this method on small naturally-occurring and model coiled coil systems have been generally confined to confirming the apparent stoichiometry of the oligomeric interaction [9, 10, 12–18]. In the course of investigations into the stability and folding properties of coiled coils we have initiated a study into the nature of the associative equilibria in a variety of synthetic model systems. Dimeric and trimeric models have been designed, synthesized and characterized by equilibrium sedimentation with the aim of not only establishing stoichiometries but also the magnitudes of the various equilibrium constants involved in the associative processes. The sequences of the model peptides are given in Table 1. We report here on a basic investigation into self-association in these models which has been carried out in order to identify suitable candidates for which a detailed characterization will be performed.

Methods

Peptide synthesis and purification

Automated and manual synthesis methods were used and, in both cases orthogonal Fmoc/tbu protection schemes and HBTU/HOBt-mediated coupling strategies were employed. Rink Amide or Rapid Amide resins that released the peptide as the C-terminal amide were used in all cases and the N-termini of the peptides were acetylated prior to side-chain deprotection and cleavage from the resin. Peptides were purified by reversed phase HPLC on C_8 columns using binary aqueous acetonitrile gradients containing 0.1% TFA. The purity of the products was confirmed by electrospray mass spectrometry and amino acid analysis. Extinction coefficients used in concentration determinations were either derived from the amino acid

Fig. 1 Representations of coiled coil structures. The repetitive sequence of FZ in the single letter code is used throughout, and subscripts indicate the position of residues within the heptad. **A & B** are helical wheel diagrams of dimeric and trimeric coiled coils, respectively. A central heptad of the sequence is shown, hydrophobic interactions are indicated by arrows and potential electrostatic interactions by dotted lines. **C** The mutual interaction of residues from two α-helices in the hydrophobic core of a parallel coiled coil is displayed schematically. Adapted from [5]

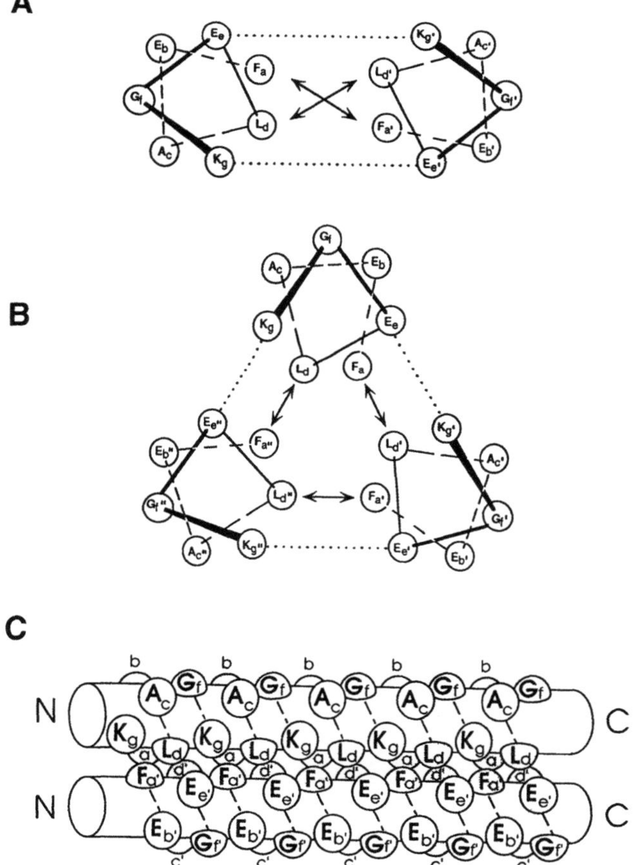

Table 1 The amino acid sequences of the model coiled coil peptides

†FZ	Ac-K	FEALEGK	FEALEGK	FEALEGK	FEALEGK–FEALEG–CONH₂
LZ	Ac-E	YEALEKK	LAALEAK	LQALEKK	LEALEHG–CONH₂
LZ12A	Ac-E	YEALEKK	LAAAEAK	LQALEKK	LEALEHG–CONH₂
LZ16A	Ac-E	YEALEKK	LAALEAK	AQALEKK	LEALEHG–CONH₂
LZ16N	Ac-E	YEALEKK	LAALEAK	NQALEKK	LEALEHG–CONH₂
LZVNV	Ac-E	YEALEKK	VAALEAK	NQALEKK	VEALEHG–CONH₂
LZVVV	Ac-E	YEALEKK	VAALEAK	VQALEKK	VEALEHG–CONH₂
Acid-LZ	Ac-E	YQALEKE	VAQLEAE	NQALEKE	VAQLEHE–CONH₂
Base-LZ	Ac-E	YQALKKK	VAQLKAK	NQALKKK VAQLKHK G–CONH₂	
GCN4-p1	Ac-R	MKQLEDKVEELLSK	NYHLENE	VARLKKL VGER–CONH₂	

† The single letter amino acid code is used. Residues at **a** and **d** positions are shown in bold type, starting with the first **a** residue. FZ, "phenylalanine zipper", LZ, "leucine zipper". Variations in the LZ sequence are indicated either by the residue number followed by the substitution, or by indicating the residues at the **a** positions in the final three heptads.
All the peptides were synthesized as the *N*-acetyl, *C*-terminal amide.

composition or calculated from the results of amino acid analysis. In general, peptides were either dialysed or desalted prior to physical studies.

Circular dichroism

CD spectra were measured on a Jasco J-500 polarimeter equipped with a thermostatted cell holder. Cells of various pathlengths were used and samples were scanned at a rate of $1 \text{ nm} \cdot \text{min}^{-1}$. Samples for the measurement of the concentration dependence of the CD signal were made by the serial dilution of a stock solution.

Sedimentation equilibrium

The apparent molecular weights of the synthetic model peptides and that of GCN4-p1 were determined by conventional sedimentation equilibrium in a Beckman XLA instrument, equipped with UV absorbance optics. Samples were run at 20 °C in aqueous 0.01 M sodium phosphate/0.09 M sodium chloride (or fluoride) over a range of concentrations of between 7 & 332 μM, and at a variety of operating speeds. Standard 12 mm double sector KelF and charcoal-filled epon centrepieces with sample column heights of 3–4 mm were used. At wavelengths of 230 nm and below corrections were made for baseline effects by running buffer *vs* buffer blanks under conditions identical to those used for the corresponding experiment. Approach to equilibrium studies indicated that equilibrium was attained within about 10 h and data collection was normally performed after a total of 20 h. At least 20 radial scans were averaged for each sample. Partial specific volumes and solvent densities were calculated in the usual manner, using the data of Laue et al. [19]. In order to provide an overview of the apparent molecular weights, whole-cell

data were fitted to the ideal single species model, using software provided by the instrument manufacturer.

Results

Choice of sequences

The peptide FZ was a modification of Hodges' extensively characterized model of tropomyosin [12, 17, 20–23] in which phenylanine replaced leucine at every *a* position. FZ was initially designed for the study of protein-protein recognition functions and for the investigation of template-directed self-assembly in peptides. The LZ sequences were based on a peptide designed by O'Neil and DeGrado [24] to test the helix forming propensities of naturally occurring amino acids in a "host-guest" approach. This peptide was itself based on the idealized model tropomyosin sequence (above), changes being made in the original sequence to enhance helix formation. One version, with serine in the guest position, termed coil-ser, has been shown crystallographically to form a three-stranded coiled coil [9]. Several variants of the basic LZ peptide in which single or multiple substitutions were made in *a* and *d* positions were also synthesized. The properties and folding kinetics of the LZ system, which was originally developed for studies on peptide-antibody recognition, have been thoroughly characterized [25, 26]. The peptides Acid-LZ and Base-LZ, which carry net charges of -7 and $+9$, respectively, at neutral pH, originated from a model of the sequences of the basic zipper transcription factors c-fos and c-jun that preferentially form heterodimeric pairs due to electrostatic repulsion in potential homomeric forms [27]. Such systems have been characterized in several investigations [13, 28, 29] and provide an ideal system in which to study two-stranded leucine zippers. As a further example of a dimeric system and as a control, the sequence of GCN4-p1 was also synthesized.

Circular dichroism

Representative CD spectra are shown in Fig. 2. FZ and all of the LZ series peptides studied had CD spectra of the form and intensity of that shown for LZ, with the exceptions of Acid-LZ, Base-LZ and an equimolar mixture of these two peptides. Standard methods of analysis indicate, with the exceptions noted above, that all these peptides are fully α-helical in aqueous buffer at room temperature at peptide concentrations above $\approx 20\ \mu M$. Peptides LZ16A, LZ16N and LZ12A all exhibited a concentration-dependent CD spectrum, a sharp decrease in ellipticity at 222 nm being seen at concentrations of below $\approx 10\ \mu M$. This is illustrated for the case of LZ16A in the inset to Fig. 2. The thermodynamic stabilities of LZ12A, LZ16A, and LZ16N determined from the CD data were all approximately equal. The CD spectra of LZ and LZVVV showed no concentration dependent behavior in the concentration range amenable to CD measurements. Studies on the urea-induced denaturation of these two peptides showed that denaturant concentrations in excess of 5 M were required to induce unfolding of LZ at peptide concentrations of the order of $2\ \mu M$, whereas LZVVV was not fully denatured by urea concentrations as high as 9 M. These results indicate that the α-helical structures formed by these peptides are extremely stable [25]. In isolation and at low peptide concentration, the peptides Acid-LZ and Base-LZ both had CD spectra indicative of a random coil structure, although there was some evidence of ordered structure in the Base-LZ spectrum at higher concentrations. The CD spectrum of an equimolar mixture of the two peptides was again reminiscent of that of an α-helix, but was lower in intensity in the 210–220 nm region of the spectrum than that of the other peptides studied, even at high peptide concentrations.

Analytical ultracentrifugation studies

Sedimentation equilibrium experiments at a variety of operational speeds and peptide concentration were performed on all the peptides studied. Typical data, the fit to it, along with the residuals to the fit are shown in Fig. 3. In order to provide an objective overview, raw data were fitted to a model that assumed that a single, ideal species was present and the statistically best fit was selected in each case. The results of this analysis for the apparent weight-average molecular mass are given in Table 2. The values obtained fell into four broad categories. Peptides Acid-LZ and Base-LZ were apparently monomeric even at relatively high peptide concentration. The slight underestimation of the mass of these two peptides, if significant may be due to either primary or secondary charge effects

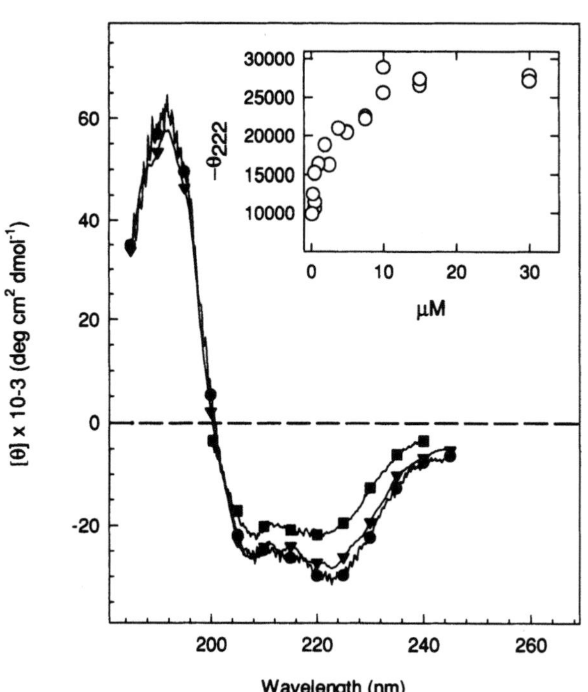

Fig. 2 The far-UV CD spectra of (●) LZ, (▼) LZ16A and (■) Acid-LZ/Base-LZ. Peptide concentration 27 μM, 100 mM phosphate buffer, pH 7.4. In the inset the dependence of the ellipticity on peptide concentration for LZ16A is shown

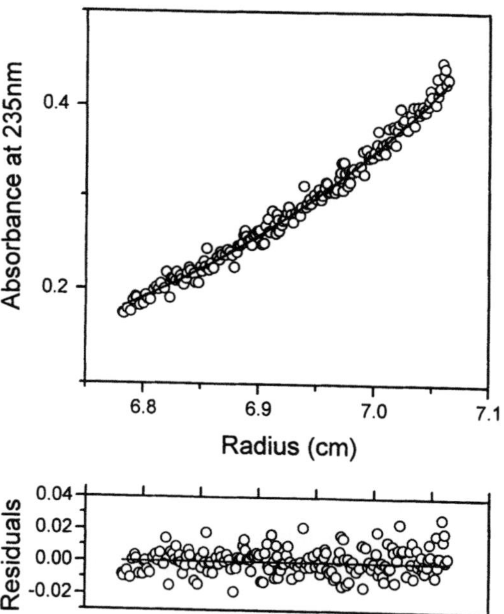

Fig. 3 Representative sedimentation equilibrium data. Peptide LZ, 33 μM, in 0.01 M sodium phosphate, 0.09 M sodium chloride, pH 7.0, 20 °C, 21 000 rpm. The raw data and the fit to it are shown in A while the residuals to the fit are shown in B

Table 2 Apparent molecular mass from sedimentation equilibrium analysis

Peptide	Conc[μM]	$\dagger M_{w,app}$[Da]	$M_{mon,c}$[Da]
LZ	33	9550 ± 180	3308
	10	9650 ± 264	
FZ	70	10770 ± 120	3933
	35	10680 ± 235	
	7	10650 ± 480	
LZ12A	100	10300 ± 240	3266
	50	11050 ± 593	
LZVVV	100	9990 ± 108	3266
LZ16A	100	8690 ± 140	3266
	50	9030 ± 106	
	20	7960 ± 84	
	11	7370 ± 100	
	7	6450 ± 301	
LZVNV	100	7420 ± 113	3281
LZ16N	100	7120 ± 99	3309
GCN4-p1	100	7250 ± 280	4038
Acid-LZ/Base-LZ	332	5510 ± 84	3407
	66	5710 ± 103	
Acid-LZ	100	3220 ± 57	3411
Base-LZ	100	3110 ± 129	3403

$\dagger$ $M_{w,app}$ values were calculated by fitting data to a single ideal species model and are shown ± S.E. $M_{mon,c}$ are the monomer masses calculated from the amino acid composition.

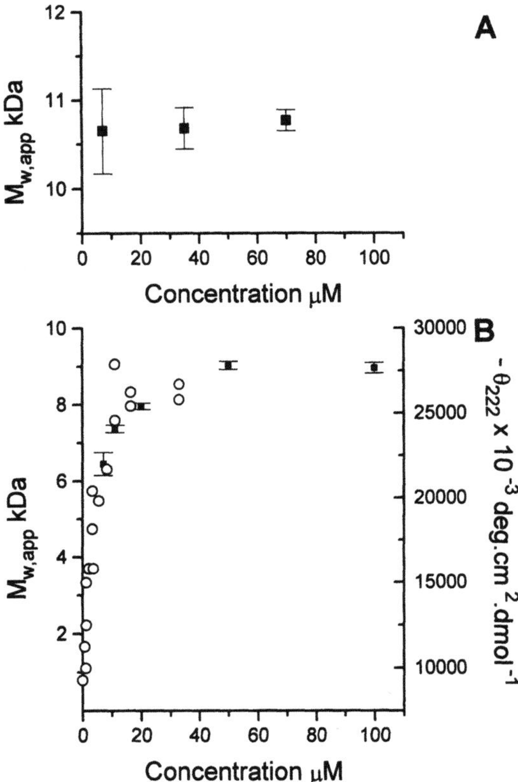

Fig. 4 A The concentration dependence of $M_{w,app}$ of FZ and B the concentration dependence of $M_{w,app}$ (■) and θ_{222} (○) for LZ16A

or both. All of the peptides in this study had a relatively high proportion of charged amino acids, although any effect would be expected to be most pronounced in these two cases. The second category comprised those systems that were apparently predominantly dimeric -GCN4-p1 and the Acid-LZ/Base-LZ mixture. This was the expected result; GCN4-p1 forms a dimer in the crystal and the Acid-LZ/Base-LZ system was predicted to form hetero-dimers, or possibly multimers of heterodimers, on the basis of charge repulsion effects. The deviation from an ideal dimer mass in both cases may indicate that monomer was present, in equilibrium, at these peptide concentrations. The weight-average masses of the third group, LZVNV and LZ16N, suggested that, while the dominant species present was probably dimer, these two peptides were also capable of forming higher-order structures. Systematic studies provided no evidence for the presence of non-specific aggregates. The fourth category, FZ, LZ and LZVVV were all clearly trimeric. There was no concentration-dependent variation in apparent mass for any of these peptides (Fig. 4A), although at high operating speed (40000 rpm) and low concentration (< 10 μM) residuals to the fit to data for peptide FZ began to show a systematic trend, indicative of the presence of material of lower mass. Peptide LZ16A seemed to fall between the two latter categories. At peptide concentration > 50 μM the majority of the material was apparently trimeric. However, at

lower concentrations the apparent mass fell, indicating that the system was dissociating. In this case the fits were statistically better at the extremes of the concentration range studied than at intermediate concentrations and there were again speed-dependent effects. The change in apparent mass mirrors the change in ellipticity with concentration for this peptide although the two effects do not appear to be fully superimposable. On the basis of the CD results discussed above, LZ12A presumably also falls into this category although it was fully trimeric at the concentrations measured.

Discussion

The origins of these studies, peptide-peptide and peptide-antibody recognition and peptide self assembly, were diverse but all started out under the assumption that the peptides that were being designed and synthesized would form stable dimeric structures. Contrary to this expectation and to the principles involved in the design process of the model sequences, all the peptides with the exceptions of Acid-LZ, Base-LZ and GCN4-p1, apparently tended to form trimeric structures that were stable to a greater or

Progr Colloid Polym Sci (1995) 99:24–30
© Steinkopff Verlag 1995

lesser degree and were all fully α-helical in aqueous solution at moderate peptide concentrations. It has been shown, in the case of the LZ-series of oligomers, that the coiled coil formed is in-register and, on the basis of fluorescence quenching studies, in a fully parallel configuration [25] and similar studies on FZ are in progress. The organization of LZ contrasts with the situation found in coil-ser, where the polarity is parallel-parallel-antiparallel due, probably, to side-chain packing effects.

In general, small, isolated α-helices are not stable in aqueous solution. Exceptions to this rule have been provided by the investigations of Baldwin and coworkers who have shown that it is possible to produce small peptides with this structure [30], although the design principles differ somewhat to those used in the construction of the model coiled coil sequences. It is generally accepted that the coiled coil sequences will only form ordered structures as a result of mutually stabilizing interactions among assemblies of monomers. In this sense the CD spectra are diagnostic for the coiled coil state. The concentration dependence of both CD intensity and molecular mass for the LZ16A peptide therefore raises questions as to the relative population of the available states (monomer, dimer, trimer) at equilibrium at low peptide concentration. The apparent non-coincidence of the dissociation process in this peptide when monitored by CD and by sedimentation equilibrium, if it can be shown to be significant, would seem to indicate that the dimeric state is populated. This observation makes the assumption that the ideal CD spectra for fully helical dimers and trimers are effectively identical and that, therefore, changes in the CD will only occur when there is a measurable contribution to the spectrum by the randomly coiled monomer. A similar view has been expressed as the result of studies on the dimer-tetramer system found in the tropomyosin analogue AcTM43 [15]. Armed with a knowledge of the stoichiometry of the associated state in these peptides, we are currently conducting a systematic series of experiments to test this assumption. A monomer-dimer-trimer system has been reported for coil-ser and its analogues coil-trp and coil-D-ala [9]. If this turns out to be a realistic scheme for the LZ system it may be that the trimeric state of the peptides LZVNV and LZ16N is only poorly stabilized relative to the dimer. The presence of higher-order association (eg dimer-of-dimers) in these peptides cannot be ruled out, although it seems unlikely as the closely related analogues LZ, LZ12A, LZVVV and LZ16A do not appear to adopt an oligomerization state of greater than three. Evidence for the presence of a relatively small amount of lighter material at equilibrium with an apparent trimer was obtained for the FZ peptide at low concentration. A reexamination of the sedimentation equilibrium behaviour of this peptide is to be made using longer columns and a wider range of operational speeds, with the aim of identifying accurately the nature of the smaller species involved.

GCN4-p1 was included in this series of investigation primarily as a control. The apparent molecular mass obtained agrees well with published data and confirms that the dominant dimeric state is in equilibrium with monomer [18, 31]. The mixed Acid-LZ/Base-LZ system associates to form a dimer. However, dimerization is incomplete even at a peptide concentration of 332 μM. The CD spectrum indicates a reduced helical component indicating that either the peptide is not fully helical but is fully associated or that association is incomplete and that unordered monomer is also present. The latter seems more probable on a number of grounds, one of which is the low apparent molecular mass. Why these peptides remain incompletely associated even at high concentration is still an open question.

Considerable attention has been focussed on the nature and strength of the stabilizing interactions and on the specificity of association in coiled coils. The primary targets of these investigations have been the hydrophobic core of the molecules, made up of residues at the a and d positions, and the electrostatic interaction between the residues at the e and g positions. Studies in which residues at these four positions have been systematically and/or combinatorially altered have shown that the oligomerization state of the coiled can be modulated by changes in any of these positions. Variations in the distribution of β-branched and non-β-branched residues in the hydrophobic core lead to the formation of either dimers, trimers or tetramers [5, 10, 12], while replacement of the negative charge at the e position by electrically neutral alanine causes a switch from dimer to tetramer formation [16]. The contribution of the present study can then be summarized. i) The sequence of FZ only differs from its parent peptide [12] by having the aromatic residue phenylalanine at all a positions. The parent peptide has leucine at all a and d positions and is reported to be dimeric. The parent peptide has leucine at all a and d positions and is reported to be dimeric. Substitution of all leucines at a positions with phenylalanine therefore causes an apparent shift from a dimeric to a trimeric configuration. However, LZ, which has a slightly different sequence but which does have leucine residues at three out of four a positions is also trimeric. ii) Substitution of leucine at a positions in LZ with valine has no effect on the trimeric state of this peptide. iii) Single substitutions can cause trimer to dimer shifts that are sensitive to the nature of the residue substituted. The last point is demonstrated by the facts that an alanine for leucine mutation at an a position in LZ16A considerably reduces the stability of the trimer and, more impressively, the substitution of asparagine for leucine or valine at the central *a* position (LZ16N and LZVNV) results in the dimer becoming the predominantly occupied state. This confirms the importance of the equivalent asparagine residue in maintaining the dimeric state in the naturally-occurring GCN4 coiled coil [8, 10], presumably by destabilizing higher-order oligomers.

There have been few reports of association constants derived from sedimentation equilibrium measurements on coiled coils [15, 16]. In order to determine values of this kind with any certainty requires an accurate model of the equilibrating system. Now that some clear information has been obtained on the FZ and LZ peptides the nature and magnitude of the associative processes are to be investigated in detail. In turn, these peptides will serve as relatively simple, well-defined and easily modulated systems for the modeling of self-association phenomena in general.

Note added in Proof Since this paper was submitted a reappraisal of Hodges' TM-43 peptide [12] has been published (Betz S, Fairman R, O'Neil K, Lear J, DeGrado W (1995) Phil Trans R Soc Lond B 348:81–88). On the basis of new experimental evidence, including determination of the partial specific volume by the $H_2O–D_2O$ difference method, this peptide is now also reported to exist predominantly as a trimer.

References

1. McLachlan AD, Stewart M (1975) J Mol Biol 98:293–304
2. Hope IA, Struhl K (1987) EMBO J 6:2781–2784
3. Landschulz WH, Johnson PF, McKnight SL (1988) Science 240:1759–1764
4. Crick FHC (1953) Acta Cryst 6:689–697
5. Adamson JG, Zhou NE, Hodges RS (1993) Curr Opin Biotech 4:428–437
6. Cohen C, Parry DAD (1990) Proteins: Struct Func Genet 7:1–153
7. Cohen C, Parry DAD (1994) Science 263:488–489
8. O'Shea EK, Klemm JD, Kim PS, Alber T (1991) Science, 254:539–544
9. Lovejoy B, Choe S, Cascio D, McRorie DK, DeGrado WF, Eisenberg D (1993) Science 259:1288–1293
10. Harbury PB, Zhang T, Kim PS, Alber T (1993) Science 262:1401–1407
11. Lupas A, Van Dyke M, Stock J (1991) Science 252:1162–1164
12. Lau SYM, Taneja AK, Hodges RS (1984) J Biol Chem 259:13253–13261
13. O'Shea EK, Rutkowski R, Stafford WF, Kim PS (1989) Science 245:646–648
14. Peterandl R, Nelson HCM (1992) Biochemistry 31:12272–12276
15. Greenfield NJ, Hitchcock-DeGregori SE (1993) Protein Sci 2:1263–1273
16. Krylov D, Mikhailenko I, Vinson C (1994) EMBO J 13:2849–2861
17. Zhu B-Y, Zhou NE, Kay CM, Hodges RS (1993) Prot Sci 2:383–394
18. O'Shea EK, Rutkowski R, Kim PS (1989) Science 243:538–541
19. Laue TM, Shah BD, Ridgeway TM, Pelletier SL (1992) In: Harding S, Rowe A, Horton J (eds) Analytical Ultracentrifugation in Biochemistry and Polymer Science. Royal Society of Chemistry, Cambridge, pp 90–125
20. Hodges RS, Semchuk PD, Taneja AK, Kay CM, Parker JMR, Mant CT (1988) Peptide Res 1:19–30
21. Zhou NE, Kay CM, Hodges RS (1992) J Biol Chem 267:2664–2670
22. Hodges RS, Saund AK, Chong PCS, St-Pierre SA, Reid RE (1981) J Biol Chem 256:1214–1224
23. Zhou NE, Zhu B-Y, Kay CM, Hodges RS (1992) Biopolymers 32:419–426
24. O'Neil KT, DeGrado WF (1990) Science 250:646–651
25. Wendt H, Berger C, Baici A, Thomas RM, Bosshard HR (1995) Biochemistry, in press
26. Wendt H, Baici A, Bosshard HR (1994) J Am Chem Soc 116:6973–6974
27. O'Shea EK, Lumb KJ, Kim PS (1993) Curr Biol 3:658–667
28. O'Shea EK, Rutkowski R, Kim PS (1992) Cell 68:699–708
29. Graddis TJ, Myszka DG, Chaiken IM (1993) Biochemistry 32:12664–12671
30. Marqusee S, Baldwin RL (1987) Proc Natl Acad Sci 84:8898–8902
31. Saudek V, Pastore A, Morelli MAC, Frank R, Gausepohl H, Gibson T, Weih F, Roesch P (1990) Prot Eng 4:3–10

Progr Colloid Polym Sci (1995) 99:31–38
© Steinkopff Verlag 1995

A. Seifert
K.D. Schwenke

Improved approach for characterizing the coalescence stability of legumin stabilized O/W emulsions by analytical ultracentrifugation

Received: 6 April 1995
Accepted: 5 June 1995

This paper is based on a poster presented at the IX. Symposium on Analytical Ultracentrifugation, Berlin-Buch, Germany, March 2/3, 1995

Dr. A. Seifert (✉) · K.D. Schwenke
Research Group Plant Protein
Chemistry (WIP)
University of Potsdam
Arthur-Scheunert-Allee 114-116
14558 Bergholz-Rehbrücke, Germany

Abstract The coalescence stability of n-decane-in-water emulsions stabilized by native and highly (>90%) acetylated legumin, the 11 S main storage protein from faba beans (**Vicia faba L.**) has been investigated by analytical ultracentrifugation. The method is based on centrifugation at constant rotor speeds (sufficiently high to produce a detectable amount of separated oil) up to (pseudo-) equilibrium values for all layers. Coalescence pressures are calculated as a measure of coalescence stability.

Acetylation of legumin results in a significant increase of the coalescence stability (about three-fold). The coalescence pressure of the emulsion with the native legumin decreases after a storage period of about 2 months. By contrast, the emulsion prepared from the acetylated legumin shows a higher coalescence pressure, even after about 8 months. This observation is in agreement with the storage stability.

Initial results from applying sedimentation velocity experiments in connection with the method for characterizing the coalescence stability have provided the following results: i) only the 11–12 S protein component in the starting native legumin solution occurred in the separated continuous phase after (incomplete) demulsification. A 15 S aggregate (11%) could no longer be detected in the separated continuous phase. ii) The 16–17 S aggregation component and a lower molecular weight dissociation product (≤ 2 S, 25–30%) present in the starting solution of acetylated legumin did not occur in the separated continuous phase, but only an aggregated 22–26 S component. This supports the conclusion for a preferred adsorption of low molecular components during preparation of the emulsion.

Key words Coalescence stability – analytical ultracentrifugation – O/W emulsions – proteins – legumin

Introduction

Plant protein isolates from grain legumes or oilseeds have been shown to be effective stabilizers of food emulsions [1]. Typical protein isolates from legume seeds are composed of two oligomeric storage proteins, a 300–360 kDa 11 S globular protein and a 150–210 kDa 7 S globulin; these are respectively named legumin and vicilin, for the case of pea and faba bean [2]. Furthermore, as shown for faba bean protein isolates, non-protein components such as lipids and polysaccharides are also present which can influence the surface functional properties of the proteins [3].

Chemical modification such as succinylation or acetylation has proved to be a powerful tool for effecting the emulsifying and foaming properties [4–7]. An increased coalescence stability of oil-in-water emulsions stabilized by faba bean protein isolates have, for example, been reported [8]. Detailed studies of such emulsions provided some hints for a preferred adsorption of some components of these isolates at oil droplets and the special role of lipid components in stabilizing surface films [3, 9]. In order to exclude the influence of different protein and non-protein components that are present in isolates and to explain the changed emulsion stabilizing properties of an acetylated faba bean protein on the basis of the physico-chemical changes induced by this chemical modification, the present study deals with isolated legumin in both the native and highly acetylated state.

Centrifugation is a widely used method for determining coalescence stability of emulsions. The use of an analytical ultracentrifuge (AUC) is particularly important as a fundamental probe onto such phenomena [10]. This is primarily because the optical system of the AUC provides a direct mean of recording phenomena occurring in the emulsions *during* the centrifugation process.

The aim of this work is to examine the influence of native and acetylated legumin on the coalescence stability of O/W emulsions using analytical ultracentrifugation. A further objective is to obtain additional information on the hydrodynamic properties and possible conformation of acetylated legumin and the role of legumin components in emulsification and demulsification using sedimentation velocity experiments. We also examine the components involved in the desorption of the proteins in demulsification and, indirectly, on adsorption phenomena in the preparation of the emulsions.

Experimental

Preparation and modification of the proteins

The legumin was obtained from beans of the variety (*Vicia faba L.* minor, var. "Fribo") by a combined salt fractionation and isoelectric precipitation according to Popello et al. [11] and as described in ref. [12]. It was found to contain ≤3% vicilin. Legumin acetylation was performed according to ref. [13]. The degree of acetylation of amino groups (>90%) was determined using a modified TNBS method [14]. All samples were dialyzed against deionized water and freeze-dried. The freeze-dried legumin samples used for sedimentation velocity experiments and preparation of the emulsions were dissolved in the same phosphate buffer solution (5.852 g K_2HPO_4,

0.871 g KH_2PO_4, 5.844 g NaCl and 0.200 g NaN_3 (p.a., Merck, Germany) per litre sol.; pH = 7.6, $I = 0.3$). The buffer was prepared with analytical grade water obtained by steam distillation of deionized water from potassium permanganate. In order to ensure the same solution conditions in the sedimentation velocity experiments and for the preparation of the emulsions no further dialysis followed. The protein content of all aqueous phases was determined using a modified micro-Biuret-method [15] employing a nitrogen to protein conversion factor of 5.7.

Sedimentation velocity investigations of the proteins

In sedimentation velocity experiments with single- and partly double-sector cells at 30 000, 35 000 and 40 000 r.p.m. using the Philpot–Svensson–Schlieren-(refractive index gradient) optics, the rate of movement of the maximum of the Schlieren-peaks was measured to obtain the sedimentation coefficients of the non-modified and starting acetylated legumin samples. This technique was also used for investigating the presence of possible impurities, and dissociation products of the proteins in the usual way [16]. For the velocity runs an initial low speed of 2000 r.p.m. was used to detect the presence of only high molecular aggregates.

Measurements were performed on a Model 3170-B analytical ultracentrifuge (Hungarian Optical Works, MOM, Budapest). All measurements were performed at 20 °C.

The optical Schlieren records were photographed and sedimentation coefficients were evaluated from the negatives directly (without making positive enlargements) i) on a magnification equipment (Carl Zeiss Jena, magnification 20-fold) or/and ii) a photographic enlarger projecting the image onto a graphics digitizing tablet using partly a new computer program for the evaluation of sedimentation velocity experiments with variable time intervals written by Cölfen [17].

Cell-loading concentrations of 3.34, 5.02 and 10.03 mg/ml and 3.85, 5.78 and 11.56 mg/ml of the native and highly acetylated legumin samples were used respectively in the phosphate buffer described.

Infinite dilution sedimentation coefficients ($s^{\circ}_{20,\,buffer}$) were attained by extrapolation using the well known equation [16]

$$1/s_{20,\,buffer} = 1/s^{\circ}_{20,\,buffer}\,(1 + k_S c)\,. \tag{1}$$

The values were not corrected to standard conditions in order to allow a direct comparison between these data and the sedimentation velocity data of the proteins in the continuous phase separated from the emulsions.

Emulsions

n-decane (Merck, for synthesis, GC > 99%, relative density = 0.729–0.730) was dispersed in protein stock solutions (protein concentration: 9.97 mg/ml and 10.00 mg/ml for the non-modified and acetylated legumin samples, respectively) with an "Ultra-Turrax" dispersing apparatus with revolution counter (Janke & Kunkel, IKA-Labortechnik, Staufen, FRG). All the emulsions were prepared by three emulsification runs, each for 1 min at 20 000 r.p.m. followed by a 1-min pause using the dispersing head S25-10G (diameter of the stator 10 mm, rotor 7.5 mm, number of teeth 8 (stator) and 2 (rotor)). The "oil" was dispersed in the protein solution (total volume in each case 5 ml) in a 10 ml lockable bottle suitable for the stator diameter choosed. The temperature of the sample during dispersing did not exceed 30 °C. The phase volume ratio (oil/water) was 70/30–30/70 and 60/40–40/60 for native and acetylated legumin, respectively.

The droplet sizes of some emulsion samples were determined by light microscopy using an automatic imaging system CUE 3 (Olympus, FRG).

The stability of the emulsions stored at temperatures of 20–28 °C was judged visually in parallel to the ultra-centrifuge investigations.

Investigation of the coalescence stability of the emulsions by analytical ultracentrifugation

The emulsions (phase volume ratio $\geqslant$ 40/60) were investigated for coalescence stability according to the "final value method" [10]. The method is based on centrifugation at constant rotor speed (sufficiently high to produce a detectable amount of separated dispersed phase) until a (pseudo-) equilibrium state was reached, for all layers. The method includes in its precise form an exact mass balance of the layers formed based on weight, density measurements and determination of the volumes of the layers with the help of a distance vs. volume calibration curve for the centrepieces in the ultracentrifuge cells. Further, the concentration of the components in the layers should be analyzed. For many practical applications it is possible to simplify the method without affecting the validity of the mean results.

The investigations of the coalescence stability were carried out with 3° and 4° single sector-shaped centre-pieces of 4–12 mm optical light path at 20 °C after various storage periods.

Coalescence pressures P_c were calculated as a measure of coalescence stability of the emulsions. The pressure is cal-

culated for sector-shaped centerpieces according to [10]

$$|P_c| = \tfrac{1}{2} \cdot \omega^2 \cdot \delta\rho \cdot \phi_{dp} \cdot (r_u^2 - r_0^2) \,, \tag{2}$$

where ω = angular velocity (rad/s), $\delta\rho$ = difference of densities of dispersed phase (ρ_{dp}) and continuous phase (ρ_{cp}) (kg/m^3), ϕ_{dp} = volume fraction of the dispersed phase in the emulsion layer, r_u, r_0 = distances from the lower (r_u) and upper (r_0) boundary of the emulsion layer to the rotation centre (m).

Investigation of the protein components in the continuous phase separated from the emulsions during centrifugation

The sedimentation behavior of the protein in the continuous (aqueous) phase separated from the emulsion samples at rotor speeds above and below the critical speed for oil separation was investigated in addition to the hydrodynamic characterization of the parent legumin samples. This included the use of sedimentation velocity experiments in the usual way to obtain information about the number of protein components, dissociation phenomena, impurities and sedimentation coefficients.

Results and discussion

Sedimentation velocity investigation of the proteins

The unmodified legumin preparation shows two clear sedimenting boundaries (Fig. 1): the major component sediments at $s^\circ_{20,\,buffer} = 12.4$ S, but there is also a minor, faster component sedimenting at $s_{20,\,buffer} \approx 18$ S. These boundaries represent legumin and an aggregation product of the protein. The concentration ratio of the 12 S and 18 S components amounts to $\sim 8:1$. High and low molecular weight impurities were not present.

The acetylated legumin sample shows a single sedimenting Schlieren peak with $s^\circ_{20,\,buffer} = 16.5$ S as main fraction (Fig. 2). A second component of lower molecular weight separates slowly from the meniscus (Fig. 2). Further runs with samples of the identical solution at higher speeds (50 000–55 000 r.p.m.) gave the same result. Corresponding to its low sedimentation coefficient, which was estimated to be ≤ 2 S, this component can be assumed to represent dissociated subunits of modified legumin. The concentration ratio of the 16 S and ≤ 2 S components can be estimated to be $\sim 3–4:1$. These results are in agreement with our former results on protein isolates [18]. There are, however, characteristic differences between the dissociation of acetylated isolates and acetylated legumin. While the high molecular weight (11–12 S) component of the isolates disappeared completely after

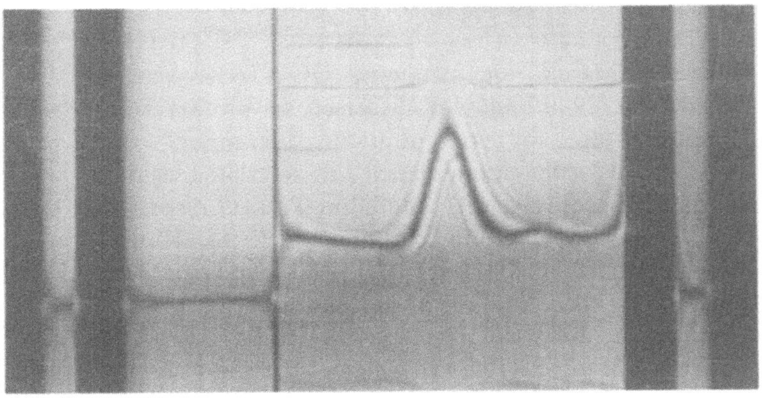

Fig. 1 Sedimentation velocity diagram of native legumin (5.02 mg/ml, phosphate buffer, pH 7.6) at 40 000 r.p.m. after 56 min of centrifugation at 20 °C

Fig. 2 Sedimentation velocity diagram of (>90%) acetylated legumin (5.78 mg/ml upper and 3.85 mg/ml lower profile) in phosphate buffer (pH 7.6) at 35 000 r.p.m. after 20 min of centrifugation at 20 °C. The sample had previously been centrifuged at 30 000 r.p.m. for 26 min

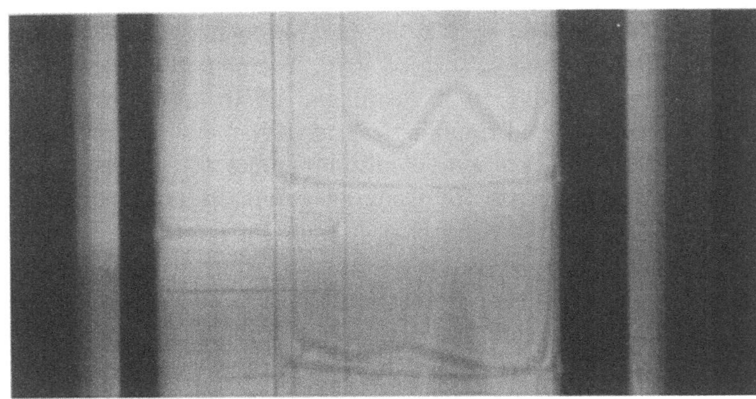

exhaustive acetylation [18] this component remained as the main fraction in legumin. This incomplete dissociation of acetylated faba bean legumin is in contrast to the corresponding homologous glycinin from soybean [19] which dissociates completely after acetylation. It is also in contrast to the succinylated legumin from pea which dissociates completely into the 3 S subunit [20]. Moreover, the increased sedimentation coefficient of the main component of acetylated legumin (from 12.4 S to 16.5 S) may indicate an aggregation of the modified legumin due to hydrophobic interactions since the modification leads to a significant increase of hydrophobicity [21]. Obviously, acetylation of legumin did not induce an unfolding of the protein as in the case of succinylation [20].

The differences in the physico-chemical properties of the acetylated protein isolate which was completely dissociated and was in a rather unfolded state [22] and the acetylated legumin may be due to structural differences of the legumin in both unmodified protein preparations. In contrast to the native legumin, the proteins in the isolate were in a non-native state which was easier to modify and to dissociate [22].

General behavior of the emulsions

All emulsions were coalescence-stable under gravity and suitable concerning their consistency and stability for filling the measuring cells with an injection syringe. In the case of the emulsion with a phase volume ratio dispersed/continuous phase of 30/70 it was difficult to introduce an aliquot of the emulsion sample in the cell because of its relatively quickly creaming under gravity. For this reason emulsions with phase volume ratios $\geqslant$ 40/60 were investigated.

The droplet sizes of the emulsions ranged from 1 to 15 μm (light microscopy). A typically average droplet size was 7 μm (acetylated sample).

The emulsion samples containing native or acetylated legumin showed oil separation at a relative centrifugal force of 2000–45 000 g (5000–25 000 r.p.m.) and 120 000–230 000 g (40 000–55 000 r.p.m.), respectively, depending on the storage period.

In comparison to emulsions stabilized by low molecular surfactants there were difficulties in identification of the phase boundaries in the legumin stabilized emulsions because of protein residues at the cell windows.

All the emulsions investigated met the condition for a sufficient constancy of the final values of the layers according to the method employed. Furthermore, no visible change occurred in the phase boundaries of the emulsions after relatively long centrifugation times below the critical rotor speed necessary for oil separation.

The emulsions both with the native and acetylated legumin formed a transparent emulsion layer under the action of a centrifugal field between the non-transparent (turbid) remaining emulsion layer and the separated oil layer.

The reversibility of this phenomenon, i.e., the formation of the transparent emulsion layer under the action of the centrifugal field and its disappearance after the run has been stopped (or the rotor speed has been decreased) has been proved for the first time on an AUC also in protein stabilized systems.

The existence of a mechanical barrier between the transparent and non-transparent emulsion layer that has previously been proved with SDS stabilized emulsions [10] could be confirmed for these emulsion systems. The phenomenon of the transparent emulsion investigated by analytical ultracentrifugation is discussed in more detail in ref. [23].

Coalescence pressures

The values of coalescence pressure calculated according to Eq. (2) are presented in Table 1. Exhaustive acetylation of legumin result in a significant increase of the coalescence stability (about three-fold). The coalescence pressure of the emulsion with the native legumin decreases after a storage period of about 2 months. By contrast, the emulsion prepared from the acetylated legumin shows a higher coalescence pressure even after about 8 months.

These findings are in agreement with the storage stability. The legumin containing emulsion was stable at least over a period of 2 months. Oil droplets in the cream

were observed after 70 and 124 days (of visual examination). At later times separation of an oil layer took place. The emulsion sample containing the acetylated legumin was stable for more than 8 months.

These examples substantiate the view that changes in the emulsion structure can be reasonably indicated by the coalescence pressures determined.

The values obtained for the coalescence pressures of the emulsions containing native and acetylated legumin are in agreement with those of emulsions stabilized by sodium dodecyl sulphate, ethoxylated alkyl phenols and mixtures of ethoxylated fatty amine/nonyl phenol [10]. Of course, it never will be possible to predict the stability of an emulsion on the day of preparation; however the investigation of its stability over a reasonable period can nonetheless give a valuable estimation of its further stability.

The results obtained with the coalescence pressures are supported by the values of the relative centrifugal force necessary for separating oil. With comparable experimental conditions these values allow an estimation of coalescence stability (cf. Tables 2 and 3).

Investigation of the protein components in the continuous phase separated from the emulsions by analytical ultracentrifugation

All the emulsions stabilized by the non-modified legumin (Table 2, samples 1–5) formed a Schlieren profile at the emulsion layer-aqueous phase boundary in centrifugation at rotor speeds below that necessary for oil separation. After increasing the rotor speed to the critical level for separating oil a more or less large and broad (but single) peak sedimented (see Fig. 3). The peak is due to the protein dissolved in the aqueous phase and that released by demulsification. The protein did not dissociate. The time of centrifugation (30–325 min) at rotor speeds below that necessary for oil separation has no influence on the apparent sedimentation coefficients of the protein component in the separated continuous phase ($s_{20,scp}$) indicating an 11–12 S component (Table 2). The emulsion sample 6 (Table 2) was centrifuged immediately at a speed higher than the critical value of 25 000 r.p.m.: an 11–12 S component is indicated also in this sample. This result means that the protein components dissolved in the separated continuous phase, in the continuous phase of the remaining emulsion and adsorbed at the oil droplets are not influenced by the centrifugal stress itself during centrifugation before and after demulsification.

After the centrifugation of emulsions at a speed just sufficient for oil separation a stepwise increase of the rotor speed to 30 000, 40 000 and 50 000 r.p.m. (after the

Table 1 Characterization of the coalescence stability of *n*-decane-in-water emulsions by analytical ultracentrifugation. (Phase volume ratio of the emulsions: dispersed/continuous phase = 40/60)

Emulsion	Legumin	Rotor sp./ r.p.m.	Coalescence pressure/ $Pa \times 10^{-5}$	Storage period emulsion/days
1	native	10 000	2.68	14
2		25 000	3.10	55
3		25 000	1.13	70
4	>90% acetylated	50 000	8.68	15
5		50 000	9.02	44
6		30 000	1.78	250

Table 2 Demulsification of native legumin stabilized *n*-decane-in-water emulsions by analytical ultracentrifugation. Centrifugation conditions before oil separation, rotor speed of oil separation and in determining the apparent sedimentation coefficient $s_{20,scp}$ in the separated continuous phase

Emulsion	Centrifug. below the speed of oil sep. Rotor speed/r.p.m. (time/min)	Oil separation Rotor speed/ r.p.m.	Centrifug. after oil sep. Rotor speed/ r.p.m.	App. sediment. coefficient $s_{20,scp}$/(S.D./) sec $\times 10^{13}$
1	2000–10 000 (225)	> 10 000 ≤ 12 000	25 000	12.31 (0.36)
2	2000–18 000 (325)	> 18 000 ≤ 25 000	25 000	12.61 (not calc.)
3	4000–8000 (35)	> 8000 ≤ 15 000	40 000	11.54 (not calc.)
4	2000–4000 (35)	> 4000 ≤ 10 000	25 000	13.20 (not calc.)
5	4000–6000 (30)	> 6000 ≤ 10 000	25 000	12.80 (0.89)
6	no centrifugation		25 000	11.62 (0.65)

- $s_{20,scp}$ = Apparent sedimentation coefficient of the protein component in the separated continuous phase after (incomplete) demulsification
- S.D. = Standard deviation (not calc. when S.D. < 0.8% of $s_{20,scp}$)
- The emulsion samples differ in their storage periods and phase volume ratio dispersed/continuous phase

Table 3 Demulsification of (> 90%) acetylated legumin stabilized *n*-decane-in-water emulsions by analytical ultracentrifugation

Emulsion	Centrifug. below the speed of oil sep. Rotor speed/r.p.m. (time/min)	Oil separation Rotor speed/ r.p.m.	Centrifug. after oil sep. Rotor speed/ r.p.m.	App. sediment. coefficient $s_{20,scp}$/(S.D./) sec $\times 10^{13}$
1	no centrifugation	40 000	40 000	25.57 (2.15)
2	2000–50 000 (250)	> 50 000 ≤ 55 000	55 000	21.88 (not calc.)

- $s_{20,scp}$ = Apparent sedimentation coefficient of the protein component in the separated continuous phase after (incomplete) demulsification
- S.D. = Standard deviation (not calc. when S.D. < 0.8% of $s_{20,scp}$)
- The emulsion samples differ in their storage periods and phase volume ratio dispersed/continuous phase

Fig. 3 Incomplete demulsification of an *n*-decane-in-water emulsion (stabilized by native legumin) at 40 000 r.p.m. after 12 min of centrifugation. The sample had previously been centrifuged at 2000–30 000 r.p.m. for 90 min. Phase volume ratio dispersed/continuous phase: 40/60, storage period: 70 days. Sedimentation coefficient of the legumin component in the separated continuous phase = 11.54 S

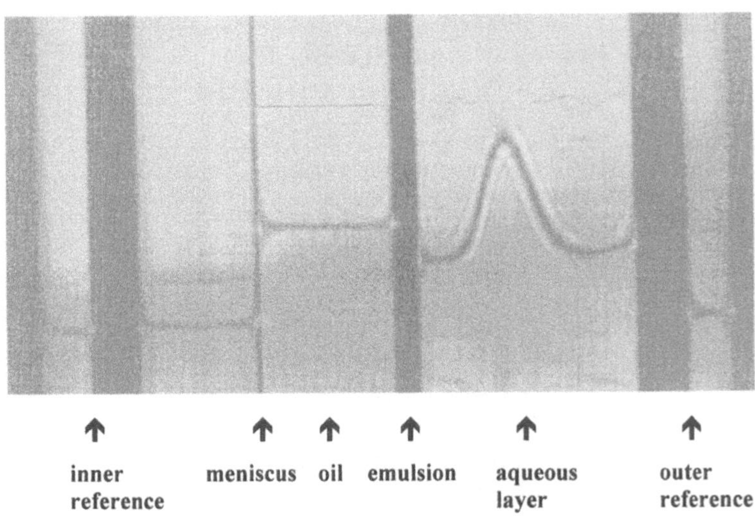

Schlieren peak had disappeared at the cell bottom) results in no additional protein peak. This might be due to a too low amount of protein released (centrepiece of 4 mm optical light path) or that the protein is left in the remaining emulsion. Therefore, a sample of the same emulsion was centrifuged first at 15 000 r.p.m. employing a 4° sector-shaped centrepiece of 12 mm optical light path. In comparison to the 3° centrepiece of 4 mm optical light path the loading emulsion volume is about five-fold. The large Schlieren peak formed showed some turbulence after increasing the speed to 25 000 r.p.m., and did not dissociate, giving a sedimentation coefficient indicating once more an 11–12 S component. All these results are an indication that, in the case of the native legumin at least, the mean component stabilizing the oil droplets is the 11 S fraction. Further, one can assume that no dissociation of this component into smaller subunits took place either during adsorption of the protein in the preparation of the emulsions or desorption after demulsification. This result is somewhat surprising as one could expect that the energy input in preparation of emulsions ("Ultra-Turrax", 20 000 r.p.m.) is high enough to induce dissociation of the legumin.

The 15 S component that was present in the starting legumin sample at a level of about 11% could not longer be detected in the separated continuous phase by sedimentation velocity. Therefore, it can be assumed that the aggregation of the 11 S component to the 15 S aggregate was reversible and it dissociated again in preparation of the emulsions.

Proceeding from the sedimentation coefficient of the protein component in the separated continuous phase shown in Fig. 3 (11.54 S) and using Eq. (1) the concentration of this component in the solution has been estimated to be ≈ 10 mg/ml. This value corresponds to the initial concentration of the native legumin solution used for the preparation of the emulsions (10.0 mg/ml). The agreement in the concentration values and the fact that the component in the separated continuous phase has been identified as an 11–12 S component supports the assumption that in the case of the native legumin stabilized emulsions only this component is adsorbed at the droplets and is not changed during the preparation procedure (adsorption) and during demulsification (desorption). Further, one can conclude that the amount of protein adsorbed at the droplets of the remaining emulsion after the final value centrifugation speed (40 000 r.p.m.) must be very small.

In the emulsions stabilized by the acetylated legumin the rotor speed could be chosen high enough to allow a fast sedimentation of the protein components in the separated continuous phase, both below and above the critical speed for oil separation.

In centrifugation of an emulsion containing acetylated legumin at 40 000 r.p.m. (Table 3, sample 1) a small amount of oil was separated and a single Schlieren peak of a highly aggregated 26 S fraction sedimented fastly. After increasing the rotor speed to 45 000 r.p.m. and 50 000 r.p.m. some more oil was separated but the amount of protein released was too small to allow determination of the sedimentation coefficient. Another emulsion (Table 3, sample 2) centrifuged at 25 000 r.p.m. (below the critical speed of oil separation) gave a sedimentation coefficient corresponding to a highly aggregated legumin (22 S). In the following centrifugation at 55 000 r.p.m. for 60 min only a small amount of oil was separated and no further visible gradient was formed.

These results support the view that in the case of the acetylated legumin the emulsification procedure itself might induce a further exposure of hydrophobic regions of the modified protein leading to an additional aggregation by hydrophobic interactions.

The starting solution of the acetylated legumin used for the preparation of the emulsions contained a relative large amount of low molecular dissociation products (25–30%). Considering that they could not be found during centrifugation, either below or above the critical speed for oil separation, the view is supported for a preferred adsorption of low molecular components in the preparation of the emulsion which aggregate to the 22–26 S fractions *after* desorption during the demulsification process. Another explanation could be that the energy input during the emulsification procedure at first results in an aggregation of the low molecular weight fractions and after that the aggregates will be adsorbed by the droplets.

Further, the results demonstrate the experimental limitations if the emulsions are both very high coalescence-stable and the degree of demulsification is not sufficient to release enough protein to enable determination of sedimentation coefficients using the Philpot–Svensson–Schlieren optical system on the ultracentrifuge.

Conclusions

By correlating characterization of the coalescence stability of emulsion with sedimentation velocity experiments, we have thus shown in this study that it is possible to identify different protein components immediately after desorption from an O/W emulsion interface. A comparison of the data of the protein components in the initial native and acetylated legumin solutions with those in the separated continuous phase after (incomplete) demulsification gives additional information. So far, this approach may be a useful tool for the approach to the conformational state of proteins at interfaces. This information can be considered as a first step to provide an answer to the question as to why the emulsions containing acetylated legumin are highly stable against coalescence. Further progress could be expected by including legumins of different degrees of acetylation and after other modifications, e.g., succinylation or enzymatic hydrolysis. Moreover, in a future study it would seem to be of great value to investigate the protein components in the separated continuous phase after removal from the measuring cell by using the more sensitive interference and UV-absorption optical systems. A comparison of these results with those obtained directly during the demulsification runs could confirm our results and provide valuable additional information. Further, data on the amount of adsorbed protein at the oil droplets obtained by independent methods would be valuable.

Acknowledgement Financial support from the Deutsche Forschungsgemeinschaft (grant No. Schw 462/3-2 and /3-3) is gratefully acknowledged.

The authors thank Dipl.-Phys. J.-P. Krause (Research Group Plant Protein Chemistry (WIP), University of Potsdam) for the helpful discussions and Dr. S.E. Harding (University of Nottingham, U.K.) for the helpful reading of the paper. The technical support of Mrs. H. Beyer, Mrs. B. Junker & Mrs. A. Krajewski is appreciated.

References

1. Kinsella JE (1976) CRD Crit Rev Food Sci Nutr 7:219–280
2. Derbyshire E, Wright DJ, Boulter D (1976) Phytochemistry 15:3–24
3. Schultz M, Schmidt G, Krause J-P, Seifert A, Schmandke H (1991) Fat Sci Technol 93:294–297
4. Schmandke H, Maune R, Schmidt G, Schultz M (1977) Nahrung 21:901–909
5. Schmandke H (1986) Nahrung 30:303–309
6. Rauschal EJ, Linow K-J, Pähtz W, Schwenke KD (1981) Nahrung 25:241–248
7. Schwenke KD, Rauschal EJ, Robowsky KD (1983) Nahrung 27:335–350
8. Muschiolik G, Dickinson E, Murray BS, Stainsby G (1987) Food Hydrocolloids 1:191–196
9. Krause JP, Buchheim W (1994) Nahrung 38:455–463
10. Strenge K, Seifert A, Progr Colloid Polym Sci (1991) 86:76–83
11. Popello IA, Suchkov VV, Grinberg VYa, Tolstogusov VB (1988) Prikl Biochem Mikrobiol 24:50–55
12. Schwenke KD, Dudek St, Seifert A, Mothes R, Staatz A (1994) Nahrung 38:559–567
13. Schwenke KD, Raab B, Pähtz W, Zirwer D, Hak KY (1989) J Food Biochem 13:321–334
14. Fields R (1972) Methods Enzymol 25:464–468
15. Itzhaki RF, Gill DM (1964) Analyt Biochemistry 9:401–410
16. Elias HG (1961) Ultrazentrifugenmethoden, Beckman Instruments GmbH, Munich
17. Cölfen H (1994) unpublished
18. Schwenke KD, Dudek St, Mothes R, Raab B, Seifert A (1993) Nahrung 37:258–268
19. Kim KS, Rhee JS (1989) J Food Biochemistry 13:187–199
20. Schwenke KD, Zirwer D, Gast K, Görnitz E, Linow K-J, Gueguen J (1990) Eur J Biochem 194:621–627
21. Schwenke KD, Dudek St, Mothes R, Seifert A, Staatz A, Knopfe C, Henning Th (1995) Proceedings of the 2nd AEP Conference on Grain Legumes – Improving Production & Utilization, July 9–13, Copenhagen, Denmark, 360–361
22. Schwenke KD, Dudek St, Mothes R, Seifert A, Raab B (1992) In: Schwenke KD, Mothes R (eds) Food Proteins – Structure and Functionality (4th Symposium on Food Proteins), Reinhardsbrunn, Germany, October 5–8, 1992), Verlag Chemie, Weinheim 1993, 154–157
23. Seifert et al., in preparation

Progr Colloid Polym Sci (1995) 99:39–44
© Steinkopff Verlag 1995

P.M. Budd

Preliminary ultracentrifuge studies of the polyelectrolyte behaviour of Welan gum

Received: 3 March 1995
Accepted: 17 May 1995

Dr. P.M. Budd (⊠)
Department of Chemistry
University of Manchester
Manchester M13 9PL, United Kingdom

Abstract Sedimentation velocity and synthetic boundary diffusion studies have been undertaken for aqueous solutions of the branched microbial polysaccharide welan gum. Centrifugation and filtration to remove insoluble material from the crude gum is shown also to remove a proportion of soluble high molar mass material. Purified welan which has been dialysed exhaustively against distilled water to remove low molar mass impurities exhibits typical polyelectrolyte behaviour in salt-free solution: an enhanced apparent diffusion coefficient and diminished sedimentation coefficient at modest polymer concentrations, and a dramatic drop in apparent diffusion coefficient at concentrations below 0.6 g dm^{-3}. In the presence of a salt, welan solutions exhibit 'weak gel' behaviour. At NaCl concentrations greater than about 0.1 mol dm^{-3}, sedimentation velocity shows no evidence of charge effects or conformational change. The solution behaviour may reflect a stable double-helical structure, stabilized by interactions between side chains and carboxylate groups in the polysaccharide backbone, coupled with weak association between double-helices.

Key words Welan gum – polyelectrolyte – sedimentation velocity – diffusion – analytical ultracentrifuge

Introduction

Welan gum is an extracellular polysaccharide produced by the bacterium *Alcaligenes* ATCC 31555. It was previously known as S-130 and has been commercialized by the Kelco Division of Merck & Company under the name Biozan. Aqueous solutions of welan exhibit a high low-shear viscosity which is maintained to high temperatures and is remarkably insensitive to pH and to the concentration of calcium and other ions [1–4]. Welan finds application in oilwell fluids (hence the name).

The chemical structure of welan (Fig. 1) comprises a backbone with the tetrasaccharide repeat

$\rightarrow$ 3)-β-D-Glc$p$-(1 $\rightarrow$ 4)-β-D-Glc$p$A-(1 $\rightarrow$ 4)

-β-D-Glc$p$-(1 $\rightarrow$ 4)-α-L-Rha$p$-(1 $\rightarrow$

and a side chain of either α-L-Rha$p$ or α-L-Man$p$ joined (1 $\rightarrow$ 3) to the 4-linked β-D-Glc$p$ [5, 6]. Approximately two-thirds of side chain units are α-L-Rha$p$ and these are randomly distributed [7]. In native welan, approximately 85% of the 3-linked β-D-Glc$p$ units have O-acetyl groups in the 2-position [8].

Fig. 1 Chemical structure of welan

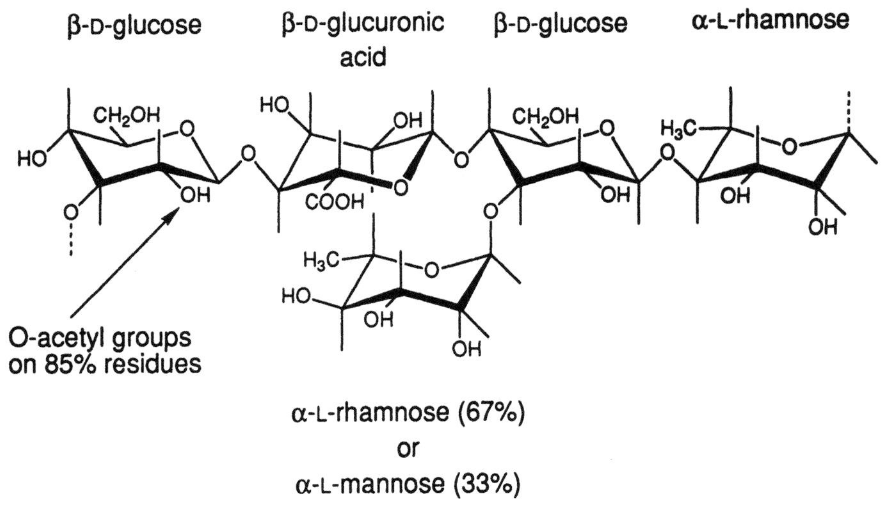

A related polysaccharide, gellan (previously known as S-60), has the same backbone as welan, but no side chains. Gellan readily forms gels in the presence of various monovalent and divalent metal ions. Under conditions which inhibit gelation, gellan exhibits a conformational transition on changing the temperature. By contrast, welan exhibits only 'weak gel' behaviour in aqueous solution and there is no evidence of a thermally induced conformational transition [3, 9]. This behaviour has been interpreted as indicating that welan exists as a disordered coil in solution [10]. However, there is evidence that welan exists in a stable, highly ordered structure in aqueous solution [3, 11–14]. An order-disorder transition may be induced by adding dimethyl sulphoxide [15].

X-ray diffraction studies suggest that the welan exists in a half-staggered, parallel, double-helix in hydrated fibres [16, 17]. It is conceivable that a double-helical structure is retained in aqueous solution.

The presence of glucuronic acid in the backbone is expected to give welan the characteristics of a polyelectrolyte. Campana et al. [13] have reported that the intrinsic pK of the carboxyl groups is abnormally low ($pK_0 = 2.2$ for welan, compared with 3.06 for gellan, which implies that welan is a stronger acid), and that the viscosity of a dilute salt-free solution increases on titration with NaOH as is typical for a polyelectrolyte. In contrast, Urbani and Brant [12] reported that potentiometric titration and the dependence of intrinsic viscosity on ionic strength suggest only very weak polyelectrolyte behaviour.

The molecular characterization of welan in aqueous solution is difficult because of its 'weak gel' behaviour at very low polymer concentrations. Molar masses have been determined by classical light scattering for samples of welan degraded by sonication (M in the range $1 \times 10^5 - 2 \times 10^6$ g mol^{-1}) [12, 13].

In this paper, we report a preliminary study of aqueous welan solutions using the analytical ultracentrifuge. The aims of the study were i) to explore the possibility of using ultracentrifuge techniques to study welan under condtions close to those in which it is used industrially and ii) to investigate polyelectrolyte effects in the transport properties of welan.

Experimental

A sample of welan gum from the Kelco Division of Merck & Company was provided by Prof AH Cervenka of Shell Research. The crude material contained many gross impurities and on dissolution in water with heating gave an extremely murky solution.

Purification of welan

Insoluble material was removed from crude welan as follows: The crude material was dissolved in water at 70–90 °C with stirring to give a 0.1% solution. The murky solution was centrifuged at 12 000 rpm for 3 h. The supernatant was carefully removed then filtered through a sintered crucible with a filter aid (Celite 545 paste). The purified polymer was recovered by freeze drying (30–45% yield).

For some experiments, low molar mass impurities were removed from the welan by exhaustive dialysis against distilled water.

Analytical ultracentrifugation

Sedimentation velocity and synthetic boundary diffusion studies [18] were carried out using a Beckman Model

E Analytical Ultracentrifuge. For sedimentation velocity, a 12 mm single-sector aluminium centrepiece was employed, the rotor speed was 40000 rpm, Schlieren optics were used and sedimentation coefficients were computed from the rate of movement of the maximum in the Schlieren image. For synthetic boundary experiments, an aluminium-filled epoxy capillary-type synthetic boundary centrepiece was employed, the rotor speed was 8000 rpm and experiments were performed with both Schlieren and interference optics.

Results

Effect of purification

An advantage of the sedimentation velocity method is that useful information may be obtained from very 'dirty' solutions. On centrifugation, gross impurities sediment rapidly and do not interfere with soluble material. For some purposes, however, it is preferable to use purified polymer, and a key question concerns the extent to which the purification process has affected the system.

Figure 2 compares the reciprocal sedimentation coefficients of crude and purified welan in aqueous NaCl (0.25 mol dm^{-3}). The area under the Schlieren image is utilised as an internal measure of the concentration of soluble species. It can be seen that the purified polymer exhibits slightly lower sedimentation coefficients than the crude polymer at comparable concentrations. This suggests that the purification procedure has removed a proportion of higher molar mass species from solution.

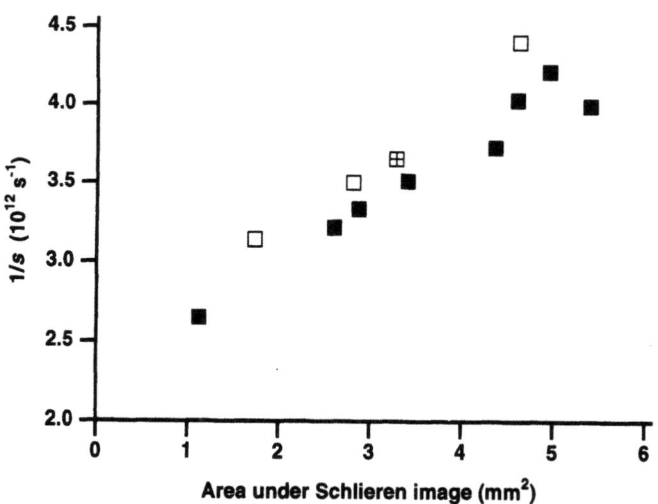

Fig. 2 Dependence of reciprocal sedimentation coefficient on polymer concentration (expressed as area under Schlieren image) for solutions in aqueous NaCl (0.25 cm dm^{-3}) of (■) crude welan, (□) purified welan and (⊞) purified and exhaustively dialysed welan

Included in Fig. 2 is the reciprocal sedimentation coefficient for a sample which was first dialysed exhaustively against distilled water to remove low molar mass impurities, before addition of NaCl. In the presence of NaCl, it is seen to behave similarly to welan which has been purified only by centrifugation and filtration. It is shown below that in the absence of added salt, dialysis has a marked effect on the solution behaviour of welan.

Diffusion and sedimentation in salt-free solution

Figure 3a shows the Schlieren image obtained after 36 min in a synthetic boundary experiment for welan purified by centrifugation and filtration then dissolved in distilled water. The boundary is seen to be unsymmetrical but fairly sharp, although the image is not very clear because the experiment has been performed under conditions which are near the limits of resolution for the optical system. A similarly-shaped boundary is obtained for welan dialysed to equilibrium against a dilute aqueous NaCl solution (0.0001 mol dm^{-3}). Figure 3b shows the Schlieren image after the same period of time for welan purified then dialysed exhaustively against distilled water; the diffusing boundary is extremely broad and includes an anomalous peak on the leading edge. Very similar behaviour has been observed for strong polyelectrolytes such as poly(styrene sulphonate) [19–21].

If the height-area method is applied to the results in Fig. 3, the diffusion coefficients obtained are 1.1×10^{-7} cm^2 s^{-1} for the sample not dialysed and 1.6×10^{-6} cm^2 s^{-1} for the dialysed sample. Whilst the height-area method is not strictly valid for unsymmetrical boundaries such as these, an order of magnitude increase in diffusion coefficient on dialysis is indicated. Enhanced diffusion in the absence of salt is typical of a polyelectrolyte and has been attributed to the "primary charge effect" or "coupled-ion diffusion" [22, 23]. Welan which has not been dialysed may be assumed to contain sufficient ionic impurities to suppress this effect to a considerable extent.

The shape of the diffusing boundary for a polyelectrolyte in salt-free solution may be interpreted in terms of a diffusion coefficient that depends markedly on concentration. If it is assumed that diffusion is one-dimensional (this is not strictly correct for a sector-shaped centrepiece), the Boltzmann–Matano procedure may be used to calculate the apparent diffusion coefficient/concentration curve over the whole concentration range of the boundary, using the results at a single, finite time t after formation of the initial boundary. For poly(styrene sulphonate) in salt-free solution, remarkably good agreement was achieved between the apparent diffusion coefficients obtained by this method

P.M. Budd
Welan gum

Fig. 3 Schlieren image obtained after 36 min in a synthetic boundary diffusion experiment for (a) welan purified but not dialysed and (b) welan purified and exhaustively dialysed against distilled water (polymer solution on right-hand side, initial concentration: $2\ \mathrm{g\,dm^{-3}}$ in distilled water)

and the fast diffusion coefficients determined in comparable dynamic light scattering studies [20]. For diffusion in the x direction, the apparent diffusion coefficient D at a concentration c_1 is given by

$$D_{c=c_1} = -\frac{1}{2t}\frac{dx}{dc}\int_0^{c_1} x\,dc \tag{1}$$

Figure 4 shows the results of this analysis for a synthetic boundary experiment carried out using interference optics for welan purified then dialysed exhaustively against distilled water. The apparent diffusion coefficient is seen to be constant (approximately $1.2 \times 10^{-6}\ cm^{-2}\,s^{-1}$) for concentrations greater than $0.6\ g\,dm^{-3}$, but to decrease markedly on dilution at lower concentrations. The analysis has not accounted for the sector-shape of the centrepiece or polydispersity of the sample, so the values obtained should be treated with some caution. However, the shape of the apparent diffusion coefficient/concentration curve is significant and is very similar to that found for other polyelectrolytes [20]. The molecular interpretation of this polyelectrolyte behaviour is open to question and further work in this area is required.

The sedimentation coefficient s for a solution of welan purified then dialysed exhaustively against distilled water (welan concentration $= 0.6\ g\,dm^{-3}$) was found to be rather low and the boundary was distorted by diffusion effects. For the same sample with NaCl added (NaCl concentration $= 0.25\ mol\,dm^{-3}$), a larger value of s was obtained ($2.74 \pm 0.04 \times 10^{-1}$ s). An increase in s on increasing the ionic strength is typical of a polyelectrolyte

and has been attributed to the "primary charge effect" [19, 20, 22, 24–26].

This work has demonstrated that exhaustive dialysis has a pronounced effect on the diffusion and sedimentation behaviour of welan, and shows that at very low ionic strengths welan behaves in the same way as other polyelectrolytes.

Sedimentation in presence of added salt

Figure 5 shows the reciprocal sedimentation coefficient as a function of welan concentration for solutions with various amount of added NaCl. Whilst the data are limited, it can be seen that the slopes of these plots vary little with increasing ionic strength. This is in contrast to the behaviour usually observed for flexible polyelectrolytes (a marked decrease in slope with increasing salt concentration) [19, 20, 22, 24–26].

Figure 6 illustrates the dependence of limiting sedimentation coefficient s_0 on NaCl concentration, both with and without correction of s to the viscosity and density of water (assuming the partial specific volume of welan to be $0.67\ cm^3\,g^{-1}$, a value estimated from the fibre density [16, 17]). It can be seen that at NaCl concentrations greater than about $0.1\ mol\,dm^{-3}$ the observed sedimentation behaviour is determined largely by the solvent properties; there is little evidence of charge effects or of conformational change.

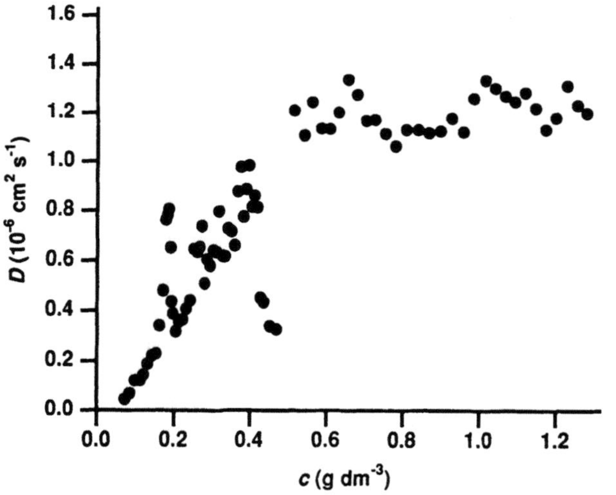

Fig. 4 Dependence of apparent diffusion coefficient on polymer concentration for welan purified and exhaustively dialysed against distilled water, determined from synthetic boundary data by the Boltzmann–Matano procedure

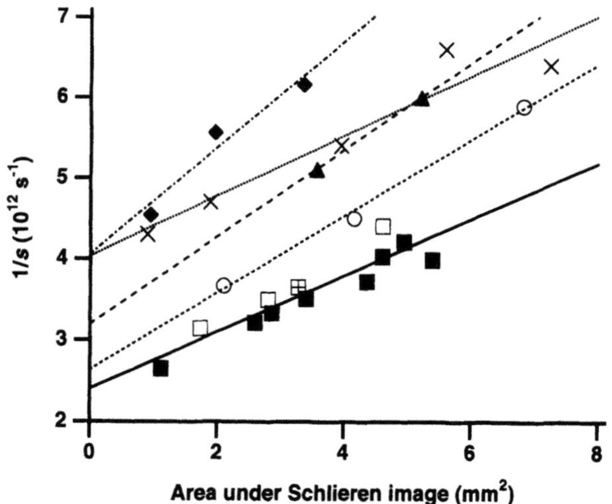

Fig. 5 Dependence of reciprocal sedimentation coefficient on polymer concentration (expressed as area under Schlieren image) for ($\times$) purified welan in distilled water with no added salt, (o) purified welan in $0.01\ mol\,dm^{-3}$ NaCl, ($\square$) purified welan in $0.25\ mol\,dm^{-3}$ NaCl, (⊞) purified and dialysed welan in $0.25\ mol\,dm^{-3}$ NaCl, (■) crude welan in $0.25\ mol\,dm^{-3}$ NaCl, (▲) crude welan in $2\ mol\,dm^{-3}$ NaCl, (◆) crude welan in $3\ mol\,dm^{-3}$ NaCl

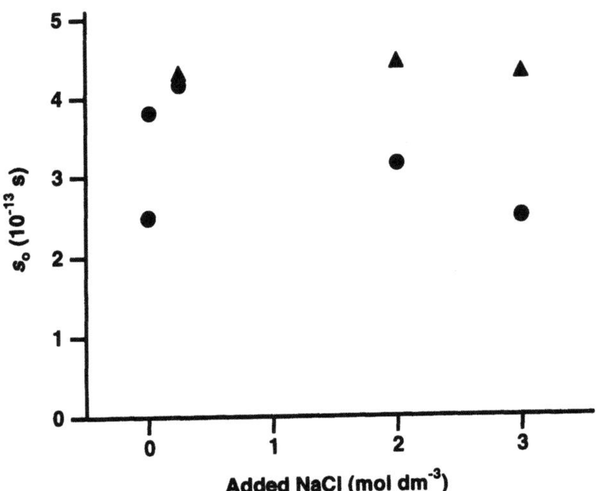

Fig. 6 Dependence of limiting sedimentation coefficient on NaCl concentration (●) as measured and (▲) after correction to the viscosity and density of water

Discussion

These preliminary results show that, whereas welan exhibits typical polyelectrolyte properties at very low ionic strength, charge effects are effectively eliminated at salt concentrations greater than about $0.1 \, mol \, dm^{-3}$. These results closely parallel the observed rheological behaviour of aqueous welan solutions [1–4].

Two factors may contribute to the observed insensitivity of solution properties to salt concentration above about $0.1 \, mol \, dm^{-3}$:

(i) Screening of the carboxylate groups through intramolecular interactions with the side chains.

(ii) Weak intermolecular association giving rise to a transient network structure.

In hydrated fibres, it has been proposed on the basis of x-ray studies that welan forms a double-helix in which the side chains fold back on the main chain to form hydrogen bonds with the carboxylate groups, and that double helices can associate with each other either through a combination of counterions and water molecules or through side chain [16, 17]. It is likely that these interactions persist in aqueous solution. The side chains serve to stabilize the double-helical structure and to inhibit direct COO^---- counterion---^-OOC interactions, such as are believed to lead to the formation of true gels in gellan solutions. The weaker intermolecular interactions that may occur in welan give rise to a tenuous network structure which dominates the transport properties at all but the lowest ionic strengths.

Acknowledgements Thanks are due to Miss Penelope Catlow, Mr L. John O'Hare and Mr Henry Chan, who contributed to this work in the course of undergraduate projects.

References

1. Kang KS, Veeder GT, Cottrell IW (1983) Prog Ind Microbiol 18:231–254
2. Sandford PA, Cottrell IW, Pettitt DJ (1984) Pure Appl Chem 56:879–892
3. Robinson G, Manning CE, Morris ER (1991) In: Dickinson E (ed) Food Polymers, Gels and Colloids, Special Publication No 82. Royal Society of Chemistry, Cambridge, pp 22–23
4. Tako M, Kiriaki M (1990) Agric Biol Chem 54:3079–3084
5. Jansson P-E, Lindberg B, Widmalm G, Sandford PA (1985) Carbohydr Res 139:217–223
6. O'Neill MA, Selvendran RR, Morris VJ, Eagles J (1989) Carbohydr Res 147:295–313
7. Jansson P-E, Widmalm G (1994) Carbohydr Res 256:327–330
8. Stankowski JD, Zeller SG (1992) Carbohydr Res 224:337–341
9. Crescenzi V, Dentini M, Coviello T, Rizzo R (1986) Carbohydr Res 149:425–432
10. Crescenzi V, Dentini M, Dea ICM (1987) Carbohydr Res 160:283–302
11. Cesàro A, Gamini A, Navarini L (1992) Polymer 33:4001–4008
12. Urbani R, Brant DA (1989) Carbohydr Pol 11:169–191
13. Campana S, Andrade C, Milas M, Rinaudo M (1990) Int J Biol Macromol 12:379–384
14. Campana S, Ganter J, Milas M, Rinaudo M (1992) Carbohydr Res 231:31–38
15. Hember MWN, Richardson RK, Morris ER (1994) Carbohydr Res 252:209–221
16. Lee EJ, Chandrasekaran R (1991) Carbohydr Res 214:11–24
17. Chandrasekaran R, Radha A, Lee EJ (1994) Carbohydr Res 252:183–207
18. Budd PM (1989) In: Allen G, Bevington JC, Booth C, Price C (eds) Comprehensive Polymer Science, Vol 1, Pergamon, Oxford, Chapter 10, pp 199–214
19. Budd PM (1989) In: Allen G, Bevington JC, Booth C, Price C (eds) Comprehensive Polymer Science, Vol 1, Pergamon, Oxford, Chapter 11, pp 215–230
20. Budd PM (1994) Progr Colloid Polym Sci 94:107–115
21. Nagasawa M, Fujita H (1964) J Am Chem Soc 86:3005–3012
22. Eisenberg H (1976) Biological Macromolecules and Polyelectrolytes in Solution, Oxford University Press, Oxford
23. Berne BJ, Pecora R (1976) Dynamic Light Scattering, Wiley, New York, Chapter 9
24. Budd PM (1985) Polymer 26:1519–1522
25. Budd PM (1988) Brit Polym J 20:33–37
26. Budd PM (1992) In: Harding SE, Rowe AJ, Horton JC (eds) Analytical Ultracentrifugation in Biochemistry and Polymer Science, Royal Society of Chemistry, Cambridge, Chapter 32, pp 593–608

Progr Colloid Polym Sci (1995) 99:45–54
© Steinkopff Verlag 1995

H. Triebel
H. Bär
R. Geuther
G. Burckhardt

Netropsin-induced changes of DNA supercoiling; sedimentation studies

Received: 5 April 1995
Accepted: 6 June 1995

Part of this paper was presented at the IX. Symposium on Analytical Ultracentrifugation, Berlin, March 2-3, 1995

Dr. H. Triebel (✉) · H. Bär
Abteilung für Biophysikalische Chemie
Friedrich-Schiller-Universität Jena
Institut für Molekularbiologie
Winzerlaer Straße 11
07745 Jena

R. Geuther
Hans-Knöll-Institut für Naturstofforschung
07745 Jena, Germany

G. Burckhardt
Abteilung für Molekulare Biochemie
Institut für Molekularbiologie
Friedrich-Schiller-Universität Jena
Winzerlaer Straße 11
07745 Jena, Germany

Abstract Sedimentation velocity techniques have been used to study the effect of the antibiotic drug netropsin (Nt) on the structural properties of different forms of closed circular DNA. Nt removes supercoils upon interaction with positively supercoiled DNA and introduces negative supercoils upon binding to relaxed DNA. Applying the classic assay of DNA unwinding by ethidium to complexes of negatively supercoiled DNA with Nt, we found that, compared with naked DNA, more ethidium is needed to convert the Nt-DNA complex into a superhelix-free form. From the shift in the critical ethidium/nucleotide binding ratio, a Nt-mediated decrease in the linking number difference and superhelix density (or increase in the number of superhelical turns) of negatively supercoiled DNA is calculated, synonymous with an increase in the linking number and a decrease in the helical repeat of the relaxed form. A significant influence of the presence of intrinsic bends in the DNA duplex on the magnitude of the drug-induced alterations in DNA supercoiling could not be detected.

Although it is unclear whether the *primary* action of netropsin consists of changes in writhe or twist of DNA, or both, the measurable effects can be described by a Nt-induced overwinding of the DNA double helix. Nt-specific DNA overwinding angles can also be derived from the (positive) linking number difference observed in literature upon relaxation of supercoiled Nt-DNA complexes by topoisomerase I and subsequent removal of Nt. A compilation of the data available from literature and our own work shows that the average Nt-mediated DNA overwinding angle exhibits a strong decrease with increasing Nt/nucleotide ratio. This finding suggests that Nt exerts the strongest effects on DNA supercoiling when interacting with DNA binding sites which have the highest affinity and, hence, primarily are occupied by Nt.

Key words Sedimentation – supercoiled DNA – netropsin – DNA interaction – DNA overwinding

Introduction

In the past, numerous studies were devoted to the binding properties of several types of non-intercalating DNA binding drugs, both with regard to possible relationships to the antibiotic activities of these drugs and to their role as tools for elucidating the molecular factors involved in the recognition of DNA base pair sequences by proteins (for reviews, see [1–4]). The vast majority of such interaction studies was carried out with natural DNAs of different base composition, synthetic polynucleotides, or short oligonucleotides of defined sequence, i.e., with linear nucleotide chains. There are only few reports in which drug-induced structural changes of closed DNA are described. Snounou and Malcolm [5, 6] have shown that positively supercoiled DNA is produced by relaxation of a netropsin-plasmid complex with topoisomerase I and subsequent removal of netropsin, indicating that netropsin increases the linking number of (relaxed) DNA.

Since nearly all naturally occurring DNA in living cells is negatively supercoiled and since the variation of superhelicity is of great importance with respect to the functional state of DNA [7], we have recently investigated the effect of different nonintercalators on the structural properties of supercoiled DNA. This was done by applying the classic s/EB titration assay of DNA unwinding by ethidium [8, 9], monitored by analytical sedimentation techniques, to complexes of superhelical DNA with various minor groove binders [10]. Although only the superhelical form of DNA was used in our experiments, numerous topological and geometric parameters of DNA structure for both the supercoiled and the relaxed form of DNA (and the complexes with minor groove binders) could be derived from the experimental data [10]. Our findings demonstrated that, for a negatively supercoiled DNA, the negative superhelix density is reduced (i.e., the absolute value is enhanced) by the interaction of DNA with certain minor groove binders which apparently induce an increase in the average duplex winding angle of DNA. The drug-induced conformational changes of supercoiled DNA were found to depend on the structure of the ligand and to be most strongly pronounced for the AT-specific DNA binding drug netropsin (Nt). Although it is not clear at all whether the *primary* local action of Nt consists of changes in winding or bending, or both [6, 10], the measurable structural changes of closed DNA can at last be described by an alteration of the average duplex winding angle. For Nt, the finding of a drug-induced overwinding of the DNA double helix (in solution!) [10] agrees with the conclusions drawn from the topoisomerase assay [5, 6, 11], but contrasts with the situation in the crystal, at least for one particular sequence. According to x-ray diffraction data, binding of Nt to a double-helical dodecamer of sequence [d(CGCGAATTCGCG)]$_2$ produces no striking systematic changes in the average duplex rotation [12].

In the present paper, we have selected the dicationic oligopeptide antibiotic Nt as a paradigm of the minor groove binders previously investigated and have expanded our experiments with Nt to the positively supercoiled and the relaxed form of DNA as well as to DNA which contained an insert with intrinsic base sequence-directed bends of the double helix. The results obtained with these DNA substrates qualitatively support our previous conclusions. Experiments with the nonintercalator distamycin A (Dst-A), structurally related to Nt, are performed for comparison.

Materials and methods

Drugs

Ethidium bromide (EB) from SERVA (Heidelberg) was of research grade and was used without further purification. EB concentrations are based on the measured molar extinction coefficient at 480 nm of 5530 cm^2/mmol which is close to the frequently used [13–15] value of 5600 cm^2/mmol. Nt was a highly purified product isolated from *Streptomyces netropsis* as described previously [16]. Dst-A was from SERVA, Novobiocin from Sigma. The structures of Nt and Dst-A are given in Fig. 1.

DNA

pGRB DNA

Halobacterium strain GRB, a gift from P. Forterre (Orsay, France), was grown in classical halophilic medium [17]. The 1.7 kb plasmid of this archaebacterium is negatively supercoiled but is positively supercoiled when cells are grown in the presence of novobiocin [18]. Novobiocin was added when the cell culture reached an optical density of

Fig. 1 Chemical structures of the minor groove binders netropsin (1) and distamycin A (2)

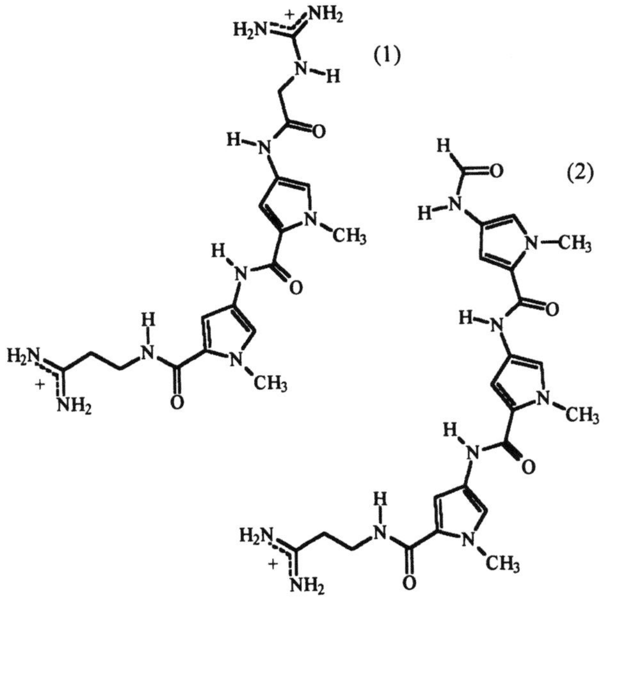

0.3 at 600 nm. The novobiocin-treated cells (and the control) from 500 ml cultures were harvested 18 h after drug addition and were lysed in 10 ml of 25 mM Tris-HCl, pH 7, 10 mM EDTA and by treatment with alkali [18]. The plasmids were purified by standard ethidium-CsCl gradient centrifugation [19]. Relaxation of pGRB was performed with topoisomerase I isolated from chicken erythrocytes. The positively supercoiled and the relaxed forms of the plasmid are designated pGRB(+) and pGRB(r), respectively. As judged by sedimentation analysis and agarose gel electrophoresis, the different preparations of pGRB(+) plasmid DNA contained about 5–10% of open circular DNA and about 90–95% of the superhelical form.

pUGD1202-3 DNA

The plasmid pUGD1202, a gift of Prof. Mizuno (Sendai, Japan), consists of the *E. coli* plasmid pUC9 (2665 bp) and an 1211 bp EcoRI fragment of the W chromosome of chicken [20]. Transformation, isolation and purification of this plasmid in *E. coli* strain TG 1 followed methods described in [19]. We found that the plasmids of all our transformants had inserts of smaller length than the original plasmid. The plasmid with the longest insert (960 bp as derived from agarose gel electrophoretic analysis) was called pUGD1202-3 and used for further experiments. The preparation of supercoiled pUGD1203 was contaminated by about 10% supercoiled dimers and about 10% nicked monomers.

Prior to analytical sedimentation experiments, the DNA stock solutions were extensively dialysed against buffer A (50 mM Tris-HCl, 1 mM Na$_2$-EDTA, pH 7.8) or, in a special case, to buffer B (30 mM NaCl, 1 mM Na$_2$-EDTA, 2 mM phosphate buffer, pH 7.8). Since, at pH 7.8, Tris is protonated to only about 60%, both buffers have the same ionic strength. DNA-P concentrations were determined from optical densities, using a molar absorption coefficient at 260 nm of 6620 cm^2/mmol (0.02 cm^2/µg).

Ultracentrifugation

Sedimentation velocity experiments were carried out as described recently [10], but with the photoelectric scanning system of the analytical ultracentrifuge (Spinco model E, Beckman Instruments, München) now being interfaced to an AT 386 personal computer for on-line data acquisition [21]. The operating and data handling software was a kind gift of Dr. C. Urbanke (Medical School, Hannover) who also gave advice concerning the hardware components. The UV absorption optics was equipped with a collimating system [22] (version K2, Schrader Feinmechanik & Optik, Braunschweig), yielding a considerable enhancement of light intensity of rotor illumination and, hence, an improvement of the signal-to-noise ratio of the photoelectric detection system. With this collimating system, the exit slit width of the monochromator and the entrance slit width of the photomultiplier could be lowered from 2 mm to 1 mm and from 0.02 mm to 0.01 mm, respectively, thus reducing the spectral band width of the monochromator and increasing the resolution of radial scanning. With its entrance slit seeing the reference hole, the photomultiplier now exhibits a dynode voltage of 60 V at 263 nm (instead of about 74 V before), indicating a higher level of light energy in the detector plane even with these lower slit widths used.

Results and discussion

Netropsin removes positive supercoils upon interaction with positively supercoiled DNA

Recently, we have shown that, for a negatively supercoiled DNA, the number of superhelical turns is enhanced by interaction with Nt [10]. For positively supercoiled DNA, the underlying mechanism should lead to an opposite effect, namely, to a reduction in the number of superhelical turns. (Note that the sign-dependent superhelix density decreases in both cases). The use of positively supercoiled DNA permits measurements without a competition with EB the presence of which was required in our experiments with negatively supercoiled pBR322 DNA [10].

Figure 2A shows that, on addition of EB, the mean sedimentation coefficient of pGRB (+) increases and approaches a constant value. This behavior proves that the DNA preparation has a positive superhelix density which is further enhanced due to the unwinding of the DNA duplex concomitant to the progressive intercalation of ethidium. Binding of Nt to the same DNA sample, however, leads to a drop in the sedimentation coefficient (Fig. 2B) which obviously is due to the removal of positive supercoils resulting from the Nt-induced overwinding of the double helix. A similar effect was seen in a gel electrophoretic experiment with pGRB(+) at a single Nt concentration [18]. In Fig. 2B, no equivalence point (i.e., no minimum of the curve) with subsequent renewed increase in *s* is attained, indicating that a conversion of the positively supercoiled into negatively supercoiled DNA by titration with Nt could not be achieved. This behavior may be due to two reasons. First, the Nt-induced duplex overwinding angle is considerably smaller compared with the ethidium unwinding angle of 26°. Second, the pGRB plasmid with its low (A +T)-content of 36 mol-%, as determined by analytical equilibrium centrifugation in a CsCl

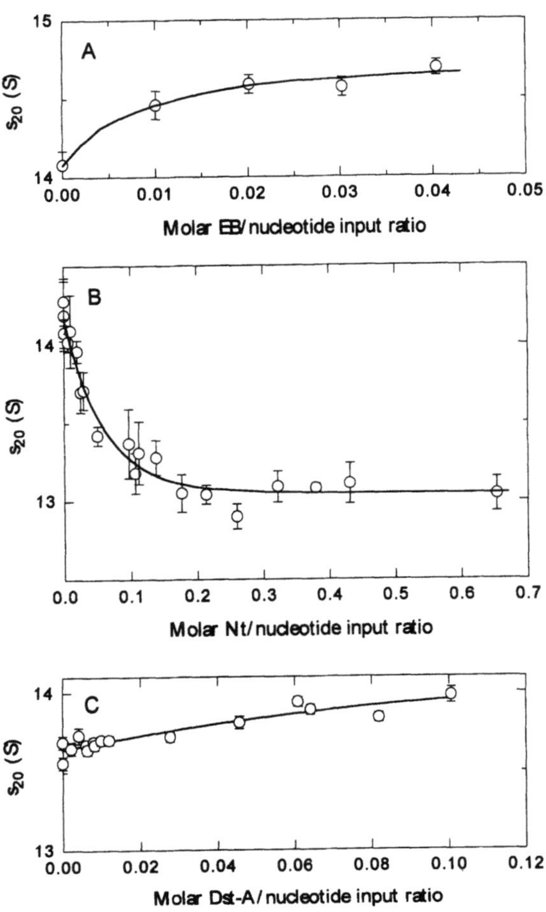

Fig. 2 Sedimentation behavior of positively supercoiled pGRB(+) DNA upon titration with EB (A), Nt (B), and Dst-A (C). In Fig. 2B, the DNA sample was the same as in Fig. 2A, whereas in Fig. 2C a DNA sample of somewhat lower positive superhelix density was used. Buffer A

density gradient (not shown), may be expected to harbor only a low number of AT-clusters suitable as strong binding sites for Nt. The nucleotide sequence of the *Halobacterium* plasmid pHSB1, namely, which likewise has an (A + T)-content of 36 mol-% and shows strong homology to the small plasmid of *Halobacterium* strain GRB [23], e.g., contains only 16 sequences consisting of at least four AT base pairs like AAAA, AAAT, AATT, etc. Apparently, both properties together lead to an early saturation of Nt binding, yielding a constant *s*-value (Fig. 2B), before the positively supercoiled DNA has attained a superhelix-free or even a negatively supercoiled state.

In contrast to the Nt-induced drop of the sedimentation coefficient (Fig. 2B), pGRB(+) DNA exhibits a small but significant increase of *s* upon addition of the structurally related antibiotic Dst-A (Fig. 2C), probably reflecting the increase in molar mass due to complex formation. From this result one is tempted to conclude that Dst-A does not alter the superhelix density of DNA and, hence,

does not affect the duplex winding angle. This conclusion would be in line with the finding that, in a topoisomerase I assay, no or only a very small effect of Dst-A on the linking number of relaxed DNA could be found [5, 6, 11].

Netropsin induces negative supercoils upon binding to relaxed DNA

According to our experience with negatively supercoiled DNA, the linking number difference of the supercoiled form with regard to the relaxed form of DNA, ΔLk, decreases ($|\Delta Lk|$ increases) upon interaction of DNA with Nt [10]. Here, we try to corroborate this effect by using a relaxed DNA substrate, pGRB(r) (with $\Delta Lk = 0$), which should be converted into a negatively supercoiled DNA ($\Delta Lk < 0$) upon binding of Nt. (Note that, in contrast to the linking number Lk which represents a topological invariant in our experiments, ΔLk may change by the binding of Nt even in the absence of breakage and religation of the DNA backbone. See [10, 24] for a discussion of this aspect).

For comparison with the sedimentation behavior of our DNA sample in the presence of the classical intercalator ethidium, we first have performed a titration experiment with EB. The strong increase of the sedimentation coefficient observed on addition of EB to pGRB(r) (Fig. 3A) reflects the generation of positive supercoils and may serve as a proof that both strands of the DNA duplex are covalently closed. The lowest *s*-value of 11.7_0 S observed slightly above zero EB concentration (Fig. 3A) corresponds to a superhelix-free DNA conformation.

Figure 3B shows the sedimentation behavior of two Nt-pGRB(r) complexes of different Nt/nucleotide input ratio as a function of increasing EB concentration. The *s*-values of both complexes are higher than that of the naked relaxed plasmid and slightly depend on the Nt/nucleotide input ratio chosen (Fig. 3B). Upon addition of EB, both Nt-pGRB(r) complexes behave like negatively supercoiled DNA of relatively low absolute superhelix density: the sedimentation coefficient first decreases, attains a minimum already at very low EB/nucleotide input ratio, and increases again. This behavior means that negative supercoils have been introduced into the relaxed DNA by interaction with Nt. Thus, the two sets of experiments with both positively supercoiled and relaxed DNA qualitatively support our previous conclusions drawn from experiments with negatively supercoiled DNA [10].

The minima of the curves in Fig. 3B represent the superhelix-free state of the complexes. At this point, the accumulated unwinding of the DNA duplex by ethidium just compensates the linking number deficit, ΔLk, which was introduced into the DNA by the binding of Nt (and, to

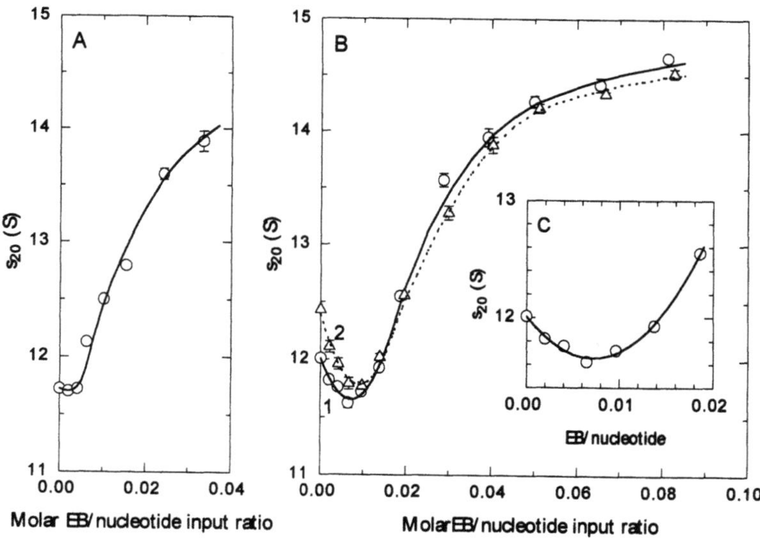

Fig. 3 Effect of EB on the mean sedimentation coefficient of relaxed pGRB(r) DNA (A) and of complexes of the same pGRB(r) sample with Nt (B). Nt/nucleotide input ratios were: 0.04 (1), 0.15 (2); buffer A. For complex 1, the second order polynomial fit of the data in the surrounding of the minimum is shown in detail (C)

a minor degree, by the difference in the thermodynamic conditions existing during the relaxation procedure and during the sedimentation experiments). ΔLk may be calculated from the equation [10]

$$\Delta Lk = -\frac{N \cdot \gamma}{180} \cdot r_c \qquad (1)$$

where N is the number of base pairs, r_c is the critical molar ethidium/nucleotide binding ratio at the curve minimum, and γ is the DNA unwinding angle of ethidium. In the Nt-pGRB complex, ethidium will bind to DNA regions uncovered by Nt. We assume that, at least at the low Nt/nucleotide input ratio of 0.044 (Fig. 3B, curve 1), ethidium has the same DNA unwinding angle of $\gamma = 26°$ (Table 8 in [25]) in the complex as in unperturbed DNA. This assumption is supported by the results of a thorough calorimetric study of local *versus* long-range effects of the Nt-DNA interaction [26], according to which the thermodynamic binding data primarily reflect local Nt-DNA interactions rather than long-range binding-induced conformational changes at regions distant from the binding site of Nt. Nevertheless, in our error analysis, for the complex γ was allowed to deviate by $d\gamma = \pm 2°$ from its DNA standard value to account for possible minor conformational changes of DNA structure which might occur outside the attachment site of Nt [10]. (Generally, in comparative measurements the reference value of 26° for DNA is treated as an error-free quantity).

A second order polynomial fitting of the sedimentation data in the vicinity of the minimum (Fig. 3C) yields a value of $(r_c)_t = (0.0072 \pm 0.0002)$ for the molar EB/nucleotide input ratio at the equivalence point for complex 1. Using the ethidium-DNA binding isotherm from a previous work [10] and the procedure reported therein, this value

may be converted to a critical EB/nucleotide binding ratio of $r_c = (0.0070 \pm 0.0005)$, valid for naked DNA. A displacement of ethidium by Nt is negligible at the very low ethidium concentration existing at the equivalence point (Fig. 3C); the corresponding input ratios of Nt and EB amount to six Nt molecules and one ethidium molecule per 70 base pairs. At a nearly tenfold higher ethidium concentration, only 5% of the ethidium molecules bound to DNA were found to be displaced by 0.044 Nt/nucleotide [10]. Therefore, the r_c value derived here for naked DNA may likewise be taken for the complex with 0.04 Nt/nucleotide. From Eq. (1), we find that this complex may be characterized by a linking number difference ΔLk of $-(1.7 \pm 0.2)$. (ΔLk may be regarded as an average for the population of topoisomers present in the DNA preparation and, therefore, does not need to be an integer). This small value is somewhat below the critical ΔLk value for the transition between the circle and figure-8 conformation of closed circular DNA [27, 28], meaning that the hydrodynamic frictional coefficient of the complex still is very similar to that one of the fully relaxed conformation. This might explain why the sedimentation coefficient of the Nt-DNA complex (ordinate axis value of curve 1, Fig. 3B) is a little higher only than the s-value of relaxed DNA. Generally, the sedimentation coefficient represents a very complex response of the Nt-mediated changes in molar mass, partial specific volume, and DNA conformation. Particularly, it is unclear how the changes of persistence length (local stiffening and curvature effects) and hydrodynamically operative contour length of DNA, well described for the interaction of Nt with linear DNA [29–31], together with a possible Nt-induced overwinding of the double helix sum up to change the linking number and the frictional coefficient of closed circular DNA.

Notably, the ΔLk value of -1.7 measured for the Nt-pGRB(r) complex cannot be fully accounted for by the action of Nt. The thermodynamic conditions under which the relaxation of pGRB($-$) DNA took place were different from those used in our sedimentation experiments. Since the superhelix density of DNA varies with temperature and ionic strength [32], a correction for the change in experimental conditions has to be made. Using the equations given in [32], we find that an average linking number difference of $\Delta Lk = -0.4$ has been introduced by the alteration in temperature and ionic strength, meaning that the 'relaxed' pGRB(r) had already adopted a very low negative superhelix density of -0.0025. Consequently, the Nt-induced effect consists only of $\Delta Lk = -1.3$, which would correspond to an average DNA overwinding angle of $\delta = (3.1 \pm 0.5)°$ in the case of complete binding of Nt. The latter figure is significantly lower than the value of $7.3°$ estimated for pBR322 under identical ionic conditions [10], probably due to the lower (A + T)-content of pGRB (36 mol-%) compared to that of pBR322 (46 mol-%), as discussed above.

Intrinsic DNA bends do not significantly alter the effect of Nt on DNA supercoiling

We were also interested to see whether the Nt-mediated effects on DNA supercoiling are influenced by sequence-directed bends of the DNA helix which can be partially abolished upon binding of Nt [31]. For this reason we have studied the interaction of Nt with a derivative of the plasmid pUGD1202 containing AT clusters (cf. Materials and methods). The sequence of the 1.2 kb insert of the original plasmid consists of tandem repeats of 59 basic units of average length 21 bp, most of which contain regularly positioned adenine or thymine clusters [20]. Due to the alternating appearance of (A_{3-5}) and (T_{3-4}) clusters at every pitch of the DNA helix this fragment, in solution, behaves as an intrinsically bent DNA molecule with strongly reduced electrophoretic mobility in 4% polyacrylamide gels [20]. In our transformants no clones were found containing the original 1.2 kb insert which obviously is not stable during DNA replication but is shortened by recombination processes. As revealed by restriction analysis, the plasmid pUGD1202-3 used in our sedimentation experiments consists of the unimpaired vector plasmid and an insert of approximately 960 bp in length. That means, the number of tandem repeats in the insert was reduced from 59 to about 46, without changing the base sequence of the repeats preserved. Therefore, despite the lower size of the insert, it will maintain the property of a bent segment. Previously, it was reported [20] that five fragments, differing in length, of the cloned 1.2 kb insert all migrate unusually slowly in 4% polyacrylamide gel compared with their migration in 2% agarose gel, showing that all fragments of the 1.2 kb sequence form bent DNA structures.

Figure 4 shows how the sedimentation coefficient of supercoiled pUGD1202-3 DNA (curve 1) and of its complexes with Dst-A (curve 2) and Nt (curve 3) varies with increasing amounts of EB added. Just as in our previous work with pBR322 [10], the curve minimum is shifted to higher EB/nucleotide input ratios by the binding of Nt, showing that, in comparison with the naked plasmid, more EB is needed to convert the supercoiled Nt-pUGD complex into a superhelix-free state. In other words, ethidium and Nt act on DNA in opposite ways: ethidium unwinds, Nt overwinds the double helix. In contrast to the behavior of the Nt-DNA complex, the titration curve of the Dst-A·DNA complex attains its minimum at the same EB concentration as the titration curve of the naked plasmid (Fig. 4).

The results presented in Fig. 4 were obtained in the presence of sodium chloride (buffer B). In Tris-buffer, sequence-directed bends of DNA are more pronounced compared with the situation in sodium chloride solution [33]. We have measured s/EB titration curves (not shown) in Tris-HCl (buffer A) and have found the same minima positions for the naked DNA and the netropsin-DNA complex (see $(r_c)_t$ values in Table 1) as in sodium chloride

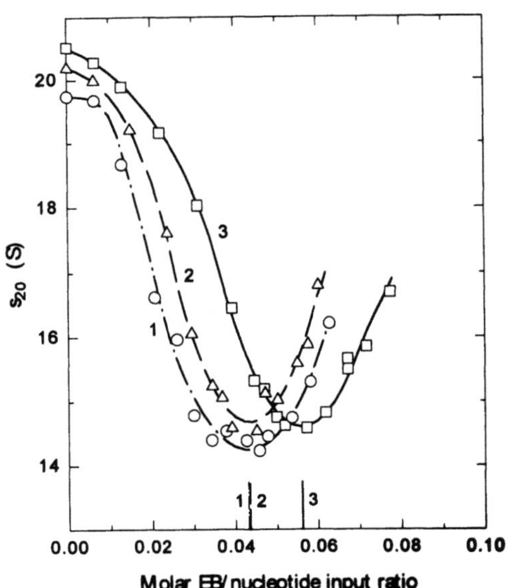

Fig. 4 Effect of ethidium bromide on the sedimentation behavior of supercoiled pUGD1202-3 DNA (curve 1) and of complexes of the same plasmid with Dst-A (curve 2) and Nt (curve 3) in buffer B. Molar ligand/nucleotide input ratios were 0.048 for Nt and 0.050 for Dst. Critical EB/nucleotide input ratios $(r_c)_t$, obtained by a second order polynomial fitting of the data in the vicinity of the minimum and indicated by vertical lines, are (0.0431 ± 0.0005) for DNA, (0.0434 ± 0.0005) for Dst-A·DNA, and (0.0561 ± 0.0004) for Nt·DNA

solution of the same ionic strength (see $(r_c)_t$ values in the legend of Fig. 4). This means that, for the system investigated, an alteration in the magnitude of local intrinsic helix bends has no significant influence both on the titratable superhelix density of the DNA nor on the Nt-mediated shift in the superhelix density and related quantities. A possible reason for this result could be that cation-induced changes of helix bending, acting on writhe Wr, and concomitant changes in the average duplex rotation angle [34], acting on twist Tw, cancel each other with regard to changes in the topological parameters (ΔLk, Lk_0) of supercoiled DNA.

A binding correction [10] of the $(r_c)_t$ values given in Table 1 yields the equivalence point binding ratios, r_c, for both the naked DNA and its complex with Nt (Table 1) which served as a starting point for the calculation of some structural parameters of DNA (see [10] for a compilation of the equations used). Although the helical repeat of the relaxed form of pUGD1202-3 DNA may be slightly affected by the existence of the AT-clusters in the insert, we have used the literature value of $h_0 = 10.6$ bp/turn [25] valid for DNA of random sequence (Table 1). The choice of h_0 is uncritical inasmuch as, for our purpose, the magnitude of the Nt-mediated changes is of essential interest instead of the true values of the parameters. Anyway, the superhelix density of DNA may differ from one DNA preparation to another due to cell cycle variation [25].

Table 1 shows that the binding of Nt to supercoiled pUGD1202-3 DNA is accompanied by a decrease in the linking number difference $\Delta Lk = Lk - Lk_0$, where Lk and Lk_0 are the linking numbers of the supercoiled and the relaxed form of DNA, respectively. The increase of $|\Delta Lk|$ suggests that, in the Nt-DNA complex, the DNA duplex behaves like a more strongly underwound DNA prior to ring closure than in the absence of Nt. A synonymous description of this finding is that the linking number of the relaxed form, Lk_0, is enhanced by the interaction with Nt (Table 1); the two DNA strands are more strongly intertwined in the relaxed complex compared with relaxed DNA. Correspondingly, a lower helical repeat, h_0, has to be assigned to the relaxed complex compared to the relaxed DNA (Table 1). The decrease of the superhelix density, σ, (increase of $|\sigma|$) upon complex formation (Table 1) indicates that Nt introduces additional negative supercoils into the DNA. All these effects may be summarized by a Nt-specific average duplex overwinding angle which amounts to $(5.2 \pm 1.4)°$ per Nt molecule added (Table 1).

Our previous experiments concerning the binding of ethidium to DNA and the displacement of ethidium by Nt [10] were performed in buffer A. With the assumption that these binding data may be used for a binding correction of the $(r_c)_t$ values obtained in buffer B of the same ionic strength (Fig. 4), we arrive at parameter values which, within the limits of error, agree with those summarized in Table 1. In this case, the DNA overwinding angle amounts to $\delta = (5.6 \pm 1.2)°$.

The minimum value for the netropsin-induced DNA overwinding angle of $(5.2 \pm 1.4)°$ derived here for pUGD1202-3 DNA in Tris buffer is somewhat lower compared to the value of $(7.3 \pm 1.1)°$ measured for pBR322 [10], although the limits of error just overlap each other. This lower value is surprising in so far as the plasmid pUGD1202-3, due to its AT-rich insert, has a higher percentage of sites suitable for a strong specific binding of Nt

Table 1 Critical molar EB/nucleotide binding ratios, r_c, of supercoiled pUGD1202-3 DNA and of a supercoiled Nt-PUGD1202-3 complex, and some topological and geometric parameters of the relaxed and superhelical forms of both the naked DNA and the complex derived therefrom. Molar Nt/nucleotide input ratio: 0.050; buffer A. Calculations are based on $N = 3625$ bp for pUGD1202-3 and $\gamma = 26°$ for the ethidium unwinding angle[1] [25]

Parameter	Symbol	DNA	Nt-DNA complex
Critical input ratio (EB$_t$/nucl.)	$(r_c)_t$	0.0439 ± 0.0009	0.0568 ± 0.0012
Critical binding ratio (EB$_b$/nucl.)	r_c	0.0419 ± 0.0016	0.0518 ± 0.0019
Helical repeat (relaxed form) (bp/turn)	h_0	10.6[1] (From literature [25])	10.44 ± 0.07
Linking number difference	ΔLk	-21.9[2] ± 0.8	-27.1 ± 2.3
Linking number (relaxed form)	Lk_0	342	347.2 ± 2.4
Linking number (supercoiled form)	Lk	320.1 ± 0.8	320.1 ± 0.8
Superhelix density (Specific linking number difference)	σ	-0.0640 ± 0.0025	-0.0781 ± 0.0067
Duplex overwinding angle (°) (Minimum value)	δ	–	5.2[3] ± 1.4

[1] Since not the "true" values of the parameters but rather the Nt-induced changes are of main interest here, we have treated h_0 and γ for naked DNA as well as N as error-free quantities. Only the propagation of our experimental error in r_c and $d\gamma = \pm 2°$ (for complexed DNA) are taken into account (cf. text).
[2] Although different topoisomers cannot be resolved in our experiments, ΔLk, Lk_0 and Lk are permitted to adopt non-integer values in our computer program to account for non-integer averages on a topoisomer population basis.
[3] Calculated with the assumption of complete binding of Nt.

than pBR322 and is expected to bind even more Nt than pBR322 under comparable conditions. On the other hand, it is known from findings with linear DNA that Nt, when bound to DNA, abolishes periodically arranged intrinsic helix bends in AT tracts [31]. In supercoiled DNA, this effect should be primarily accompanied by changes in the writhe *Wr* of DNA which might partially compensate an Nt-induced alteration in twist, *Tw*.

Nt excerts the strongest effects on DNA supercoiling upon interaction with DNA binding sites of highest Nt affinity

The magnitudes of the effects of Nt on the supercoiling of the plasmids pGRB and pUGD1202-3 described under 1.-3. are smaller than expected on the basis of the data of Snounou and Malcolm [5, 6] and of our previous experience with pBR322 [10] and pBR313 (unpublished). Although it seems possible to attribute these discrepancies to the low AT content of pGRB or to the structural peculiarities of the A-tracts in the insert of pUGD1202-3, these differences prompted us to compile all data hitherto available about Nt-mediated changes in DNA supercoiling. In some previous papers dealing with the influence of supercoiling on the mechanism of restriction endonucleases [35] or on the transcriptional activity [36], Nt was used as a tool to generate positively supercoiled DNA by a topoisomerase I assay. From the details concerning total Nt and DNA concentrations, linking number differences, or superhelix densities given in [5, 6, 11, 35, 36] we have calculated minimum values for the average Nt-induced overall DNA overwinding angle δ, simply using

$$\delta = \frac{\Delta Lk \cdot 360}{Z_t} \qquad (2)$$

Here, in the case of the topoisomerase I-mediated relaxation assay, ΔLk is the (*positive!*) change in linking number measured for naked DNA *after removal of Nt* and enzyme from the originally 'relaxed' Nt-DNA complex (with $\Delta Lk = 0$), and Z_t is the total number of Nt molecules added per plasmid molecule. To avoid confusion, it should be recalled that negative ΔLk (or $\Delta(\Delta Lk)$) values are obtained from experiments performed with relaxed (or supercoiled) DNA *in the presence of Nt* [10]. In this case, a negative sign has to be added to equation (2).

Figure 5 shows the overall Nt-specific DNA overwinding angles of Nt as a function of increasing Nt/nucleotide input ratio for various DNAs (see legend). Most strikingly, the δ values exhibit a strong decrease with increasing concentration of Nt added to DNA. The overwinding angles represent minimum values inasmuch as they are related to the number of Nt molecules, added, and not to the number of Nt molecules bound. Any binding correc-

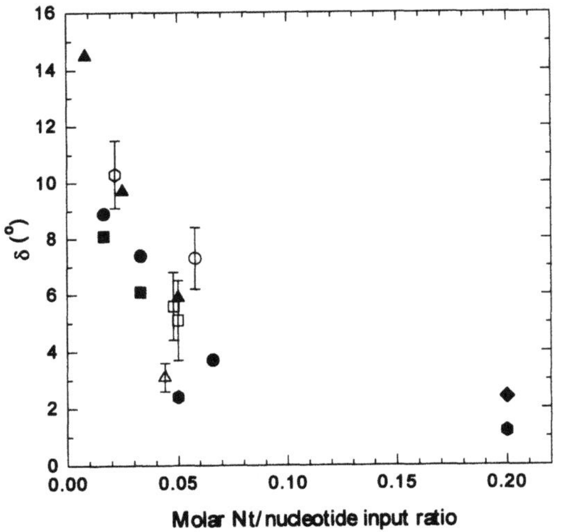

Fig. 5 Average Nt-mediated DNA overwinding angles, δ, calculated with the assumption of complete binding of Nt (*cf.* text). *Filled symbols*: data refer to linking number differences, ΔLk, or superhelix densities, σ, reported for positively supercoiled DNA which was produced by Nt in a topoisomerase-mediated DNA relaxation assay. DNAs were: (filled circle), pAT153 [6]; (filled squares), ΦX174 [6]; (filled triangles), pBR322 [35]; (filled diamond), pAT78/pAA82 [36]; (filled hexagons), pUC18-3,4 [11]. For the test point of pAT153 (circles) at an Nt/nucleotide input ratio of 0.066, no ΔLk value is given in [6]. We have deduced this test point from the fact that, in Fig. 1a of paper [6], no further change in ΔLk occurs upon increasing the Nt concentration from 10 to 20 μg/ml. *Open symbols*: data were derived from sedimentation experiments following the unwinding of Nt-DNA complexes by ethidium. DNAs were: (open circle), pBR322 [10], (open squares), pUGD1202-3 (this work); (open triangle), pGRB (this work); (open hexagon), pBR313 (unpublished)

tion would shift the data to lower abscissa and, concomitantly, to higher ordinate values. Due to the lack of reliable binding data and the complex properties of the Nt-DNA systems, we are unable to perform a meaningful binding correction, the more so as the data compiled in Fig. 5 differ in DNA base composition (and sequence!) and in the ionic conditions used in the experiments. Furthermore, from extensive viscometric studies on the interaction of Nt with linear DNAs of different base composition (see [29, 31] and literature cited therein) it is known that i) distinct DNA base-sequence dependent binding modes can be resolved at low degree of Nt binding (≤ 0.05 Nt/nucleotide), ii) the DNA conformational changes per bound Nt molecule decrease from mode to mode, iii) the different successive binding modes considerably differ in the association constant, iv) the respective binding sites of different binding strength are AT clusters of different sequence, v) for DNAs with an (A + T) content ≥ 50 mol-% and Nt/nucleotide input ratios below about 0.05, the concentration of free Nt is negligible compared to that of bound Nt, indicating that the apparent overall association constant is larger than 10^6 l/mol within this region. The latter

Progr Colloid Polym Sci (1995) 99:45–54
© Steinkopff Verlag 1995

finding is also supported by the fact that from quantitative footprinting data obtained at Nt/nucleotide input ratios ≤ 0.1, an *average* association constant of about 10^7 l/mol could be derived for the strong binding sites on a 139 base pair fragment of pBR322 [37]. From i) and iii) it follows that the data shown in Fig. 5 cannot be treated with any single association constant. Particularly, data corrections on the basis of association constants around 10^5 l/mol, previously derived by Scatchard analysis of binding experiments at Nt/nucleotide input ratios above 0.05 [38], are no longer meaningful at low levels of Nt binding. From (v), however, we may conclude that, perhaps with the exception of pGRB DNA with its low AT content of only 36 mol-%, in the range below 0.05 Nt/nucleotide the necessary binding corrections would be small and, above all, would not alter the important fact that δ strongly decreases with increasing amount of Nt bound (Fig. 5).

With the results summarized under (i)–(v) in mind, we have to conclude from Fig. 5 that the essential effect of Nt on supercoiled DNA originates from the interaction of Nt with the DNA binding sites of highest affinity, which primarily are occupied when Nt is added to DNA in solution. The preference of Nt for AT-rich sequences along the minor groove of B-DNA is a well-established fact [2, 29, 39, 40]; footprinting studies clearly reveal that the binding sites of Nt all contain runs of at least four A and T nucleotides [37, 41, 42]. Sequences consisting of five or four consecutive (AAAA) or alternating (ATAT) A·T pairs seem to be strongly preferred [2, 31]. These binding sites exhibit comparable and extremely high affinities to Nt [43, 44]; calorimetric binding studies in which poly(dAT)·poly(dAT) and poly(dA)·poly(dT) served as DNA host duplexes for the association of Nt yielded binding free energies, ΔG^0, of -12.7 kcal/mol and -12.2 kcal/mol at 25 °C, respectively [43]. The corresponding binding constants amount to $2.1 \cdot 10^9$ l/mol for poly(dAT)·poly(dAT) and $8.9 \cdot 10^8$ l/mol for poly(dA)·poly(dT). The sequence AATT, e.g., is expected to be correlated with a binding mode occurring at higher Nt/nucleotide ratios [31]; for the binding of Nt to the central AATT core of the decamer duplex [d(GCGAATTCGC)]$_2$, an association constant of $2.8 \cdot 10^8$ l/mol was obtained at 25 °C [26]. From Fig. 5 it follows that the "later" the site is occupied upon addition of increasing amounts of Nt (and the lower the affinity of the DNA binding site), the smaller is the alteration in the duplex winding angle of supercoiled DNA per Nt molecule bound to that site. Since, in our calculation of δ, the total change in the winding angle always is related to the total number of Nt molecules added up to a certain Nt level, independent of whether the Nt molecules bound do or do not contribute to a change in DNA conformation, a decrease of δ with rising Nt concentration (Fig. 5) is an obligatory consequence.

We do not intend to discuss the origin of the considerable scattering of the data shown in Fig. 5 in detail. Besides the fact that different DNAs under distinct ionic conditions have been used by different authors, also different levels of error may be characteristic for the topoisomerase assay and the s/EB titration method. In the case of the classic DNA unwinding assay based on sedimentation measurements, a question hitherto unanswered is whether the chemical equilibrium of ethidium-DNA binding may be noticeably affected by possible surface adsorption processes connected with synthetics centerpieces (and windows) of the ultracentrifuge cell. This problem merits study. An analytical ultracentrifuge equipped with a fluorescence detection system [45] should be a suitable tool to directly measure the fraction of ethidium bound to DNA within an ultracentrifuge cell, in comparison with separate usual spectroscopic measurements of the EB/DNA binding isotherm.

Acknowledgement The authors are indebted to Dr. Claus Urbanke (Hannover) for his advice concerning the interfacing of the scanning system of the Spinco E ultracentrifuge to a personal computer and for providing the operating and data handling software, as well as to Dipl.-Ing. W. Schütze (Jena) for technical realization of this on-line coupling. We wish to thank Prof. P. Forterre (Orsay, France), Prof. S. Mizuno (Sendai, Japan), and Dr. H. Simon (Jena) for generous gifts of *Halobacterium* strain GRB, plasmid pUGD1202, and topoisomerase I from chicken erythrocytes, respectively. The critical reading of the manuscript by Prof. Ch. Zimmer, Dr. K.E. Reinert and Dr. A. Walter is gratefully acknowledged. Part of this work was supported by the Deutsche Forschungsgemeinschaft, project Nr. Fr 876/1-1.

References

1. Hahn FE (1975) In: Corcoran JW, Hahn FE (eds) Antibiotics III. Mechanism of action of antimicrobial and antitumour agents. Springer, Berlin-Heidelberg-New York, pp 79–100
2. Zimmer Ch, Wähnert U (1986) Prog Biophys Molec Biol 47:31–112
3. Zimmer Ch, Luck G (1992) In: Hurley LH (ed) Advances in DNA Sequence Specific Agents, Vol 1. JAI Press London, pp 51–88
4. Zimmer Ch, Luck G, Burckhardt G, Krowicki K, Lown JW (1986) In: Chargas C, Pullman B (eds) Molecular Mechanisms of Carcinogenic and Antitumor Activity. Int Symp at the Pontifical Acad Sci 70, Vatican Press, pp 339–363
5. Malcolm ADB, Snounou G (1983) Cold Spring Harbour Symp Quant Biol 47:323–326

6. Snounou G, Malcolm ADB (1983) J Mol Biol 167:211–216
7. Bates Ad, Maxwell A (1993) DNA Topology. IRL Press, Oxford, pp 83–105
8. Crawford LV, Waring MJ (1967) J Mol Biol 25:23–30
9. Bauer W, Vinograd J (1968) J Mol Biol 33:141–171
10. Triebel H, Bär H, Walter A, Burckhardt G, Zimmer Ch (1994) J Biomol Struct Dyn 11:1085–1105
11. Störl K, Burckhardt G, Lown JW, Zimmer Ch (1993) FEBS Lett 334:49–54
12. Kopka ML, Yoon C, Goodsell D, Pjura P, Dickerson RE (1985) J Mol Biol 183:553–563
13. Waring MJ (1965) J Mol Biol 13:269–282
14. Krugh TR, Wittlin N, Cramer StP (1975) Biopolymers 14:197–210
15. Suh D, Sheardy RD, Chaires J (1991) Biochem 30:8722–8726
16. Zimmer Ch, Reinert KE, Luck G, Wähnert U, Löber G, Thrum H (1971) J Mol Biol 72:329–348
17. Sioud M, Forterre P, de Recondo AM (1987) Nucl Acids Res 15:8217–8234
18. Sioud M, Baldacci G, de Recondo AM, Forterre P (1988) Nucl Acids Res 16:1379–1391
19. Sambrook J, Fritsch EF, Maniatis D (1989) Molecular Cloning (A Laboratory Manual), Second Edition. Cold Spring Harbor Laboratory Press, 1.23–24
20. Saitoh Y, Saitoh H, Ohtomo K, Mizuno S (1991) Chromosoma 101:18–25
21. Ebel C, Altekar W, Langowski J, Urbanke C, Forest E, Zaccai G (1995) Biophys Chem 54:219–227
22. Floßdorf J (1980) Makromol Chem 181:715–724
23. Hackett NR, DasSarma S (1989) Can J Microbiol 35:86–91
24. Cozzarelli NR, Boles TC, White JH (1990) In: Cozzarelli NR, Wang JC (eds) DNA Topology and Its Biological Effects. Cold Spring Harbor Laboratory Press, pp 139–184
25. Bauer W (1990) In: Landoldt-Börnstein New Series Group VII, Vol. 1, 'Nucleic Acids', Subvolume d'Physical Data II' Springer-Verlag Berlin-Heidelberg-New York-London-Paris-Tokyo-Hongkong-Barcelona, pp. 1–30
26. Breslauer KJ, Ferrante R, Marky LA, Dervan PB, Youngquist RS (1988) In: Sarma RH, Sarma MH (eds) Structure and Expression, Vol 2: DNA and Its Drug Complexes. Adenine Press, pp 273–290
27. Le Bret M (1984) Biopolymers 23:1835–1867
28. Schlick T, Olson WK (1992) J Mol Biol 223:1089–1119
29. Reinert KE, Stutter E, Schweiß H (1979) Nucl Acids Res 7:1375–1392
30. Reinert KE, Geller K, Burckhardt G (1991) J Biomol Struct Dynamics 3:537–552
31. Reinert KE (1993) J Biomol Struct Dynamics 6:973–990
32. Bauer W (1978) Ann Rev Biophys Bioeng 7:287–313
33. Dieckmann S, Wang JC (1985) J Mol Biol 186:1–14
34. Anderson P, Bauer W (1978) Biochemistry 17:594–601
35. Snounou G, Malcolm, ADB (1984) Eur J Biochem 138:275–280
36. Brahms JG, Darggouge O, Brahms S, Ohara Y, Vagner V (1985) J Mol Biol 181:455–465
37. Dabrowiak JC, Goodisman J, Kissinger K (1990) Biochemistry 29:6139–6145
38. Luck G, Triebel H, Waring M, Zimmer Ch (1974) Nucl Acids Res 1:503–530
39. Kopka ML, Larsen TA (1992) In: Probst CL, Perun TJ (eds) Nucleic acid targeted drug design. Marcel Dekker, New York, pp 303–374
40. Dabrowiak JC, Stankus AA, Goodisman J (1992) In: Probst CL, Perun TJ (eds) Nucleic acid targeted drug design. Marcel Dekker, New York, pp 93–149
41. Portugal J, Waring MJ (1987) Eur J Biochem 167:281–289
42. Bailly C, Colson P, Houssier C, Houssin R, Mrani D, Gosselin G, Imbach JL, Waring MJ, Lown JW, Hènichart JP (1992) Biochemistry 31:8349–8362
43. Marky LA, Breslauer KJ (1987) Proc Natl Acad Sci USA 84:4359–4363
44. Marky LA, Kupke DW (1989) Biochemistry 28:9982–9988
45. Schmidt B, Rappold W, Rosenbaum V, Fischer R, Riesner D (1990) Colloid Polym Sci 268:45–54

Progr Colloid Polym Sci (1995) 99:55–62
© Steinkopff Verlag 1995

E.V. Karpova
T.N. Osipova
V.I. Vorob'ev

Sedimentation studies of specific association of oligonucleosomes from sea urchin sperm chromatin

Received: 3 March 1996
Accepted: 3 July 1995

Dr. E.V. Karpova (✉) · T.N. Osipova[1]
V.I. Vorob'ev
Institute of Cytology of the Russian
Academy of Sciences
Tikhoretsky pr. 4
St. Petersburg 194064, Russia

[1]*Present address*:
T.N. Osipova
Biological Institute of the
St. Petersburg University
St. Petersburg 198904, Russia

Abstract It is known that oligonucleosomal chromatin fragments from chicken erythrocytes enriched in inactive genes can self-associate into pseudo-higher-order structures. In this paper, we demonstrate the similar aggregation phenomenon for another type of repressed cells, sea urchin spermatozoa. Like the inactive part of erythrocyte chromatin, sea urchin sperm chromatin is characterized by the long linker DNA (100 base pairs) and the presence of specific histones typical for the repressed state of chromatin. It seems that these two factors determine the specific association of oligonucleosomes. The aggregates of sea urchin sperm oligonucleosomes resemble those of the erythrocyte chromatin in the following features: the range of the sizes of oligonucleosomes that are able to associate into pseudo-higher-order structures is limited; such self-association takes place at a low extent of the chromatin digestion by micrococcal nuclease; the aggregates undergo salt-induced compaction. Similarity in the properties of chromatin aggregates from two types of terminally repressed cells makes it possible to consider the specific association into pseudo-higher-order structures as a characteristic feature of chromatin from the inactive part of the genome.

Key words Sedimentation – chromatin structure – sea urchin sperm chromatin – sperm-specific histones – active and inactive genes

Introduction

Salt-dependent oligonucleosome aggregation is known to be one of the intrinsic properties of isolated fragments of chromatin. Usually it is observed as the appearance of bi-modal sedimentation profiles of unfractionated nuclease-digested chromatin [1–4]. Later on it became clear that there exist two different phenomena [5]: 1) nonspecific aggregation due to redistribution of linker histones of the H1 family from ends of very short fragments to larger oligomers of various sizes resulted in heterogeneous aggregates that survive exposure to low ionic strength; 2) association of two or more fragments forming rather homogeneous pseudo-higher-order structures without continuity of DNA in them. In both cases the ionic strength serves as a factor that supports aggregation. Formation of aggregates of the first type is usually induced by a high level of nuclease digestion of nuclear chromatin. It originates from the redistribution of histones H5 and H1 due to their preference for higher-order chromatin structure [6]. Such aggregation seems to be an artifact having little to do with the native chromatin structure. It was observed for various types of chromatin such as the chromatin from rat liver nuclei [1, 4], bovine lymphocyte nuclei [3] and chicken erythrocyte nuclei [4, 5].

On the contrary, appearance of aggregates of the second type was shown only for chicken erythrocyte oligonucleosomes isolated after a very mild nuclease digestion to prevent redistribution of linker histones [5, 7]. The integrity of the aggregates is supported by an increased ionic strength. It seems that such type of aggregation is characteristic of a more stable higher-order structure of the transcriptionally repressed chromatin and results from the presence of the erythrocyte specific histone H5. Moreover, the aggregates are enriched in the genes that are inactive in erythrocytes [8–10]. Presumably, the chromatin of active genes is unable to self-associate into pseudo-higher-order structures because of its slight depletion of histones H5 and H1 compared with inactive genes [10].

In this paper, we demonstrate the aggregation phenomenon of the second type for oligonucleosomes from another type of transcriptionally inactive cells, sea urchin spermatozoa. Sperm cells with their completely repressed small heterochromatic nuclei are a good model to study inactive chromatin. Sea urchin spermatozoa represent a rare example of ripe sperm cells that preserve the nucleosomal organization of chromatin in the course of spermatogenesis. Besides having the highest nucleosomal DNA repeat size ever observed (248 base pairs (bp)), the chromatin from the sea urchin (*Strongylocentrotus intermedius*) sperm has unusual sperm-specific histones spH1 and spH2B (two subfractions) characterized by a higher molecular weight and a very high arginine content [11, 12]. The sequence analysis of sperm-specific histones from other sea urchin species showed that the increase of molecular weight of these histones was due to elongation of their *N*-terminal domains as well as the *C*-terminal domain of spH1 by the repetitive peptides Ser-Pro-(Thr)-Basic-Basic [13]. Previously, we estimated the amount of histone H1 in sea urchin sperm chromatin as two spH1 molecules per nucleosome [14], which is inside the range of the H1 content for various inactive chromatins (1.3–2.4 molecules per nucleosome) [15]. Thus chromatin from the sea urchin sperm, like the erythrocyte chromatin, is characterized by a long linker DNA and the presence of specific histones characteristic of repressed chromatin. It seems that these two factors determine the formation of the second type aggregates of oligonucleosomes.

Materials and methods

We used sea urchins *Strongylocentrotus intermedius*, obtained at the Marine Biological Station of the Institute of Marine Biology of the Russian Academy of Sciences (Vladivostok, Japanese sea). The procedures used for the preparation of oligonucleosomal fragments of chromatin from the sea urchin spermatozoa are described in detail by Osipova et al. [11].

Briefly, the sea urchins were stimulated by injection of 0.5 M KCl. The spermatozoa were collected into the filtered sea water and their maturity was checked microscopically. The sperm cells were collected by centrifugation at 5000 rev./min and washed three times with buffer A (0.25 M sucrose, 10 mM Tris-HCl, pH 7.5, 3 mM CaCl$_2$, 0.1 mM PMSF). A concentrated suspension of spermatozoa was stored in buffer A with 50% glycerol at $-20\,°C$.

Nuclei were purified in buffer A with 0.5% Triton X-100, washed twice with buffer A and treated with micrococcal nuclease. The extent of chromatin digestion was estimated by the amount of acid-soluble products. Just after lysis into 5 mM Tris-HCl pH 7.5, 4 mM EDTA, 0.1 mM PMSF, the total soluble chromatin was fractionated in a linear 5–20% (w/v) sucrose gradient containing 5 mM Tris-HCl, pH 7.5, 1 mM EDTA, 0.1 mM PMSF. Centrifugation proceeded at 25 000 rpm for 16–17 h. To achieve the desired ionic strength for analytical ultracentrifugation 2 M NaCl was added to the chromatin samples 16–24 h before sedimentation. Sedimentation coefficients of oligonucleosomal fractions, the weight-average size of their DNA, the protein content, and the protein-to-DNA weight ratio were determined as described elsewhere [14].

To separate the aggregated and unaggregated material, several combined fractions from the first sucrose gradient were subjected to another centrifugation, at 27 000 rpm for 3.5 h in a 10–50% (w/w) linear sucrose density gradient in the same buffer. In this case the fractionation started not sooner than 40 h after the chromatin solubilization.

Results and discussion

Oligonucleosomes from sea urchin sperm chromatin are able to self-associate into pseudo-higher-order structures

We have started studying the salt-induced compaction of sea urchin sperm chromatin fragments as an example of sedimentation behavior of the chromatin with long linker DNA [11, 14]. For this purpose we used the isolated oligonucleosomal fractions obtained in a linear sucrose gradient (5–20%) at low ionic strength (5 mm Tris-HCl, pH 7.6, 1 mM EDTA). Surprisingly, the sedimentation boundaries besides the component corresponding to unaggregated oligonucleosomes showed the presence of rapidly sedimenting material. A typical example of the two-component sedimentation boundary is presented in Fig. 1. Both components are observed within the whole range of ionic strengths from 5 to 85 mM and the sedimentation coefficient $s_{20,\,w}$ of the fast-sedimenting component is elevated from 70 S to 180 S with increase of the ionic strength

Progr Colloid Polym Sci (1995) 99:55–62
© Steinkopff Verlag 1995

(Fig. 2). The distinct, well-resolved sedimentation boundary of the fast component and a sharp increase of its sedimentation coefficient with increasing ionic strength indicate that it is a product of association of oligonucleosomes. We have never observed similar association for other types of chromatins, such as rat thymus chromatin and chromatin from pigeon brain cortical neurons [14, 16].

Comparison of the heights of the sedimentation boundaries corresponding to the aggregated and unaggregated components has allowed us to estimate the proportion of the aggregates. It has been found that the proportion of aggregated material is determined by the extent of chromatin digestion. About 40–60% of oligonucleosomes in the aggregated state are observed when the extent of hydrolysis of the nuclear chromatin by micrococcal nuclease is lower than 0.7% of acid soluble products. The increase of the extent of digestion up to 1.4% of acid soluble products decreases the amount of aggregated material two times. The fast-sedimenting material practically disappears when the amount of acid soluble products reaches 3–4%. Thus, in spite of the fact that aggregates of sea urchin sperm oligonucleosomes survive the exposure to low ionic strength, the strong reverse dependence of the amount of aggregates on the extent of nuclease digestion supports the idea that this process is self-association of oligonucleosomes into pseudo-higher-order structures.

As linker DNA is more accessible to the micrococcal nuclease action than nucleosomal DNA, our above data suggest the critical role of maintenance of intact terminal linker fragments during the nuclease treatment of nuclear chromatin for the ability of chromatin fragments to form

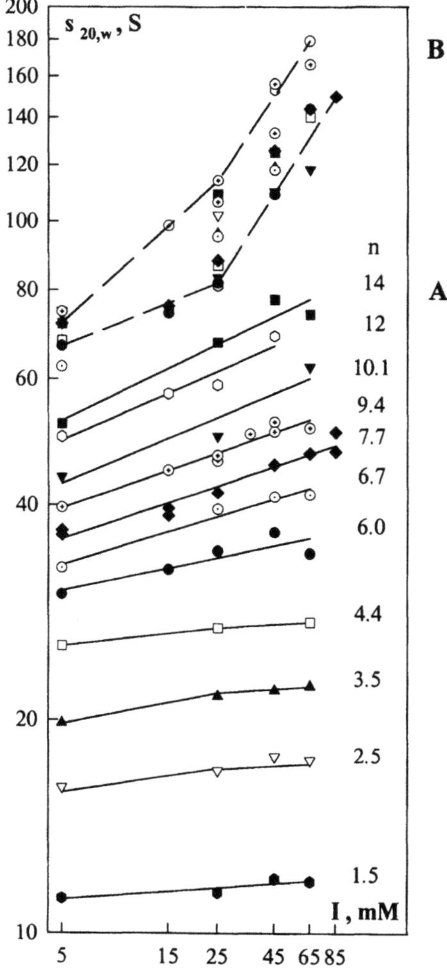

Fig. 2 Ionic strength dependence of sedimentation coefficients for both components, present in analytical sedimentation experiments: A) unaggregated oligonucleosomal fractions of the sea urchin sperm chromatin; n is the weight-average size of chromatin fragments, expressed as the number of nucleosomes; B) their aggregates. The same symbols in A and B parts indicate the slow and fast components of the same experiment

Fig. 1 Two-component sedimentation boundaries obtained in a typical sedimentation experiment performed at $I = 5$ mM with the oligonucleosomal fragment of the sea urchin sperm chromatin containing the weight-average number of nucleosomes $n = 10.1$. The extent of hydrolysis of chromatin by micrococcal nuclease is 0.9% of acid soluble products. The arrow denotes the direction of sedimentation from left to right; the speed of rotation is 29 500 rpm. The experimental curves, taken in 4-min intervals, show A_{260} against distance from the axis of rotation. The boundaries of the components are sufficiently well resolved to permit measurements of both s-values

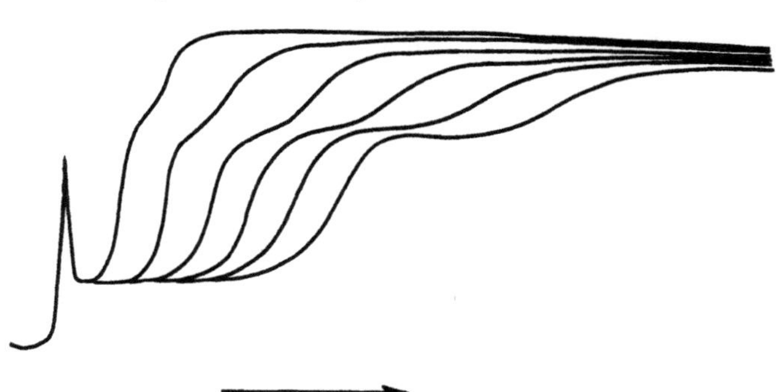

the pseudo-higher-order structures. Some additional facts to support this suggestion will be presented in the paper devoted to the biochemical properties of the aggregates that we are preparing now. The same mechanism of aggregate formation was demonstrated by Grigoryev et al. [9] for chicken erythrocyte chromatin. Earlier, we observed a similar mechanism of association for the short fragments (n is between 2.5 and 4 nucleosomes) of DNase I digested pigeon erythrocyte chromatin [17]. The specificity of DNase I action results in 20 bp elongated terminal linker. And the presence of elongated linker was crucial for the formation of aggregates in that case, too.

Figure 2 demonstrates ionic strength dependence of sedimentation coefficients of oligonucleosomal fragments of the sea urchin sperm chromatin (part A) and their aggregates (part B). The oligonucleosomes were taken from chromatin samples obtained at different levels of nuclease treatment. That is why in some cases, when the lower amounts of fast-sedimenting material were present, it was not possible to get reliable values of sedimentation coefficients of aggregates.

Ionic strength dependent condensation represents a traditional approach to the studies of the formation of higher order structure by isolated chromatin fragments [18]. Oligonucleosomal fragments of sea urchin sperm chromatin show the typical character of ionic strength induced compaction determined by the size of the fragment (Fig. 2A). Our previous papers were devoted to the detailed description of this process [14, 16]. In short, the fragments containing less than 6 nucleosomes undergo only the first stage of compaction (up to 25 mM) demonstrating inability to form any higher order chromatin structure. In our opinion, this stage may be related to some additional wrapping of a part of linker DNA over the histone octamer. The longer chromatin fragments show a power-law dependence of sedimentation coefficient on ionic strength. The value of the exponent reflects the degree of compactness of the structure. Thus, starting from six nucleosomes the chromatin fragments can form higher order structure and the compactness of the structure increases for the oligonucleosomes longer than 10–11 nucleosomes. The data presented in Fig. 2A are in good agreement with the similar data of Thomas et al. [19] for the sperm chromatin from another sea urchin species.

The sedimentation coefficients of the aggregates that comigrate with chromatin fragments containing from 2.5 to 14 nucleosomes are presented in Fig. 2B. It is to be noted that even for the fragment with $n = 1.5$, we observed the fast-sedimenting component, but of somewhat lower size ($s_{20,w}$ is about 50 S at 5 mM). The values of sedimentation coefficients of the aggregates are independent on the size of comprised fragments. Presumably, the existence of the upper limit of the aggregate sizes for each ionic strength results from the fact that the larger aggregates cannot withstand hydrodynamic shear force. The value of the highest $s_{20,w}$ of aggregates may be dependent on the amount of linker histones. This point will be discussed in more detail later.

The occurrence of association for the oligonucleosomes containing less than six nucleosomes contrasts with the results obtained in sedimentation studies of chromatin fragments from the chicken erythrocyte chromatin [5]. Other authors observed the specific aggregation for small nucleosome oligomers (2.5–5.5 nucleosomes) in electrophoresis of the chicken erythrocyte chromatin [9] and in sedimentation experiments with DNase I treated pigeon erythrocyte chromatin [17]. But in both studies the factors are present that favor aggregate formation, namely, con-

centration of the material at the starting zone of electrophoresis and the elongated terminal linker, produced by DNase I. Two facts assure us of the ability of small fragments from the sea urchin sperm chromatin to self-associate: 1) we did not observe an increase in the amount of fast-sedimenting relative to slow-sedimenting material with increasing oligomer size from 2.5 to 6 nucleosomes; 2) two-dimensional DNP-DNA electrophoresis demonstrates the presence of small oligonucleosomes in the aggregates (unpublished data).

The inability of short chromatin fragments (less than six nucleosomes) to self-associate into pseudo-higher-order structure would support the suggestion that the self-association occurs due to the formation of higher order chromatin structure. But this is not the case, at least, for sea urchin sperm oligonucleosomes. It seems that the association of chromatin fragments at low ionic strength is determined by intermolecular interactions, which are demonstrated for N-terminal tails of spH1 and spH2B [20]. And then associated, presumably end-to-end, fragments undergo salt-induced folding into higher order structure similar to continuous ones.

Sedimentation coefficients of aggregates demonstrate a sharp increase at the ionic strength higher than 25 mM. Two phenomena may be responsible for this. In this range of ionic strengths (between 15 and 35 mM NaCl), Clark and Thomas [21] observed change in the character of protein–protein and/or protein–DNA interactions in spH1-DNA complexes resulting in growth of sedimentation coefficients of the complexes and in change of their morphology (from thin filaments containing two DNA molecules to rod-like complexes consisting of thin filaments packed side-by-side). On the other hand, long continuous fragments of sea urchin (*Echinus esculentis*) sperm chromatin that have s-values at 5 mM close to those for aggregates (about 70–75 S) demonstrate a jump in the ionic strength dependence of sedimentation coefficients between 45 and 55 mM [19]. The authors consider that the jump reflects the ability of long chromatin fragments to fold into more stable higher order structure that can better (than shorter ones) withstand hydrodynamic shear forces. In our case, both processes may take place and result in the sharp increase of sedimentation coefficients of aggregates: the additional aggregation and the formation of more stable and compact higher order structure by "sewed" chromatin fragments as it was demonstrated for continuous ones, if the end-to-end association occurred at 5 mM.

Kinetics of the formation of aggregates

The presence of the second fast-sedimenting component in analytical sedimentation experiments of the oligonucleo-

Progr Colloid Polym Sci (1995) 99:55–62
© Steinkopff Verlag 1995

somes, size-fractionated previously in the linear sucrose density gradient, appears somewhat unexpectedly. In both cases, sedimentation was carried out under the same ionic conditions (5 mM Tris-HCl pH 7.5, 1 mM EDTA). The aggregated material should have been separated from the unaggregated one in the course of preparative gradient centrifugation. It seems unlikely that sedimentation of more dilute oligonucleosome solutions in the analytical ultracentrifuge can induce self-association. A more reasonable suggestion is that the formation of aggregates is a slow process that is still not completed by the end of the sucrose gradient centrifugation.

Usually, we started the sucrose gradient fractionation of total soluble chromatin just after its solubilization, and the centrifugation lasted for 16 h. The amount of aggregates that accompany oligonucleosomal fractions significantly decreases if the time interval between solubilization and size-fractionation of chromatin is 24 h. In 40–48 h after solubilization, the process of aggregate formation is completed and the entire aggregated material is separated from the oligonucleosomal fragments that are not able to self-associate in the course of the sucrose gradient centrifugation.

Taking into account such kinetics of the aggregate formation, we subjected the fractions obtained from the first sucrose gradient to a second separation in a linear (10–50%) sucrose density gradient, the first centrifugation beginning just after the solubilization of chromatin and the second one, not sooner than 40 h after the first one. We have chosen the fraction of the first sucrose gradient with the weight-average number of nucleosomes $n = 5.5$ (Fig. 3b). Corresponding fractions from three gradients, run in parallel, were combined and loaded on the top of the second (10–50% w/w) sucrose density gradient either under the same ionic conditions ($I = 5$ mM) or after addition of NaCl up to 45 or 65 mM. The profiles of the gradients are presented in Fig. 3a. It demonstrates a typical bimodal sedimentation profile. The increasing ionic strength induces the growth of sedimentation coefficients of both aggregated and unaggregated material without significant redistribution of material between these two peaks. A certain total loss of material due to precipitation was observed at 65 mM.

Comparison of the histone content of the separated fractions of associated oligonucleosomes and the free oligonucleosomes from the gradient run at 5 mM (denoted further as ASS-5 and ON-5) (Fig. 3c) using densitometry of SDS/15% polyacrylamide gels stained with Coomassie blue R revealed a 20–30% depletion of ON-5 in spH1. Although it is difficult to measure such a small difference in histone content using Coomassie blue staining, we reliably observe it in five different chromatin preparations. As the aggregates in consideration contain the oligonucleosomes with $n = 5.5$, such value of depletion corresponds well to

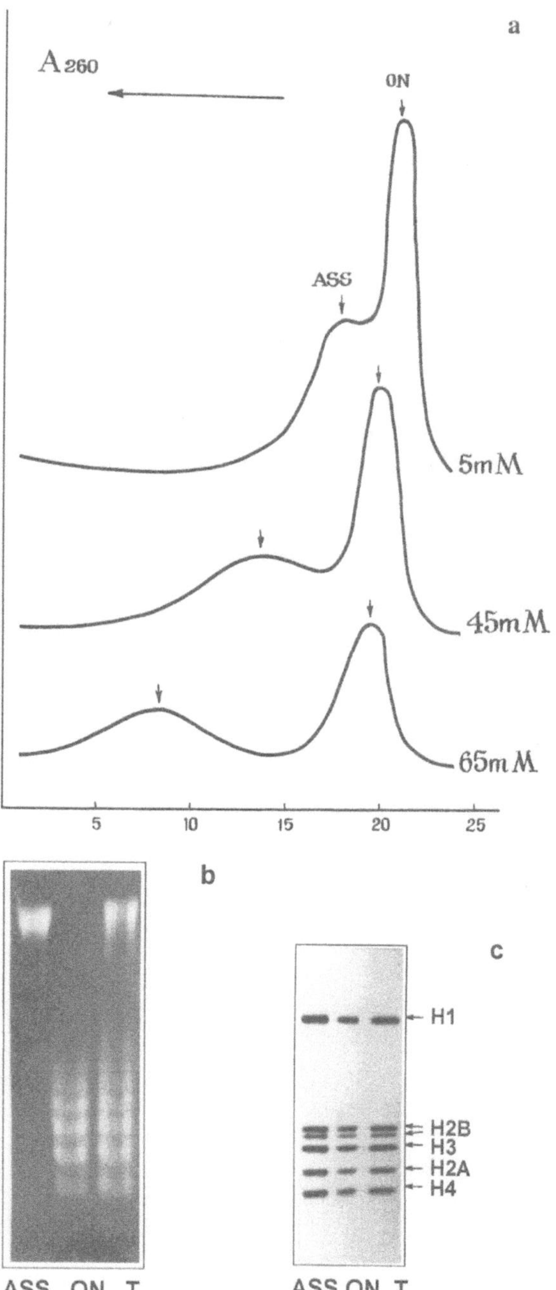

Fig. 3 a) Bimodal distribution of associated (ASS) and non-associated (ON) forms of oligonucleosomes (the weight-average size is about 5.5 nucleosomes), isolated at the ionic strength 5 mM, upon sedimentation in 10–50% linear sucrose gradient at different ionic strengths. The arrow indicates the direction of sedimentation. b) and c) The DNA and protein contents of the material in the fast (ASS) and slow-sedimenting (ON) peaks and of the total material (T), loaded on the top of the sucrose gradient, run at 5 mM

the suggestion that oligonucleosomes are unable to self-associate at $I = 5$ mM because they have lost the H1 histone (perhaps one of two molecules H1 per nucleosome) from the ends of the chromatin fragment, as was shown in [9] for chicken erythrocyte oligonucleosomes.

Reversible unfolding of aggregates of oligonucleosomes from sea urchin sperm chromatin

For refolding experiments ASS-5 and ON-5 fractions from the gradient run at 5 mM were divided into two parts. Half were dialyzed into 0.2 mM EDTA, pH 7.5, in order to evaluate stability of aggregates and reversibility of low ionic strength induced unfolding. Then the sedimentation coefficients of both parts were measured in a series of increasing ionic strengths. The samples for analytical sedimentation were diluted up to $A_{260}^{1\,cm} = 0.5$. The corresponding material from the gradient run at 65 mM (ASS-65 and ON-65) was subjected to a similar study with the only exception: the sample was dialyzed into 5 mM Tris-HCl, pH 7.5, 1 mM EDTA. Figure 4 shows the obtained results.

Fig. 4 Ionic strength dependence of sedimentation coefficients of the associated and non-associated material obtained after the sucrose gradient fractionation, presented in Fig. 3. The solid lines refer to the material from 5 mM sucrose gradient (ASS-5 and ON-5) and the dashed lines – to the material from 65 mM sucrose gradient (ASS-65 and ON-65 for "secondary" oligonucleosomes). The solid symbols correspond to the material, subjected to dialysis and the open symbols – to the not subjected one. The rectangles show the sedimentation coefficients of "secondary" aggregates arising at 45 mM and 65 mM in ON-5 fraction. For details see the text

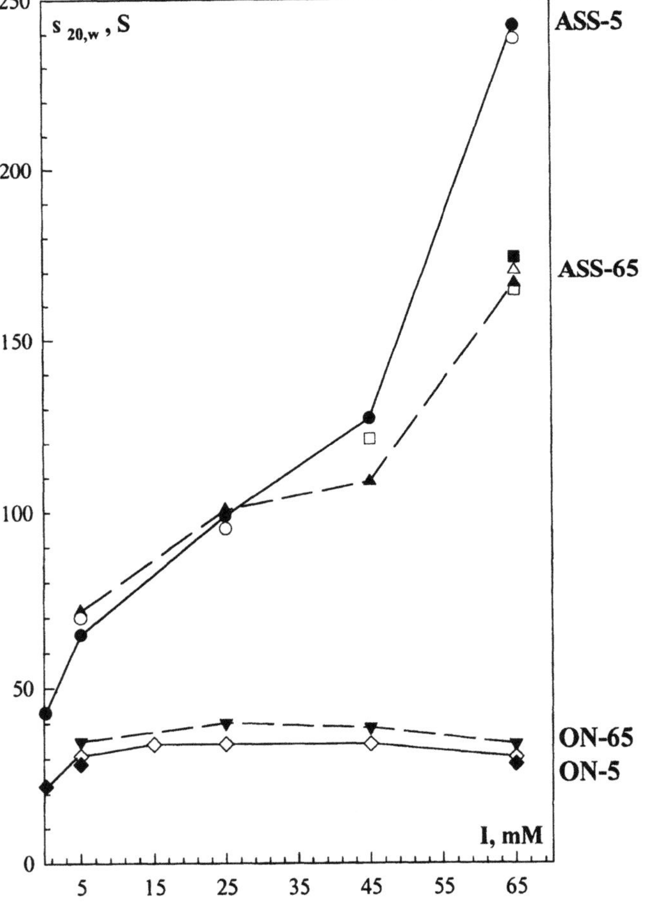

ASS-5 demonstrates striking stability: at the ionic strength of 0.2 mM the unfolded oligonucleosomes are still in associated form. Such unfolding is completely reversible. The difference between sedimentation coefficients of ASS-5 subjected and not subjected to the dialysis does not exceed the error of the experiment in the whole range of the ionic strengths used. Both samples of ASS-5 contained about 30% of unaggregated oligonucleosomes with sedimentation coefficients close to those for ON-5 (the data are not presented in Fig. 4 because they superimpose on the points for ON-65). The amount of unaggregated material was not increased after dialysis. The presence of such material results from a partial dissociation of aggregates due to the dilution of the samples for analytical ultracentrifugation. The portion of such "secondary" oligonucleosomes is dependent on the ionic strength (Table 1).

ASS-65 show similar reversible unfolding. But after dialysis well defined slow-sedimenting material, comprising about 60% of the total material, appeared. The s-values for the newly separated oligonucleosomal fragments are presented in Fig. 4 and denoted ON-65. The s-values for unaggregated oligonucleosomes, obtained after separation in sucrose gradient run at 65 mM, are not presented in Fig. 4. The experimental points lie between those for ON-5 and ON-65.

ON-5, as well as ON-65, separated from ASS-65 after dialysis, demonstrate the type of salt-induced compaction normal for such short chromatin fragments [14, 16]. Thus,

Table 1 Influence of ionic strength on the dilution induced association–dissociation of sea urchin sperm oligonucleosomes

Sucrose gradient fractions	Ionic strength mM	The percentage of "secondary" forms appearing after dilution of sucrose gradient fractions up to $A_{260}^{1\,cm} = 0.5$	
ASS-5	0.2	32	"secondary"
	5	29	oligonucleosomes
	25	25	
	45	0–10	
	65	0–10	
ASS-65	5	62	"secondary"
	25	54	oligonucleosomes
	45	46	
	65	32	
ON-5	0.2	no	"secondary"
	5	no	aggregates
	25	no	
	45	26	
	65	56	
ON-65	5	no	"secondary"
	25	no	aggregates
	45	no	
	65	no	

a slight increase of the sedimentation coefficients for ON-5 and ON-65 are observed only up to 25 mM. At 45 mM and 65 mM in both dialyzed and undialyzed oligonucleosomes ON-5 fast-sedimenting material appears (denoted by the rectangles in Fig. 4). The sedimentation coefficients of these "secondary" aggregates are close to the values of $s_{20, w}$ for ASS-65 (Fig. 4). As was mentioned above, ON-5 are slightly depleted of H1 compared with the oligonucleosomes forming ASS-5. Presumably, they have a shorter terminal linker as a result of trimming. That is the reason for the onset of self-association for ON-5 in 45 mM. Unfortunately, we have no measurements at 35 mM, and this ionic strength is considered to be a transition point from one type of spH1-DNA complexes to the other [21]. We have never observed the "secondary" aggregates for ON-65. They, and to a lesser extent ASS-65, are depleted of spH1 as a result of sucrose gradient centrifugation at 65 mM and the precipitation of a part of material enriched in spH1 (unpublished data).

All the data concerning ionic-strength dependent formation of "secondary" forms by associated and non-associated oligonucleosomes separated in sucrose gradient are summarized in Table 1. At each ionic strength a readjustment equilibrium between associated and non-associated forms takes place. Taking into account the fact that the amount of spH1 decreases in the order ASS-5 > ON-5 > ASS-65 > ON-65, one can see that specific association of sea urchin sperm oligonucleosomes is determined by the same factors as the formation of H1-DNA complexes, namely, by the sinergic action of three factors: the DNA length (an analogue of the linker DNA length), ionic strength, and the type and the amount of the linker histones [21].

The character of the ionic strength dependence of sedimentation coefficients for ASS-5 and ASS-65 (Fig. 4) is similar to that presented in Fig. 2B. Describing Fig. 2B, we discussed the possible mechanisms of the sharp increase of $s_{20, w}$ of aggregates at 25 and 45 mM. The same considerations are applicable to ASS-5 and ASS-65. The s-values of ASS-65 and "secondary" aggregates of ON-65 are close to those of aggregates presented in Fig. 2B. Of special interest are the high s-values of ASS-5 at 65 mM. The character of sedimentation boundary shows that it is not the onset of unspecific aggregation. In our opinion, much higher s-values are determined by higher spH1 content in ASS-5 compared with ASS-65 and ON-5 (and their "secondary" aggregates). As presented in Fig. 2B, aggregates have lower s-values at 65 mM because they are in equilibrium with non-aggregated oligonucleosomes and, altogether, they have lower spH1 content. Among two mechanisms presumably underlying the growth of sedimentation coefficients of the aggregates, at least the specific aggregation seems to be strongly dependent on the spH1 content.

Whatever the mechanism, either intrafilament or interfilament condensation, our data show that the chromatin, enriched in histone H1, is able to form a more tightly folded structure. This seems to be the first direct experimental demonstration of this widely accepted suggestion (for discussion see [25]).

Conclusion

Self-association of oligonucleosomes into the pseudo-higher-order structures as a characteristic feature of the inactive part of the genome

Up to now, the specific self-association of oligonucleosomes has been known only for avian erythrocyte chromatin [5, 7, 8–10]. The connection between the ability to form pseudo-higher-order structures and the repressed state of the genome was clearly demonstrated [5, 8–10].

The aggregates of sea urchin sperm oligonucleosomes described here are similar to those of erythrocyte chromatin in the following features: 1) the aggregates are observed only for oligonucleosomes from the transcriptionally inactive chromatin; 2) they exist only at a low extent of chromatin digestion by micrococcal nuclease; 3) the aggregates undergo salt-induced compaction; 4) the range of the sizes of oligonucleosomes that are able to associate into pseudo-higher-order structures is limited.

The essential distinction is an exceptional stability of aggregates from sea urchin sperm chromatin. They survive the exposure not only to a very low ionic strength (0.2 mM), but also to the traditional procedure of DNA isolation (Fig. 3b). The presence of protamine-like tetrapeptides in the tails of the sperm-specific histones spH1 and spH2B may account for such unusual stability. The exceptional ability of spH1 [22] and spH2B [23] to form giant intermolecular associates that contrast with the properties of somatic variants of these histones was demonstrated for their complexes with SV-40 DNA. In the study of the H1-DNA complexes with short linear DNA, it was shown that the sperm-specific histone spH1 as well as the avian erythrocyte-specific histone H5 (but not the somatic H1) were able to interact cooperatively with linear DNA at low ionic strength (5 mM) [21]. But comparing the two chromatins, it is to be noted that H5 is present in a much lower amount (0.9 molecule per nucleosome [24]) in erythrocyte chromatin than spH1 in sea urchin sperm chromatin (about two molecules per nucleosome [14]). Thus, the model experiments support the possibility of existence of more stable aggregates in the case of the sea urchin sperm chromatin. Such stability may reflect the necessity of dense and tight DNA packaging in spermatozoa.

The similarity in the properties of chromatin aggregates from two types of terminally repressed cells, avian erythrocytes, and sea urchin spermatozoa makes it possible to consider the specific association into pseudo-higher-order structures as a characteristic feature of chromatin from the inactive part of the genome.

Acknowledgements The authors thank Sergey Konditerov for his technical help in the sedimentation experiments. This work was supported by the Russian Program "Frontiers in Genetics" and by the Russian Basic Research Foundation.

References

1. Renz M (1979) Nucl Acids Res 6: 2761–2767

2. Ruiz-Carillo A, Puigdomenech P, Eder G, Lurz R (1980) Biochemistry 19: 2544–2554

3. Stratling W, Klingholz R (1981) Biochemistry 20:1386–1392

4. Puigdomenech P, Ruiz-Carrillo A (1982) Biochim Biophys Acta 696: 267–274

5. Thomas JO, Rees C, Pearson E (1985) Eur J Biochem 147:143–151

6. Thomas JO, Rees C (1983) Eur J Biochem 134:109–115

7. Muyldermans S, Lasters I, Hamers R, Wyns L (1985) Eur J Biochem 150: 441–446

8. Weintraub H (1984) Cell 38:17–27

9. Grigoryev S, Spirin K, Krasheninnikov I (1990) Nucl Acids Res 18:7397–7406

10. Kamakaka RT, Thomas JO (1990) EMBO J 9:3997–4006

11. Osipova TN, Karpova EV, Konditerov SV, Vorob'ev VI (1990) Molec Biol 24: 69–78

12. Zalenskaya IA, Pospelov VA, Zalensky AO, Vorob'ev VI (1981) Nucl Acids Res 9:473–486

13. Von Holt C, Strickland WN, Brandt WF, Strickland MS (1979) FEBS Lett 100:201–218

14. Vorob'ev VI, Karpova EV, Osipova TN (1994) Biorheology 31:221–234

15. Van Holde KE (1989) Chromatin, Springer-Verlag, New York, pp 355–356

16. Osipova TN, Karpova EV, Vorob'ev VI (1990) J Biomol Structure & Dynamics 8:11–22

17. Osipova TN, Karpova EV, Ramm EI, Svetlikova SB, Pospelov VA (1986) Molec Biol 20:853–861

18. Butler PJG, Thomas JO (1980) J Mol Biol 140:505–529

19. Thomas JO, Rees C, Butler PJG (1986) Eur J Biochem 154:343–348

20. Hill CS, Rimmer JM, Green BN, Finch JT, Thomas JO (1991) EMBO J 10:1939–1948

21. Clark DJ, Thomas JO (1988) Eur J Biochem 178:225–233

22. Osipova TN, Triebel H, Bar H, Zalenskaya IA, Hartmann M (1985) Molec Biol Rep 10:153–158

23. Osipova TN, Vorob'ev VI, Bottger M, Von Mickwitz C-U, Scherneck S (1982) Molec Biol Rep 8:71–75

24. Bates DL, Thomas JO (1981) Nucl Acids Res 9:5883–5894

25. Kamakaka RT, Thomas JO (1990) EMBO J 9:3997–4006

Progr Colloid Polym Sci (1995) 99:63–68
© Steinkopff Verlag 1995

J. Behlke
O. Ristau
A. Marg

Complex formation and crystallization of adrenodoxin-reductase with adrenodoxin

Received: 4 April 1995
Accepted: 18 May 1995

J. Behlke
Humboldt-University Berlin
c/o Max Delbrück Center for
Molecular Medicine
Robert-Rössle-Str. 10
13122 Berlin

Prof. Dr. J. Behlke (✉) · O. Ristau · A. Marg
Max Delbrück Center for Molecular
Medicine
Robert-Rössle-Straße 10
13122 Berlin, Germany

Abstract The interaction of bovine adrenodoxin reductase with adrenodoxin was investigated by means of sedimentation equilibrium technique using the analytical ultracentrifuge XL-A. From the radial concentration distributions recorded at sedimentation equilibrium the partial concentrations of the free proteins and the complex were obtained using the program Polymol. Independent of the ratios of proteins used in the experiments a 1:1 complex was formed. The association constant amounts to about $2 \cdot 10^6$ M^{-1} at low phosphate concentration (10–50 mM). The value drops to about 10^3 M^{-1} at 500 mM phosphate. For successful crystallization the complex was stabilized by cross-linking experiments with carbodiimide to prevent dissociation at higher ionic strength.

Key words Adrenodoxin – adrenodoxin-reductase – sedimentation equilibrium – association constant – crystallization

Introduction

Cytochrome P450-dependent monooxygenases are observed in microsomal and mitochondrial membranes of mammalian cells. The membranes contain different electron transfer chains which enable P450 enzymes to metabolize exogenous drugs and toxins as well as endogenously produced steroids, vitamin D, prostaglandins, biogenic amines and bile acids. Whereas the P450 enzymes found in the endoplasmic reticulum (type II enzymes) receive electrons by NADPH through a single FAD-containing flavoprotein, termed reductase (or adrenodoxin reductase, ADR), the mitochondrial (type I) enzymes obtain electrons from NADPH by two intermediates. First ADR transfers electrons to the second intermediate, a small iron sulfur protein, called adrenodoxin (Adx.). This shuttle protein donates the electrons to P450 [1]. ADR and Adx are synthesized on cytoplasmic ribosomes as larger precursors and transported into the mitochondrial matrix and inner membrane, respectively, where they are processed post-translationally into the mature forms [2–4]. While Adx and ADR are dissolved or loosely associated with the inner mitochondrial membrane, P450 is integrated in this membrane ([5] see also Fig. 1).

It is assumed that during the electron transfer Adx cannot form a ternary complex simultaneously with ADR and P450 [6, 7]. Furthermore, in site-directed mutagenesis experiments it was demonstrated that Arg 239 and Lys 243 in ADR are candidates for charge pair interactions with Asp 76 and Asp 79 in Adx [8].

After successful crystallization of Adx [9] and ADR [10], for x-ray crystal structure analysis we have tried to cocrystallize also the complex between both proteins. Therefore, it was necessary to study the complex formation between the proteins. This was of interest because already published association constants determined by diverse methods [11–16] differ up to several orders of magnitude. In contrast to the spectroscopic, electrochemical or kinetic methods we have analyzed the complex formation between

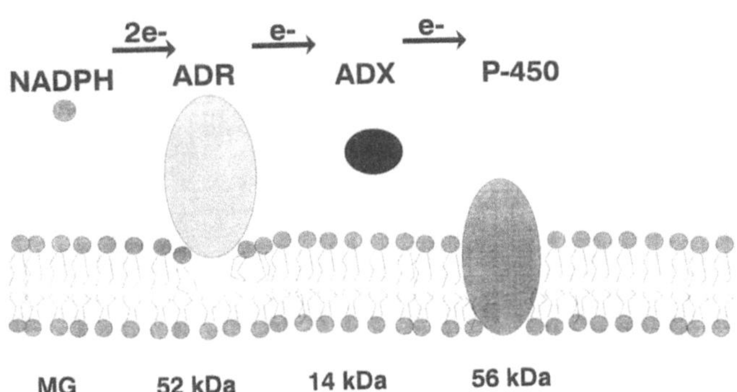

Fig. 1 Schematic model for the electron transfer in the mitochondrial cytochrome P450 system

ADR and Adx by the sedimentation equilibrium technique which considers the partial concentrations and masses of reactants and complex(es). The method and results obtained will be communicated in this paper. Furthermore, we want to report on conditions leading to a successful cocrystallization of ADR-Adx complexes.

Material and methods

ADR and Adx were isolated according to Nonaka [17] or Sakihama [18], respectively, with slight modifications by Marg [19]. Both proteins were characterized electrophoretically as pure substances and showed the expected activity [19]. Furthermore, the spectral properties given by the adsorption ratio $A_{272\,nm}/A_{450\,nm} = 7.7$ for ADR

and $A_{415\,nm}/A_{450\,nm} = 0.90$ for Adx confirm the purity of both proteins. The concentrations of the reactants were determined using adsorption coefficients of $11.3\ mM^{-1} \cdot cm^{-1}$ at 450 nm for ADR and $9.8\ mM^{-1} \cdot cm^{-1}$ at 415 nm for Adx, respectively. Molecular mass determinations of the proteins and the complex formation were analyzed from sedimentation equilibrium runs using the XL-A analytical ultracentrifuge (Beckman) with six-channel cells. About 100 μl of the samples, extensively dialyzed against K-phosphate buffer, pH 7.4, in the isolated state or different mixtures were filled in each compartment and centrifuged two hours at 24 000 rpm (overspeed technique) and then about 24 h at 20 000 rpm (equilibrium speed).

The radial concentration distributions for all samples were recorded at three different wavelengths, either 378, 415 and 450 nm or, at low initial protein concentration, at 230, 235 (see Fig. 2) and 240 nm, respectively.

As is well known, the concentration (c) distribution given for a single macromolecule with the molar mass M at sedimentation equilibrium can be described by the formula (1)

$$c_{(r)} = c_0 \cdot e^{M \cdot A} \tag{1}$$

with

$$A = [(1 - \rho\bar{v}_i)\omega^2(r^2 - r_0^2)]/2RT \tag{2}$$

where ρ means solvent density, $\bar{v}_i$ partial specific volume of component i (0.7438 ml/g for ADR and 0.7265 ml/g for Adx, respectively), ω angular velocity, R the gas constant and T the absolute temperature. In the case of two reacting macromolecules termed as receptor protein (R) and ligand (L) according to [20] we can describe the concentration

Fig. 2 Simultaneous registration of the radial concentration distribution curves of ADR, ADR-Adx mixture and Adx at wavelengths 230 and 235 nm using a six-channel cell

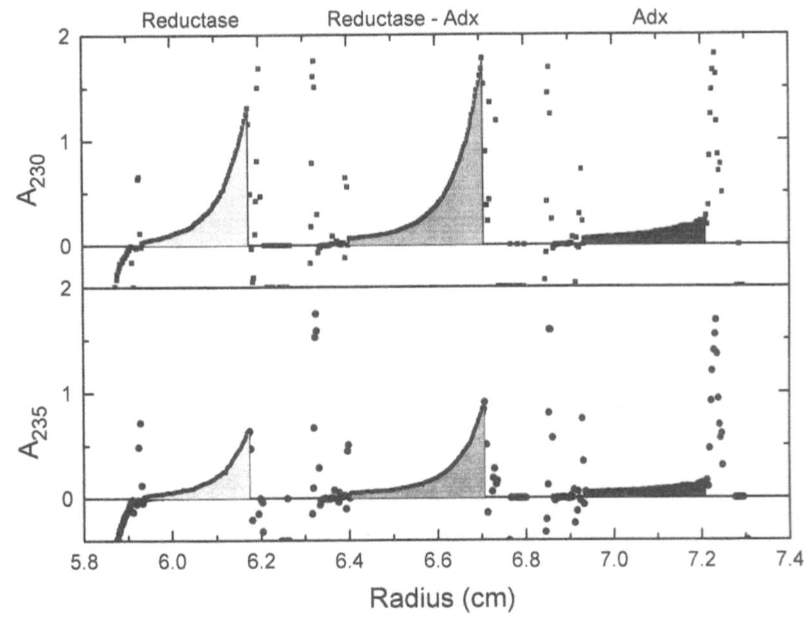

distribution as radial absorbance by equation (3):

$$Abs_r = \varepsilon_R * c_{OR} * e^{F_R * M_R * A} + \varepsilon_L * c_{OL} * e^{M_L * A}$$

$$+ c_{OR} * \sum_{j=1}^{n} \frac{1}{n^j} \binom{n}{j} * (\varepsilon_R + j\varepsilon_L) * (c_{OL} * K_1)^j$$

$$* e^{(F_R * M_R + jM_L) * A} \tag{3}$$

This equation represents the portions of free receptor protein and ligand (first two items) together with that of the complexes. ε_R and ε_L mean the extinction coefficients of both reactants and F_R considers the deviation in buoyancy between R and L (in our system R = ADR, L = Adx.) To reduce the number of estimated parameters in Eq. (3) the molecular masses of the reactants have to be determined in separate experiments. Furthermore, the initial concentrations of the reactants c_R and c_L as total concentrations can be analyzed by integration of the areas below the concentration distribution curve (Eq. 3). The total concentration of R for six channel cells means:

$$c_{Rt} = \frac{c_{OR}}{(r_b - r_m)} \int_{r_m}^{r_b} e^{F_R * M_R * A} \, dr$$

$$+ \frac{c_{OR}}{(r_b - r_m)} \int_{r_m}^{r_b} \sum_{j}^{n} \frac{1}{n^j} \binom{n}{j} * (c_{OL} * K_1)^j$$

$$* e^{(F_R * M_R + jM_L) * A} \, dr \tag{4}$$

The total concentration of L is given by:

$$c_{Lt} = \frac{c_{OL}}{(r_b - r_m)} \int_{r_m}^{r_b} e^{M_L * A} \, dr$$

$$+ \frac{c_{OR}}{(r_b - r_m)} \int_{r_m}^{r_b} \sum_{j}^{n} \frac{j}{n^j} \binom{n}{j} * (c_{OL} * K_1)^j$$

$$* e^{(F_R * M_R + jM_L) * A} \, dr \tag{5}$$

where $K_1 * c_L$ is substituted by x and c_L is expressed by x/K_1, we can eliminate the parameter c_R using the mass balance Eqs. (4) and (5). So we have to estimate only one free parameter from fitting Eq. (3) to the experimental curve, the product of free ligand and association constant K_1 for the binding of the first ligand.

A further possibility to analyze the stoichiometry of a reaction between ligands and a receptor molecule is based on usage of the following model function (6)

$$Abs_r = \varepsilon_R \cdot c_{OR} \cdot e^{F_R \cdot M_R \cdot A} + \varepsilon_L \cdot c_{OL} \cdot e^{M_L \cdot A}$$

$$+ c_{OR} \cdot (c_{OL} \cdot K)^j \cdot (\varepsilon_R + j\varepsilon_L) \cdot e^{(F_R \cdot M_R + j \cdot M_L)A} \tag{6}$$

This equation can be employed in experiments for measuring under ligand excess. It considers only one type of complex. In analogy to Eqs. 4 and 5 a mass balance is included also for this program. By fitting Eqs. (6) to the radial concentration distribution we can treat the occupa-

tion number j as a normal parameter with arbitrary values and then the estimation can be performed in the usual manner.

The total protein concentrations were obtained by numerical integration of the three recorded absorption profiles taking into account the known extinction coefficients valid for the wave lengths used in the experiments.

From the known association constant

$$K_a = \frac{[ADR - Adx]}{[ADR][Adx]} \tag{7}$$

or

$$K_a = \frac{[x]}{[a - x][b - x]} \tag{8}$$

with x = ADR − Adx, a = total ADR and b = total Adx we can estimate the amount of complex (x) in dependence on the concentration of reactants (a, b) and the association constant K_a according to Eq. (9)

$$x = \left[\frac{1}{2}\left(a + b + \frac{1}{K_a}\right)\right] - \sqrt{\left[\frac{1}{2}\left(a + b + \frac{1}{K_a}\right)\right]^2 - ab} \tag{9}$$

To get a high yield of ADR-Adx complex both proteins dissolved in 25 mM K-phosphate buffer were mixed in a 1:1 ratio. To stabilize the formed complex it was cross-linked by carbodiimide. After purification on Sephadex G75 or Adrenodoxin-Sepharose the complex with a molecular mass of about 65 000 was obtained with activity in reduction of cytochrome c. The cross-linked complex was crystallized in the presence of $(NH_4)_2SO_4$ using the dialysis method.

Results

To analyze the interaction between ADR and Adx by sedimentation equilibrium runs using the program Polymol, we have first determined the molecular masses of both reactants. When applying six channel cells we were able to analyze simultaneously the isolated proteins as well as their mixture. The radial concentration distributions recorded at the wavelengths 378, 415 or 450 nm, respectively, resulted in three absorbancy curves. From these data the molecular masses were calculated using Eqs. (1) and (2) or the first and second item of Eq. (3). From the data presented in Fig. 3 the molecular mass of ADR was determined to be 51 800 ± 500. The corresponding value for Adx (Fig. 4) amounts to be 14 000 ± 400. That means the results obtained are in good agreement with molecular masses calculated from the amino acid composition, indicating no proteolytic degration or considerable aggregates occur. The above mentioned values were taken to analyze the concentration distribution of ADR-Adx mixtures by

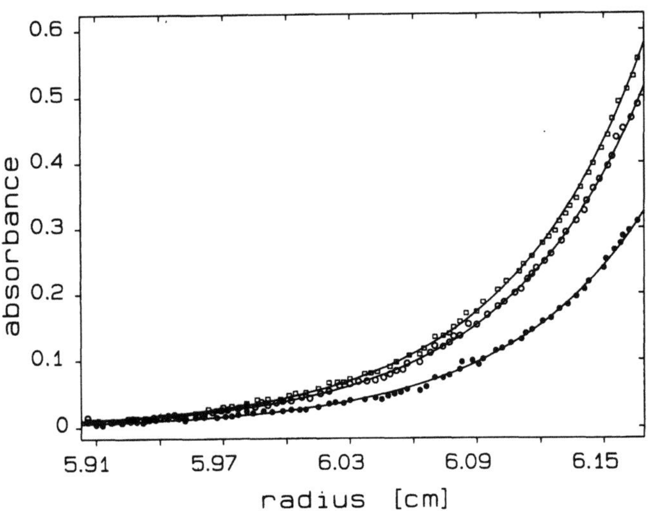

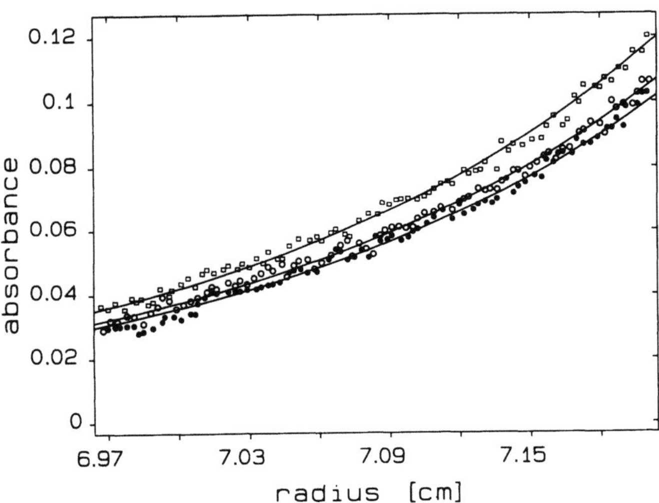

Fig. 3 Radial absorbance distribution curves of ADR at sedimentation equilibrium. Registration was at 378 nm (○) or 415 nm (●) or 450 (□), respectively. The curves are fitted to the experimental data

Fig. 4 Radial absorbance distribution curves of Adx at sedimentation equilibrium recorded at 378 nm (○), 415 nm (□) or 450 nm (●), respectively. The curves represent fits to the experimental data

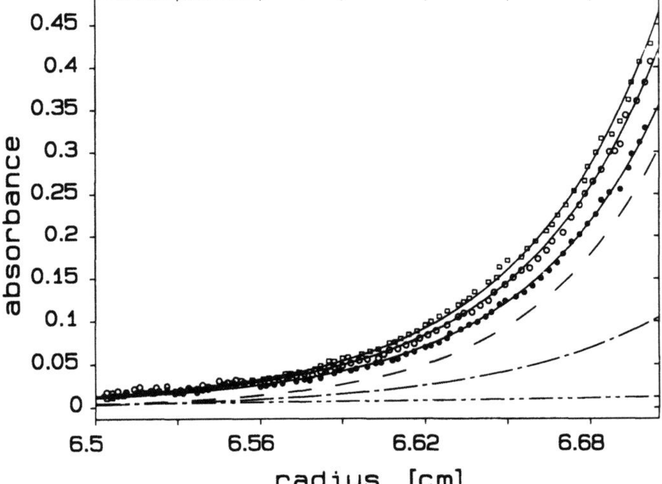

Fig. 5 Radial absorbance distribution curves of the reacting system ADR-Adx measured at 378 nm (○), 415 nm (●) and 450 nm (□). The curves for the free reactants: ADR (– – –), Adx (– – – –) as well as for the complex (– –) are due to the experimental curve at 378 nm. Both proteins were mixed in an equimolar ratio of 6.5 μmolar. Buffer: 25 mM K-phosphate, pH 7.4. Temperature 10 °C. The complex formation corresponds to an association constant $K_a = 1.9 \cdot 10^6 \, \text{M}^{-1}$

fitting Eq. (3) to the experimental data (Fig. 5). The three absorbancy curves were adjusted using three pairs of extinction coefficients for the two reactants and deconvolved into curves reflecting the 1:1 complex, free ADR and free Adx. To be sure whether other kinds beside of the 1:1 complex were formed too, other stoichiometric compositions were checked by the fitting procedure but with less success. When fitting the radial concentration distribution by Eq. (6) the occupation number j was estimated to be 1.04. From the partial concentrations obtained in the

ADR-Adx mixtures in 50 mM K-phosphate an association constant of about $2 \cdot 10^6 \, M^{-1}$ was calculated.

Besides an equimolar mixture of ADR and Adx also other compositions were tested starting with a threefold excess of ADR over Adx until to a fourfold excess of Adx over ADR. For all systems affinity and stoichiometry of the complexes were analyzed. In spite of different ADR/Adx mixtures only the 1:1 complex was formed retaining a larger amount of free reactant which was added in excess. The association constant on the average was not influenced using different ADR/Adx ratios (see Fig. 6).

Of greater interest was the influence of ionic strength on complex formation. As expected, increasing phosphate concentration in the buffer solution leads to a decrease of the association constants. As can be seen from Fig. 7 the values drop by about three orders of magnitude when the phosphate concentration exceeds 0.25 M. When using protein concentrations of about 10 μmolar only few percent of the reactants form complexes.

Most sedimentation equilibrium runs were performed at 10 °C. Experiments at 20 °C yield a slightly smaller association constant than at 10 °C (not shown) when using phosphate concentrations up to 100 mM. At higher ionic strengths both reactants are destabilized and partly aggregated during sedimentation equilibrium experiments so that association constants could not be determined.

To get ADR-Adx crystals it was necessary that the complexes have to be formed at higher protein concentrations.

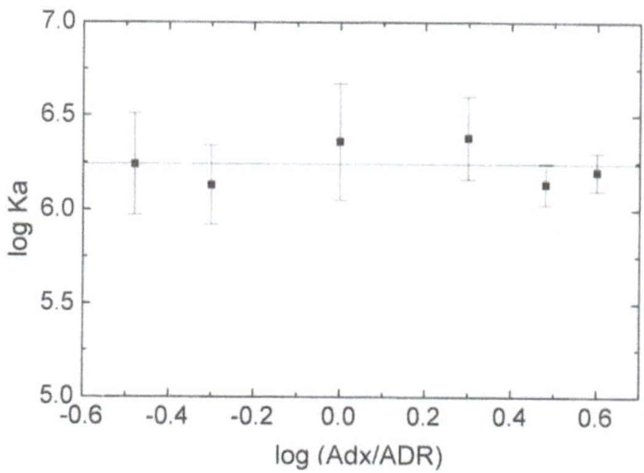

Fig. 6 Association constants for the reaction ADR and Adx obtained from different molar ratios of both proteins. Total protein concentration: 13 μmolar. Buffer: 25 mM K-phosphate. Temperature: 10 °C

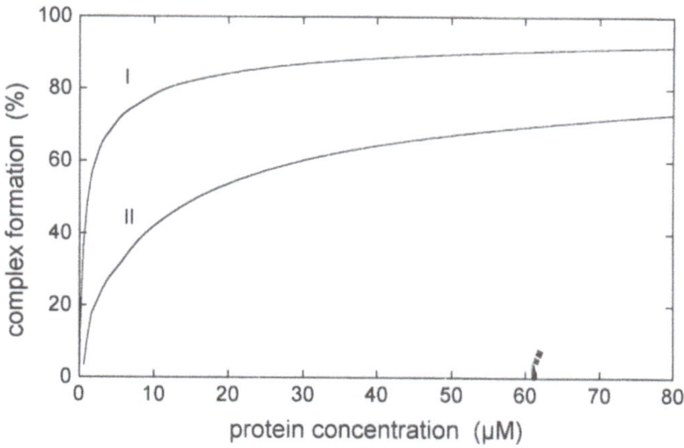

Fig. 8 Predicted concentration dependent complex formation ADR-Adx in 50 mM (|) and 100 mM (‖) phosphate buffer calculated by Eq. (9) and the association constants given in Fig. 7

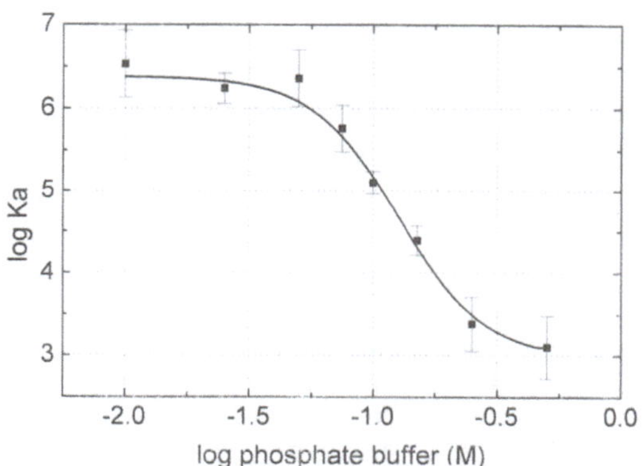

Fig. 7 Ionic strength dependent change of association constants for the interaction of ADR and Adx in an equimolar ratio (6.5 μmolar). Conditions: K-phosphate buffer, pH 7.4, temperature 10 °C

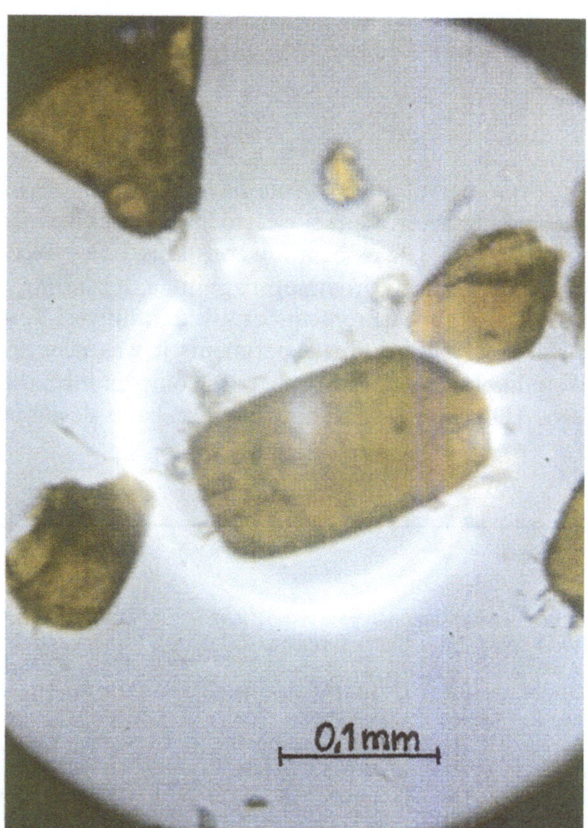

Fig. 9 Crystals of the cross-linked ADR-Adx complex

Since the association constants depend strongly on the ionic strength we have calculated the expected complex formation for 50 mM or 100 mM phosphate buffer considering the association constants and different protein concentrations using Eq. (9). As can demonstrated by Fig. 8 we may not use buffer concentrations higher than 50 mM to get about 90% ADR-Adx complex from 60–80 μmolar protein solutions. For crystallization experiments performed in solutions of $(NH_4)_2SO_4$, it was necessary to stabilize the complex by prior cross-linking with carbodiimide. At least from the purified complex coloured crystals were obtained (Fig. 9) which are different in view and shape from those one obtained for the isolated proteins [9, 10]. Furthermore the existence of a covalently bound complex also in crystals was demonstrated electrophoretically (Fig. 10).

Discussion

Existence of complex formation between ADR and Adx is an important step to realize the electron transfer in the P450 monooxygenase system. In order to characterize this

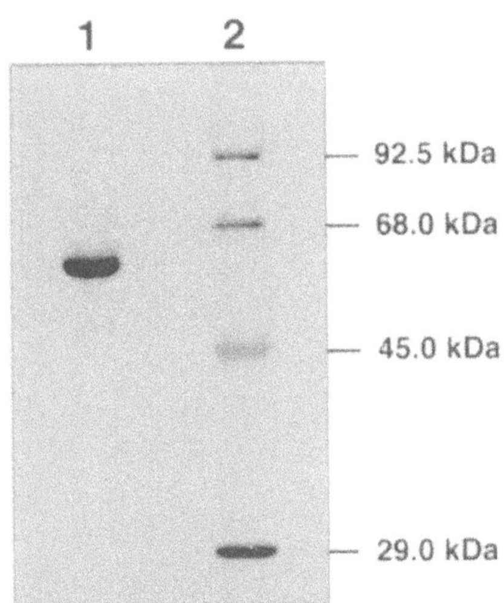

Fig. 10 Polyacrylamide gel electrophoresis of cross-linked ADR-Adx complex in the presence of sodium dodecyl sulphate (line 1) and molecular mass market proteins (line 2) given in kDa

a procedure to calculate the partial concentrations of reactants and complexes based on molecular mass determinations. By this method we were able to analyze the stoichiometry of the reaction as well as the association constant. Comparing our results with those given in the literature we agree with the data obtained by Sakamoto et al. [12] using circular dichroic measurements and low ionic strength. The corresponding association constant (transformed from the dissociation constant) published by Hamamoto et al. [16] considering kinetic methods in the presence of pyridoxal 5'-phosphate (P5P) seems to be too high by two orders of magnitude when taking into account low ionic strength. Until now it is not clear whether P5P acts as an effector to increase the affinity between ADR and Adx. However, circular dicroic measurements and sedimentation equilibrium runs of ADR and Adx without P5P result in agreeable association constants and complex formation within the expected range as found for the cross-linked complex. The proposed charge pair interactions between Arg 239 and Lys 243 in ADR and Asp 76 or Asp 79 in Adx [6–8] and their weakness at higher ionic strength has made it necessary to crosslink the complex and in this way to stabilize them as a prerequisite for successful crystallization.

system a number of experiments have been done earlier using different spectroscopic, electrochemical and kinetic methods [11–16]. The association constants, however, differ considerably up to orders of magnitude depending on direct or indirect measurements or given additives. To be successful in crystallization experiments it was necessary to have reliable results with respect to complex formation between ADR and Adx. Therefore, we have developed

Acknowledgements This contribution was presented at the IX Symposium on Analytical Ultracentrifugation March 2/3 1995 in Berlin. The authors are indebted to Mrs. Bärbel Bödner for skillful technical assistance. Work was supported by the WIP program B 002999.

References

1. Omura T, Sanders E, Estabrook RW (1966) Arch Biochem Biophys 117:660–673
2. Nabi N, Omura T (1980) Biochem Biophys Res Commun 97:680–686
3. Kramer RE, Dubois RN, Simpson ER, Anderson CM, Kashiwagi K, Lambeth JD, Jefcoate CR, Waterman MR (1982) Arch Biochem Biophys 215:478–485
4. Nabi N, Omura T (1983) J Biochem (Tokyo) 94:1529–1538
5. Hanukoglu I, Suh BS, Himmelhoch S, Amsterdam A (1990) J Cell Biol 111:1373–1381
6. Coghlan VM, Vickery LE (1991) J Biol Chem 266:18606–18612
7. Coghlan VM, Vickery LE (1992) J Biol Chem 267:8932–8935
8. Brandt ME, Vickery LE (1993) J Biol Chem 268:7126–17130
9. Marg A, Kuban RJ, Behlke J, Dettmer R, Ruckpaul K (1992) J Mol Biol 227:945–947
10. Kuban RJ, Marg A, Resch M, Ruckpaul K (1993) J Mol Biol 234:245–248
11. Chu JW, Kimura T (1973) J Biol Chem 248:5183–5187
12. Sakamoto H, Ichikawa Y, Yamano T, Tagaki T (1981) J Biochem (Tokyo) 90:1445–1452
13. Hamamoto I, Ichikawa Y (1984) Biochim Biophys Acta 786:32–41
14. Lambeth JD, Pembers S (1983) J Biol Chem 258:5596–5602
15. Massey V, Strickland S, Mayhew SG, Howell LG, Engel PC, Mathews RG, Schuman M, Sullivan PA (1969) Biochem Biophys Res Comm 36:891–897
16. Hamamoto I, Kurokohchi K, Tanaka S, Ichikawa Y (1993) J Steroid Biochem Molec Biol 46:33–37
17. Nonaka Y, Aibara S, Sugiyama T, Yamano T, Morita Y (1985) J Biochem (Tokyo) 98:257–260
18. Sakihama N, Hiwatashi A, Miyatake A, Shin M, Ichikawa (1988) Arch Biochem Biophys 264:23–29
19. Marg A (1994) Thesis, Humboldt-University, Berlin
20. Behlke J, Ristau O, Knespel A (1994) Progr Colloid Polym Sci 94:40–45

Progr Colloid Polym Sci (1995) 99:69–73
© Steinkopff Verlag 1995

B. von Rückmann
E. Huber
P. Schuck
D. Schubert

Studying heterologous associations between membrane proteins by analytical ultracentrifugation: Experience with erythrocyte band 3

Received: 21 March 1995
Accepted: 7 June 1995

B. von Rückmann · E. Huber
P. Schuck[1] · Prof. Dr. D. Schubert (✉)
Institut für Biophysik
JWG-Universität
Theodor-Stern-Kai 7, Haus 74
60590 Frankfurt am Main, Germany

[1] *Present address*:
National Institute of Health
Laboratory of Biochemical Pharmacology
Bethesda, MD 20892, USA

Abstract Applying procedures originally suggested for water-soluble proteins (Osborne JC, Powell GM, Brewer HB (1980) Biochim Biophys Acta 619:559–571), our group has studied the association of the intrinsic membrane protein band 3 (from human erythrocytes) with other proteins for which band 3 represents a binding site. In all cases, the association was studied by sedimentation equilibrium analysis on detergent-solubilized band 3 and dye-labeled ligand proteins, under conditions where the proteins show ideal sedimentation behavior. During these studies, a standard step-by-step procedure has developed which allows the determination of the stoichiometry of the predominant heterologous complex and the detection of complexes of different band 3 or ligand content. Application of the procedure to aldolase, erythrocyte band 4.1 protein and hemoglobin as ligand proteins is demonstrated. It is shown that, in all cases studied, the band 3 tetramer represents the predominant or even the exclusive binding site. The number of ligand proteins bound per band 3 tetramer can reach 4 for aldolase and hemoglobin or nearly 8 for band 4.1.

Key words Analytical ultracentrifugation – sedimentation equilibrium – heterologous associations – erythrocyte band 3 protein

Introduction

Sedimentation equilibrium analysis in the analytical ultracentrifuge undoubtedly represents the best available method for studying heterologous protein-protein associations. Its application to that field is, however, far from being trivial, due to the large number of unknown parameters in the equations which have to be applied in analyzing the experimental data. In previous reports, we have presented some of our own efforts in the field [1–3]. The purpose of the present paper is to summarize the general approach which has been developed in our laboratory during the last few years and which, for us, represents a scheme to proceed when starting another study. The data used to illustrate the procedure are taken from refs. [3–5].

The main subject of our studies is an intrinsic membrane protein, the band 3 protein of the human erythrocyte membrane. Its molar mass is 100 000. Besides functioning as the membrane's anion transporter, this protein serves as a binding site for a variety of other proteins of the erythrocyte membrane or its cytoskeleton, respectively, and of the cytoplasm. These proteins include the cytoskeletal proteins band 2.1 (ankyrin) and band 4.1, the glycolytic enzymes aldolase, glyceraldehyde-3-phosphate dehydrogenase and phosphofructokinase, and hemoglobin [6, 7]. The primary aim of our studies is to determine the stoichiometry of these associations. In this respect, a highly interesting aspect is added by the oligomeric state of the band 3 protein: In solutions of suitable nonionic detergents and, most probably, also in the intact membrane, band 3 exists as a mixture of monomers, dimers, and tetramers linked in an association equilibrium [8, 9]. This strongly increases the number of possible aggregates and adds the question for the "functional unit" of ligand binding, i.e., the smallest band 3 oligomer which can serve as a binding site for a given ligand [10]. We have isolated the band 3 protein in solutions of the nonionic detergent nonaethyleneglycol lauryl ether ($C_{12}E_9$) [8], added the ligand protein and measured the sedimentation equilibrium distribution of the particles in the analytical ultracentrifuge. Afterwards, the oligomeric state of the heterologous aggregates formed was determined by mathematical analysis of the sedimentation equilibrium data [2, 3, 11]. A discussion of the biological relevance of this approach and experimental details can be found in previous reports from our group [2, 8, 10].

General strategy

A look at the equations describing the sedimentation equilibrium distribution of mixtures of interacting proteins [2, 11, 12] immediately shows that, even under conditions of ideal sedimentation behavior (virial coefficients equal to zero) and knowing the molar masses of the complexes, a reliable determination of the numerous unknown parameters involved from the experimental absorbance-versus-radius data $A(r)$ will not be feasible. Thus, the number of parameters to be determined has to be reduced, by focusing on those parameters which are of primary interest. A well-established way to achieve this is to label one of the components (favorably that with the simpler association behavior) with a dye and to monitor the sedimentation equilibrium distributions at a wavelength where only the dye will absorb [2, 11, 12]. If present, an intrinsic dye label could be used instead of an extrinsic one, e.g., the heme group in hemoglobin [3]. The scheme to be described below makes use of both variants. Of course, by disregarding some of the aggregates other useful information, e.g., concerning association constants and thus the thermodynamics of the interactions, will be lost. An alternative procedure would be to record data from the full absorbance-versus-wavelength-and-radius space and thus to use the full informational content available [13]. An application of this more sophisticated and laborious method has already been described [13, 14]. As for any method applying extrinsic labels, it has to be established that the labeled protein shows the same association behavior as the unlabeled one.

Special problems arising from the presence of detergents

The detergent necessary for keeping intrinsic membrane proteins in aqueous solution may have a twofold influence on the measured absorbance profiles $A(r)$: 1) The values of M and $\bar{v}$ characterizing the sample's sedimentation behavior will be those of the protein/detergent complex, M_c and $\bar{v}_c$. 2) If the protein-free detergent micelles show a significant absorbance at the wavelength applied for detection (intrinsic or due to apolar absorbing buffer components or impurities), the detergent will contribute its own Boltzmann distribution to $A(r)$. For detergent densities between 1.0 and 1.1 g/ml, "density matching" (i.e. adjusting the solvent density to that of the detergent by addition of D_2O [11, 15]) will cancel both effects. Another possibility to cope with problem (1) is to base the calculations on an experimental value for $M_c(1 - \bar{v}_c\rho_0)$, for a uniform protein/detergent complex. E.g., for the erythrocyte band 3 protein a stable dimeric form can be prepared [8, 9, 14]; the figure of $M_c(1 - \bar{v}_c\rho_0)$ measured with this oligomer in solutions of $C_{12}E_9$ forms the basis of all evaluations described below (assuming that all band 3 oligomers show virtually the same protein/detergent weight ratio). Problem (2) can be disregarded when protein complexes of molar masses $> 250\,000$ are studied in solutions of $C_{12}E_9$: At the relatively low rotor speeds used, the sedimentation equilibrium distribution of the detergent micelles is virtually parallel to the baseline and can be included into the latter. In other cases, the micelle term may have to be taken into account [13]. Detailed studies on this problem can be found in two recent theses [16, 17].

First step of the analysis: Determination of the predominant binding site

The aim of this step is to find out which of the different oligomers of a protein α (in our case: the band 3 protein) represents the main binding site for a protein β (the dye-labeled ligand). The analysis should be based on experimental data obtained at low molar ratio of the proteins β and α (see below) and at low concentrations of the ligand protein β (to ensure that this protein is present in only one oligomeric form). In addition, all protein concentrations should be kept low enough to ensure ideal sedimentation behavior.

In the first round of calculations, the equation which seems to be the most useful contains only two terms: that for the unliganded protein β and that for a complex of one molecule of protein β and one of the oligomers of protein α. The absorbance (or concentration) of the ligand is treated as a free parameter, whereas the absorbance of the complex is kept fixed during one fit but varied between successive fits [3]. A demonstration of the results obtained with band 3 protein (as protein α) and rhodamine-labeled aldolase (as protein β) is given in Fig. 1. The figure shows a semilogarithmic plot of the dependence of the sum of the squared residuals, σ, in fits to experimental data collected at 557 nm, on the assumed type and amount of the constrained complex. The latter was assumed to consist of one aldolase tetramer plus a band 3 monomer (A), dimer (B) or tetramer (C). It is obvious that by far the best fit is obtained when it is assumed that the binding site for aldolase is the band 3 tetramer. In this case, the differences between the experimental and fitted data are within the uncertainty of the former ones and are statistically distributed along the r-axis (data not shown), suggesting that other complexes may not contribute significantly to $A(r)$ [4]. It should, however, be noted that, with other data sets obtained under seemingly identical experimental conditions, the differences between curves like those shown in Fig. 1 are less marked, mainly due to differences in the contributions from noise. Thus, a careful evaluation of a considerable number of data sets seems to be appropriate. Another caveat seems to be necessary: As the simulated data in Fig. 2 illustrate, a band 3 dimer as the predominant binding site could be easily suggested in a situation where both the band 3 monomer and tetramer serve as binding sites, in this case at a molar ratio of 3.5, and where noise is significant. The ligand protein β was assumed to have a molar mass of 66 000 (corresponding to that of band 4.1). In the simulation, the band 3 dimer did not represent a ligand binding site; nevertheless, it leads to a minimum of σ. However, the misleading result suggested by the figure can be corrected in part (see below).

An alternative, though related approach would be to perform a series of fits which uses again two terms: one for the unbound ligand, and a second one for a complex of fixed molar mass M^*. This mass is systematically varied between that of the ligand and that of the largest complex conceivable. What is optimized is the absorbance, at a fixed radial position r_0, of both particles. The procedure can, of course, be automated. The result of interest is the dependence of σ on M^*. A typical result, as obtained for a fit to experimental data on a band 3/band 4.1 mixture (using rhodamine-labeled band 4.1) is shown in Fig. 3. The rather well-defined minimum value of σ corresponds to a molar mass virtually identical to that of a complex of one band 4.1 molecule and the band 3 tetramer, which suggests that this complex is the predominant one [5]. Heterogeneity of the complexes formed can be inferred when the minimum of curve does not correspond to the mass of a well-defined heterologous aggregate, or if the residuals of the fit are not distributed statistically.

Second step: Evaluation of possible contributions by other binding sites

In this step, the calculations performed during the first step are extended, using the same experimental data. The absorbances (at a fixed radial position, r_0) of the uncomplexed ligand and of the predominant complex identified in step 1 are now used as the parameters to be optimized, and the absorbance (at r_0) of one of the other possible

Fig. 1 Associations between band 3 and rhodamine-labeled aldolase: Dependence of the sum of the squared residuals of the fit, σ, on the assumed type and amount of complex present (all other parameters floating). The complexes were assumed to consist of one aldolase tetramer plus either the band 3 monomer (A), dimer (B) or tetramer (C). Rotor speed: 8000 rpm [4, 10]

Fig. 2 A fit, analogous to that of Fig. 1, to simulated $A(r)$-data on a mixture of band 3 and a ligand protein β of molar mass 66 000. The dye-labeled particles contributing to $A(r)$ were unliganded protein β and complexes of one molecule of protein β with a band 3 monomer and a tetramer, at initial absorbances of 0.2, 0.1 and 0.029 OD, respectively. Noise of 0.01 OD (rms) was added to the curve [13]. In fitting the data it was assumed that the binding site for protein β is the band 3 monomer (A), dimer (B), or tetramer (C). Assumed rotor speed: 10 000 rpm

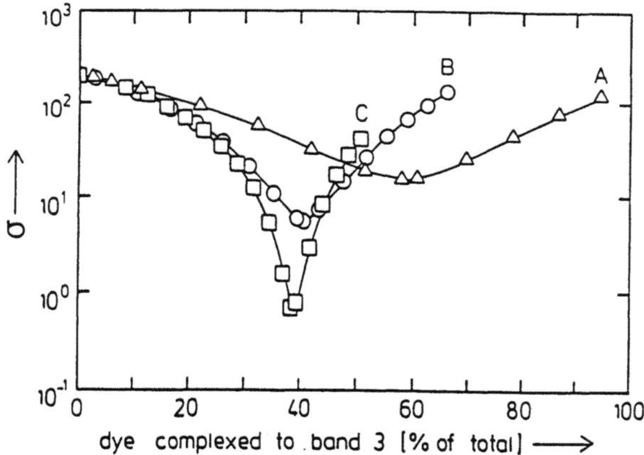

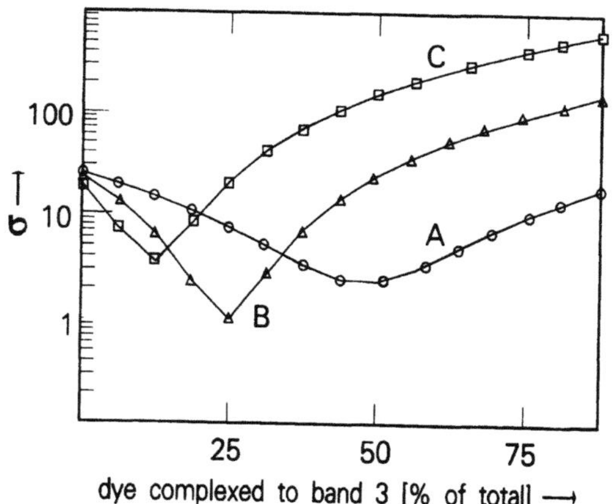

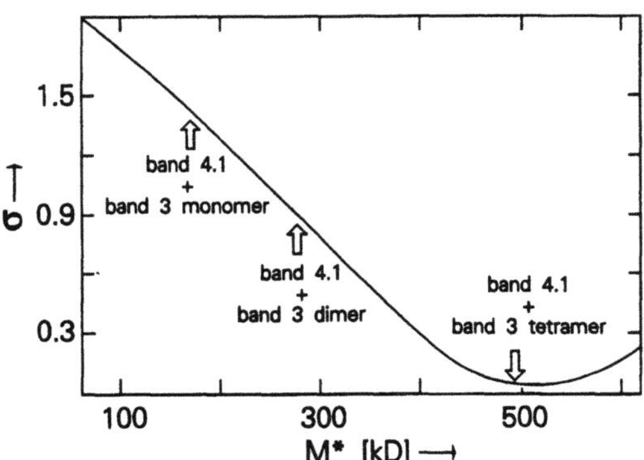

Fig. 3 Determination of the molar mass M^* of that complex which, together with the unbound ligand, allows the best fit to the experimental data on a band 3/band 4.1 mixture (low molar band 4.1/band 3 ratio, rhodamine-labeled band 4.1): Dependence of σ on M^*. Rotor speed: 10 000 rpm [5]

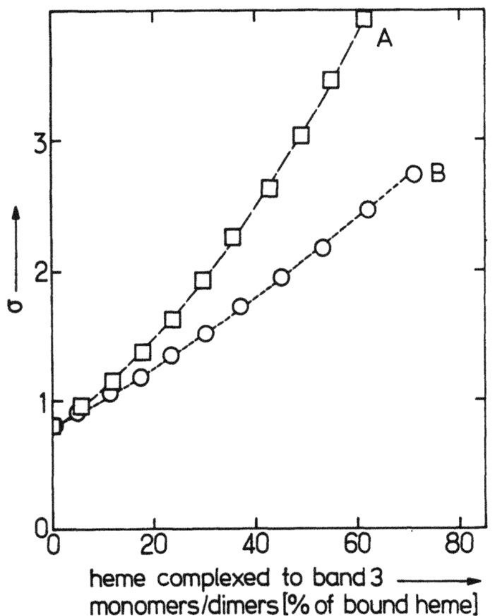

Fig. 4 Possible contributions of ligand binding sites other than the band 3 tetramer in band 3/oxyhemoglobin associations: Changes of σ which follow the addition of fixed terms for complexes of one hemoglobin tetramer and either the band 3 monomer (A) or dimer (B), in fits to experimental data at low molar hemoglobin/band 3 ratio. Rotor speed: 10 000 rpm [3]

complexes (or the corresponding amount of complex in the cell) serves as a fixed parameter. By this procedure, the possible contributions of additional complexes to $A(r)$ can be estimated, without increasing the number of free parameters in the fit beyond that limit where the reliability of the analysis becomes questionable. Results from a study on the band 3/oxyhemoglobin association are shown in Fig. 4. The free parameters were the concentrations of uncomplexed hemoglobin (dimers and tetramers) and of a complex consisting of one band 3 tetramer and a hemoglobin tetramer (according to the results of step 1). Com-

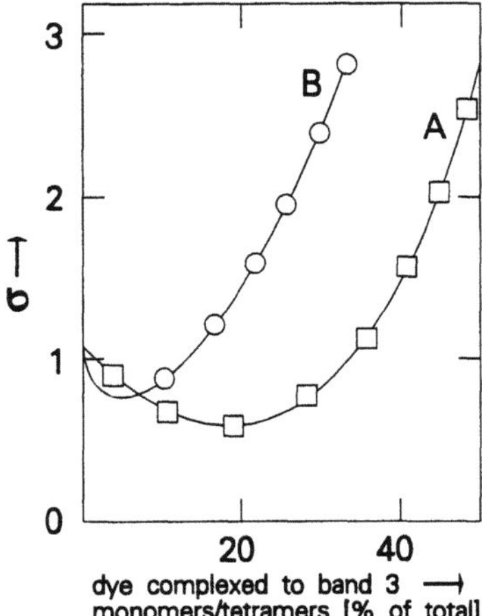

Fig. 5 Evaluation, analogous to that of Fig. 4, which extends that of the simulated data of Fig. 2: Influence of terms characterizing an additional complex of protein β and either the band 3 monomer (A) or tetramer (B), assuming that the band 3 dimer is the predominant ligand binding site

plexes of a hemoglobin tetramer and either the band 3 monomer or the band 3 dimer were characterized by fixed terms. The results clearly show that any added complex containing band 3 monomers or dimers reduces the quality of the fit [3]. Smaller slopes of the curves than those in Fig. 4 would, of course leave the possibility that part of the ligand protein may be bound to other oligomers of protein α than the one identified in step 1. Of course, in the general case the minima of the curves will be located at nonzero instead at zero complex concentration.

The procedure described, or an equivalent one based on the alternative approach of step 1, may also allow to correct the misleading result of Fig. 2. By adding to the equations a term characterizing a complex smaller or larger than the one determined before, σ may distinctly decrease (Fig. 5), which suggests that the simple picture suggested by Fig. 2 has to be modified. However, in all cases which are characterized by the presence of more than one heterologous complex and a relatively high noise level, it may be quite difficult to find the correct answer.

Third step: Determination of the number of ligand binding sites

Whereas steps 1 and 2 use experimental data collected at low molar ratio of proteins β and α, step 3 will have to use data from runs at high molar β/α ratios. In this step, which is analogous to the alternative procedure of step 1, again only two different terms should be considered: one characterizing the unbound ligand, and another one characterizing a complex with fixed molar mass larger than that determined in step 1. The mass of the latter complex is

Progr Colloid Polym Sci (1995) 99:69–73
© Steinkopff Verlag 1995

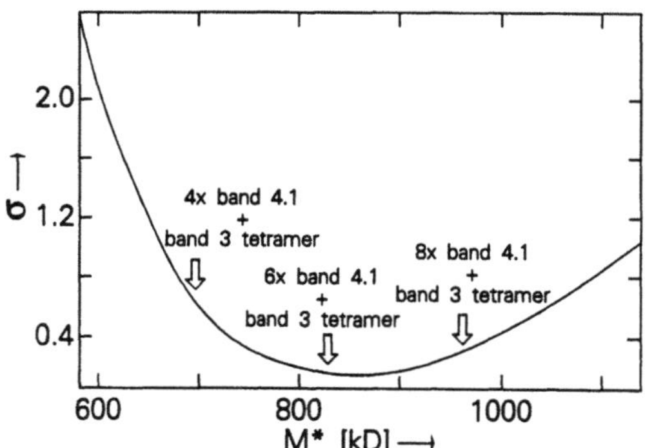

Fig. 6 Determining the maximum number of binding sites for band 4.1 on band 3: An analogue to the study of Fig. 3, using much higher molar ratios of band 4.1 and band 3 [5]

varied in a series of fits. The fits will yield an *average* complex mass which represents the best equivalent for the mass distribution probably present. Again, this procedure can be automated. An example concerning the band 3/band 4.1 association is shown in Fig. 6; it extends the data of Fig. 3 which were collected at a low molar band 4.1/band 3 ratio. At the much higher concentration of band 4.1 used in the experiment of Fig. 6, the minimum of σ is strongly shifted to a higher molar mass $M*$ and points to a ratio of 6–7 band 4.1 molecules per band 3 tetramer, indicating that more than one binding site for band 4.1 may exist on the band 3 protomer [5].

examples shown represent relatively simple cases, due to the fact that, for all associations of band 3 investigated by us until now, there was one predominant binding site. If there is no such site, the results of the analysis will probably be much more ambiguous. It should be noted that even the simple cases described are not without problems. The most obvious one concerns the band 3/aldolase and the previously studied band 3/ankyrin association. In these cases, the molar mass of the ligand is very near to that of the band 3 dimer. As a consequence, the molar mass of the complex between one band 3 tetramer and the ligand is virtually identical to that of a complex consisting of the band 3 dimer and two ligand proteins. Thus, there is no direct way to discriminate between the two types of complexes. What can solve the problem is a study of the dependence of complex concentration on the concentration of the unbound ligand, at an exess of band 3: in the former case, complex concentration should increase with the first power of the concentration of the free ligand, in the latter case with the second power. Another finding which could help to resolve the ambiguity concerns the maximum molar mass of the complex. Both arguments clearly favor the band 3 tetramer as the binding site [1, 4]. Thus, at least in relatively simple cases the methodology available for studying heterologous protein–protein associations is apparently sufficient to answer most of the questions raised.

Additional remarks

We have described the scheme we follow when studying heterologous protein-protein associations. Of course, the

Acknowledgements We thank L. Ehrhardt, M. Ernst, and Dr. Ph. Wood for helpful discussions, and the Deutsche Forschungsgemeinschaft for financial support (SFB 169).

References

1. Mulzer K, Petrasch P, Kampmann L, Schubert D (1989) Stud Biophys 134:17–22
2. Mulzer K, Kampmann L, Petrasch P, Schubert D (1990) Colloid Polym Sci 268:60–64
3. Schuck P, Schubert D (1991) FEBS Lett 293:81–84
4. Huber E, Bäumert HG, Spatz-Kümbel G, Schubert D (1995) manuscript in preparation
5. von Rückmann B, Jöns T, Dölle F, Drenckhahn D, Schubert D (1995) manuscript in preparation
6. Passow H (1986) Rev Physiol Biochem Pharmacol 103:61–203
7. Salhany JM (1993) Erythrocyte band 3 protein. CRC Press, Boca Raton, pp 27–177
8. Pappert G, Schubert D (1983) Biochim Biophys Acta 730:32–40
9. Schubert D, Boss K, Dorst H-J, Flossdorf J, Pappert G (1983) FEBS Lett 163:81–84
10. Schubert D, Huber E, Lindenthal S, Mulzer K, Schuck P (1992) Prog Cell Res 2:209–217
11. Schubert D, Schuck P (1991) Prog Colloid Polym Sci 86:12–22
12. Osborne JC, Powell GM, Brewer HB (1980) Biochim Biophys Acta 619:559–571
13. Schuck P (1991) Prog Colloid Polym Sci 94:1–13
14. Schuck P, Legrum B, Passow H, Schubert D (1995) Eur J Biochem 230:806–812
15. Tranford C, Reynolds JA (1976) Biochim Biophys Acta 457:133–177
16. Ehrhardt L (1995) Ph D thesis, JWG University, Frankfurt am Main
17. Ernst M (1995) Diploma thesis, Department of Physics, JWG University, Frankfurt am Main

Progr Colloid Polym Sci (1995) 99:74–81
© Steinkopff Verlag 1995

Chemical cross-linking and analytical ultracentrifugation study of the histone-like protein HBsu: Quaternary structure and DNA binding

C. Timmermann
J. Behlke
O. Ristau
H. Gerst
U. Heinemann

Received: 8 May 1995
Accepted: 19 May 1995

C. Timmermann · J. Behlke · O. Ristau
H. Gerst · Dr. U. Heinemann (✉)
Max-Delbrück-Zentrum für
Molekulare Medizin
Robert-Rössle-Straße 10
13122 Berlin, Germany

Abstract By chemical cross-linking and analytical ultracentrifugation the dimer is shown to be the predominant quaternary structure of HBsu, the major histone-like protein from *Bacillus subtilis*, at protein concentrations up to 1 mM. At low ionic strength between 0 and 100 mM NaCl, higher-order structures up to the hexamer are also found. A double-stranded 10-basepair DNA oligomer is demonstrated to prevent the aggregation of HBsu molecules as does an increase in ionic strength. The complex formed between HBsu and the DNA decamer is determined to consist of two protein dimers and the DNA. The association constant is about $10^6\,\mathrm{M}^{-1}$.

Key words Histone-like proteins – quaternary structure – chemical cross-linking – sedimentation equilibrium – DNA binding – association constant

Introduction

HBsu is the major histone-like protein in *Bacillus subtilis* (for a review on histone-like proteins see [1]). It is the *B. subtilis* homologue of the protein HU, the major histone-like protein of *Escherichia coli*. Both belong to the family of HU proteins, small basic DNA-binding proteins with a highly conserved primary structure, which can be isolated from several bacterial nucleoids. While the *E. coli* protein is composed of two different but highly homologous subunits, for HBsu, as for the HU proteins of other Gram-positive bacteria, only one primary structure is found.

Both HU and HBsu have been shown to introduce negative superhelical turns into circular double-stranded DNA in the presence of topoisomerase *I* [2, 3]. They are believed to be involved in the condensation of DNA by promoting the generation of nucleosome-like structures, analogous to the eukaryotic histones. Specific DNA target sequences are not required for the binding. *In vitro* a significant influence of HU can be observed on the initiation of replication [4, 5], the integration of the λ phage genome [6, 7] and the transposition of the phage Mu [8]. HU has been shown to mediate and to stabilize DNA bends [9, 10]. Deletion mutants in both genes coding for the HU subunits exhibit deficiencies in cell division, DNA supercoiling and recombination [11, 12]. *B. subtilis* cells carrying a deletion mutation in the *hbs* gene coding for the HBsu protein are not viable. Placing the gene under the control of an isopropyl thiogalactoside (IPTG) inducible promoter produces a phenotype showing IPTG dependence in sporulation and altered cell morphology [13].

HBst, the HU protein of *Bacillus stearothermophilus*, exists in crystals as a dimer of two closely interlocked monomers [14]. For HBsu, calorimetric and spectroscopic data suggest a concerted two-state model of unfolding from a native dimer to denatured monomers [15–17].

To explain the DNA-condensing activity of HU proteins different models have been suggested where the DNA is wrapped around a core built from HU protein [1, 14, 18]. These models are based on the assumption that each molecule covers a length of about 10 basepairs of DNA and that larger aggregates of at least 8 to 10 dimers are formed. From spectroscopic experiments using HBsu protein encoded by a synthetic gene [19, 20] a binding site size of 3.5 basepirs was calculated, and no cooperativity of binding could be found.

Here we describe experiments using two different techniques to determine the quaternary structure of HBsu protein and the stoichiometry of its DNA complexes. Chemical cross-linking is chosen as one method to characterize higher-order structures of HBsu, as shown earlier for *E. coli* HU [21, 22]. Analytical ultracentrifugation is used to determine the molecular mass of HBsu in solution and to derive its native quaternary structure. Furthermore, this technique permits a quantitative analysis of heterologous interactions [23] between HBsu and DNA.

Materials and methods

Preparation of HBsu

The preparation of HBsu from recombinant *E. coli* TG-1 (pHBsu1) was described in detail previously [20]. Protein concentrations were estimated from the absorption at 58 nm based on an absorption coefficient for the HBsu monomer of $751\ M^{-1}\ cm^{-1}$ as determined by Groch et al. [20].

Chemical cross-linking

Cross-linking was performed using the bifunctional imidoester dimethyl suberimidate (DMS, purchased from Sigma), as described by Carpenter and Harrington [24] and by Losso et al. [21]. DMS generates cross-links between primary amino groups as present in lysine sidechains and the at *N*-terminus of a protein. The HBsu primary structure contains 12 lysine residues per monomer.

The reaction was carried out in 50 mM triethanolamine, pH 8.5, on samples containing 80 to 430 μM protein and the cross-linking reagent in a 20-fold molar excess. Ionic strength was varied from 2.5 mM NaCl to 202.5 mM NaCl. The mixture was stirred for 60 min at 25 °C, and the reaction was stopped by addition of 0.1 vol. of 1 M Tris-HCl, pH 7.2, and 0.125 vol. of a 4 M NH_4Cl solution. The products were analysed by electrophoresis on SDS containing 15% polyacrylamide gels, according to Laemmli [25]. The slots were filled with aliquots of the reaction mixture containing 30 μg protein. Gels were stained with Coomassie Blue.

Ultracentrifugation

Sedimentation equilibrium experiments were performed in a Spinco E analytical ultracentrifuge (Beckman) equipped with Rayleigh interference and Schlieren optics system. The short-column high-speed technique with a double-sector or six-channel cell according to Yphantis [26] was used. The columns were filled with 100 μl of a HBsu solution in 50 mM Tris-HCl pH 7.5, 1 mM EDTA, 50 mM NaCl. Before starting an experiment the sample was extensively dialysed against this buffer. To determine the composition of HBsu complexes with DNA a variable amount of a double-stranded 10-basepair DNA oligomer was added to the sample (sequence of the oligomer: CGCACACACG/CGTGTGTGCG). The sedimentation equilibrium was reached after 20–24 h, and the resulting concentration gradients were recorded employing Rayleigh-interference or Schlieren optics for the higher concentrated samples. Photographs were scanned with a MIT CCD 72 camera (resolution 760 * 512 * 8 bits) using an analogue contrast adjustment device as part of the camera system. By special computer programs on a VAX station 4000/60 the fringe displacements were transformed in radial concentration distributions.

Determination of molecular weight and complex composition

Concentration gradients at sedimentation equilibrium are described by a simple function.

$$C_r = C_0 \cdot e^{M \cdot A} \tag{1}$$

with

$$A = [(1 - \rho \bar{v}_i)\omega^2(r^2 - r_0^2)]/2RT . \tag{2}$$

Here, c_0 is the concentration at the reference radius r_0, M the molecular mass, ρ the solvent density, $\bar{v}_i$ the partial specific volume of the polymer component i, ω the angular velocity, R the gas constant and T the absolute temperature. The area below the exponential function (Eq. (1)) is proportional to the initial concentration of the substance. During the experiment the concentration remains constant at one radial position (inflection point) to reach the sedimentation equilibrium. If all parameter values of Eq. (2) are known we can determine the molecular mass directly from the concentration distribution (Eq. (1)) by non-linear regression using the commercially available program Sigma Plot (Jandel) or our program POLYMOL [23] which fits the different values $c = f(r)$ given as fringe displacements by a damped least-squares procedure, in the version of Wynne and Wormell [27].

In the case of only one substance, here, the HBsu protein, the molecular mass and the concentration can be determined unambiguously. However, in reacting systems the concentration distribution at sedimentation equilibrium is represented by a sum of exponential functions with more free parameters (molecular masses and concentration of at least three components). To get reliable data from such experiments we have to reduce the number of free parameters by obtaining the molecular masses of reactants from separate experiments.

According to the equation

$$k_i = \frac{[NL_i]}{[NL_{i-1}] * [L]} \tag{3}$$

which reflects the interaction of ligand (L), or HBsu protein, with the oligonucleotide (N), from the statistical point of view and assuming *equal* binding sites for all (maximal n) bound ligands we can formulate the following relation [28] between the binding constants of n ligands

$$k_i = \left(\frac{n+1-i}{n \cdot i}\right) * k_1 . \tag{4}$$

In this equation k_1 is the binding constant of the first ligand and $i = 1, n$.

The overall binding constant

$$K_j = \frac{[NL_j]}{[N] * [L]^j} \tag{5}$$

which corresponds to the interactions between the oligonucleotide duplex and ligands can be expressed because of the validity of

$$K_j = k_1 * k_2 \cdots * k_j \tag{6}$$

in the following way:

$$K_j = k_1^j * \frac{1}{n^j}\binom{n}{j} \tag{7}$$

with $j = 1, n$ (according to [28]).

The model function which describes the fringe displacements as concentration distribution at sedimentation equilibrium can be formulated by means of the binomial coefficients and without considering the virial coefficients in the following form:

$$f_r = F_c * c_N * e^{F_N * M_N * A} + c_L * e^{M_L * A}$$

$$+ c_N * \sum_j^n \frac{1}{n^j}\binom{n}{j} * (c_L * k_1)^j * (F_c + j * M_L/M_N)$$

$$* e^{F_N * M_N + j * M_L) * A} \tag{8}$$

The factors F_c and F_N in equation (8) consider the deviating fringe displacement or buoyancy of the nucleotide (N) in comparison to the ligand (L), here the HBsu protein. Whereas the molecular masses of the nucleotide and protein are known or obtained from individual measurements, the parameters c_N, c_L and k are estimated fitting Eq. (8). As could be demonstrated earlier [29] virial coefficients have

to be considered only when macromolecular concentrations exceed 2 mg/ml.

The maximal coordination number (n) can be estimated from the radial concentration distribution curve only for a model with equal, but not for different binding sites because of the larger number of free parameters. The POLYMOL program allows to reduce this number based on the fact that the known initial concentrations of the two reacting components are reflected in the area below the model exponential functions (Eq. (8)) which can be obtained by integration. The total oligonucleotide concentration corresponds to:

$$C_{Nt} = \frac{c_N}{r_b - r_m}\left[\int_{r_m}^{r_b} e^{F_N * M_N * A} dr \right.$$

$$\left. + \int_{r_m}^{r_b} \sum_j^n \frac{1}{n^j}\binom{n}{j} * (c_L * k_1)^j * e^{(F_N * M_N + j * M_L) * A} dr \right] \tag{9}$$

For the concentration of ligands we can write:

$$C_{Lt} = \frac{1}{r_b - r_m}\left[c_L * \int_{r_m}^{r_b} e^{M_L * A} dr + \frac{c_N * M_L}{M_N} \right.$$

$$\left. * \int_{r_m}^{r_b} \sum_j^n \frac{j}{n^j}\binom{n}{j} * (c_L * k_1)^j * e^{(F_N * M_N + j * M_L) * A} dr \right], \tag{10}$$

where r_b and r_m are the radial positions at the bottom and the meniscus of the cell. The integration by a numerical method has to be executed at every iteration step. When replacing the value c_L by the new parameter X/k_1, the model function (8) as well as Eqs. (9) and (10) contain two linear parameters, c_N and $1/k_1$, and one non-linear parameter, X. By means of the Eqs. (9) and (10) c_N and $1/k_1$ can be eliminated in the model function Eq. (8). This enables us to retain only the unknown parameter X as a product of the free ligand concentration and the association constant k_1 for the binding of the first ligand.

Depending on the quality of the obtained experimental curves we can estimate the association constant and the partial concentrations of reactants with a relative error of about 10% by this fitting procedure. Furthermore, from the best fit using different occupation numbers we can derive information about the stoichiometry of the reaction.

Results

Results of the DMS cross-linking of HBsu are shown in Fig. 1. Among the reaction products four main species from monomer to tetramer (1–4) can be clearly distinguished. A small amount of pentamer and hexamer is also recognizable.

Progr Colloid Polym Sci (1995) 99:74–81
© Steinkopff Verlag 1995

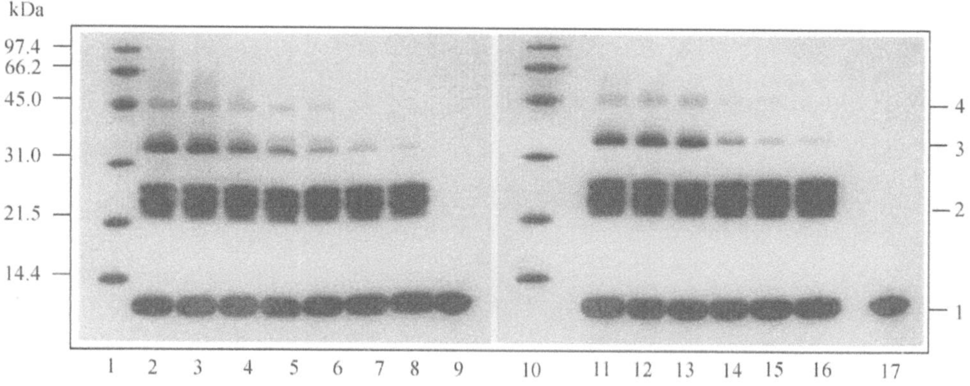

Fig. 1 Cross-linking of HBsu with dimethyl suberimidate (DMS). 15% SDS-PAGE, Coomassie-stained. Lanes (1) and (10) marker proteins, (9) and (17) HBsu without cross-linking reagent. From lane (2) to (8) increasing ionic strength:(2) 2.5 mM NaCl, (3) 12.5 mM, (4) 52.5 mM, (5) 77.5 mM, (6) 102.5 mM (7) 152.5 mM, (8) 202.5 mM. From lane (11) to (16) increasing concentrations of a double-stranded 10-basepair DNA oligomer, 50 mM NaCl. Lane (11) no DNA (12) 5 μM, (13) 10 μM, (14) 20 μM, (15) 40 μM, (16) 80 μM

The amount of residual monomer is not affected by changes in protein concentration (not shown), ionic strength and DNA concentration. This suggests that it is determined mainly by cross-linking efficiency. The predominant reaction product is the dimer. It is present in four distinct bands, presumably representing dimerization products differing in the sites of cross-linker attachment and hence in gel mobility.

Conversely, the formation of trimers, tetramers and higher-order structures is clearly affected by ionic strength. At NaCl concentrations above 100 mM the yield of reaction products larger than the dimer becomes almost negligible. The formation of those aggregates can be assumed to be mediated by electrostatic interactions involving the amino groups crucial for DMS cross-linking reactions. For the dimer formation electrostatical interactions do not seem to play that role: Its yield is not influenced by changes in ionic strength.

A similar effect can be observed for the titration of HBsu with the double-stranded 10 bp DNA oligomer CGCACACACG/CGTGTGTGCG. Added in low concentrations up to 10 μM to a sample containing HBsu at a concentration of 160 μM it slightly increases the yield of trimers and tetramers. As higher DNA concentrations are added their formation is prevented. Some of those amino groups connected by the cross-linking reagent seem to be involved in DNA binding. Binding of DNA does not increase the HBsu protein's tendency towards oligomerization.

Analytical ultracentrifugation demonstrates the dimer to be by far the predominant solution structure of HBsu. Between initial protein concentrations of 30 μM (Fig. 2) and 160 μM (not shown) no significant difference is found in the results obtained from ultracentrifugation at sedimentation equilibrium. The apparent molecular masses

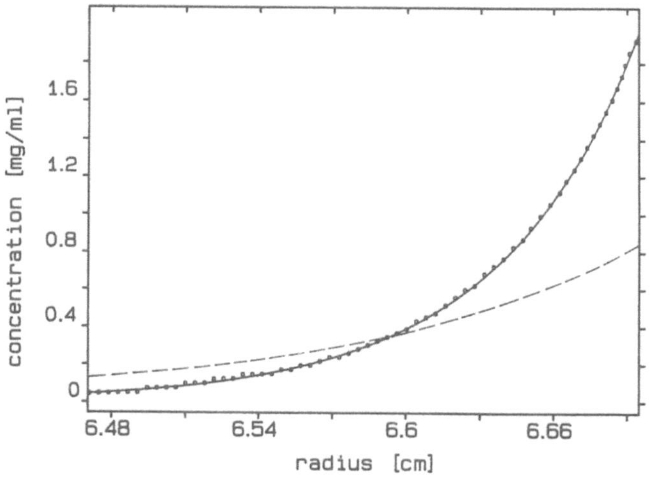

Fig. 2 Radial concentration distribution (o) of 30 μM HBsu, dissolved in 50 mM Tris-HCl, pH 7.5, 1 mM EDTA, 50 mM NaCl at sedimentation equilibrium, (speed:32 000 rpm, temperature 14.3 °C). The molecular mass determined from these data is 20950. The dotted line was calculated by Eq. (1) for a molecular mass of 9882 using the same conditions. A concentration of 1 mg/ml corresponds to 101 μM HBsu

calculated for both concentrations, 21.0 ± 0.6 and 22.0 ± 0.5 kDa, are slightly higher than the 19764 Da of the dimer derived from the amino acid sequence of HBsu. This indicates that the dimer is not the only compound present at the chosen conditions (50 mM NaCl). At least one larger complex must be assumed to contribute to the apparent molecular weight in the solution. The range of protein concentrations covered by the centrifugation method is wide: The bottom concentration in a sample cell at sedimentation equilibrium amounts to about 0.19 mM for the lower and about 1 mM for the higher initial concentration (not shown), while the protein concentration

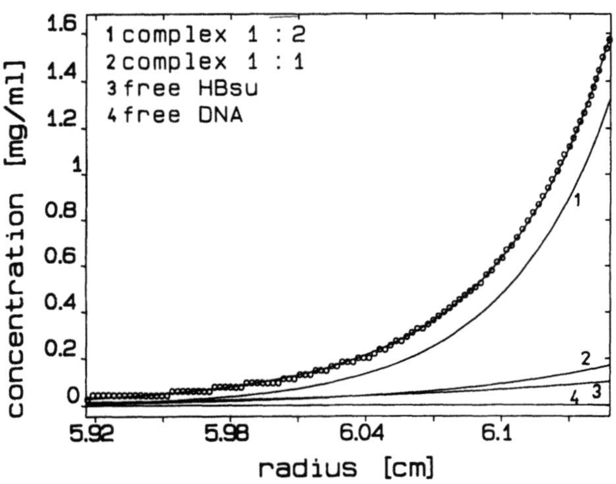

Fig. 3 Radial concentration distribution (○) of a solution of 7.6 μM oligonucleotide and 15 μM HBsu (dimer) in 50 mM Tris-HCl, pH 7.5, 50 mM NaCl and 1 mM EDTA at sedimentation equilibrium (speed:24 000 rpm), temperature 14.3 °C. The curves below the experimental curve correspond to the 2:1 complex (1), the 1:1 complex (2), free HBsu (3) and free oligonucleotide (4). A concentration of 1 mg/ml corresponds to 50.5 μM HBsu (dimer)

remaining at the meniscus becomes negligibly low. In this range, a protein-concentration dependence of formation of higher-order quaternary structures is not observed.

The radial concentration distributions of sedimentation equilibrium runs obtained from different mixtures between HBsu and the double-stranded DNA oligomer in a range from the fivefold molar excess of the oligonucleotide over HBsu (dimer) to the fivefold molar excess of protein over DNA are screened using the program POLYMOL. Figure 3 represents the radial concentration distribution of a mixture of DNA and the twofold molar excess of dimeric HBsu including the portions of free reactants and complexes. The optimal fit to the experimental curve is obtained for a mixture of 2:1 and 1:1 complexes (two or one HBsu dimers per DNA). Stoichiometries with higher protein-DNA ratios are theoretically possible but can be excluded because of the poor fit obtained when assuming occupation ratios higher than 2:1 (see Fig. 4). In addition to the residuals as measure for the quality of radial fits (Fig. 4) the lowest value for the standard deviation or variance (Fig. 5) favors the binding of two HBsu dimers

Fig. 4 Fits to the experimental radial concentration distribution and residuals for the data given in Fig. 3 assuming occupation numbers 1, 2, 3, or 4

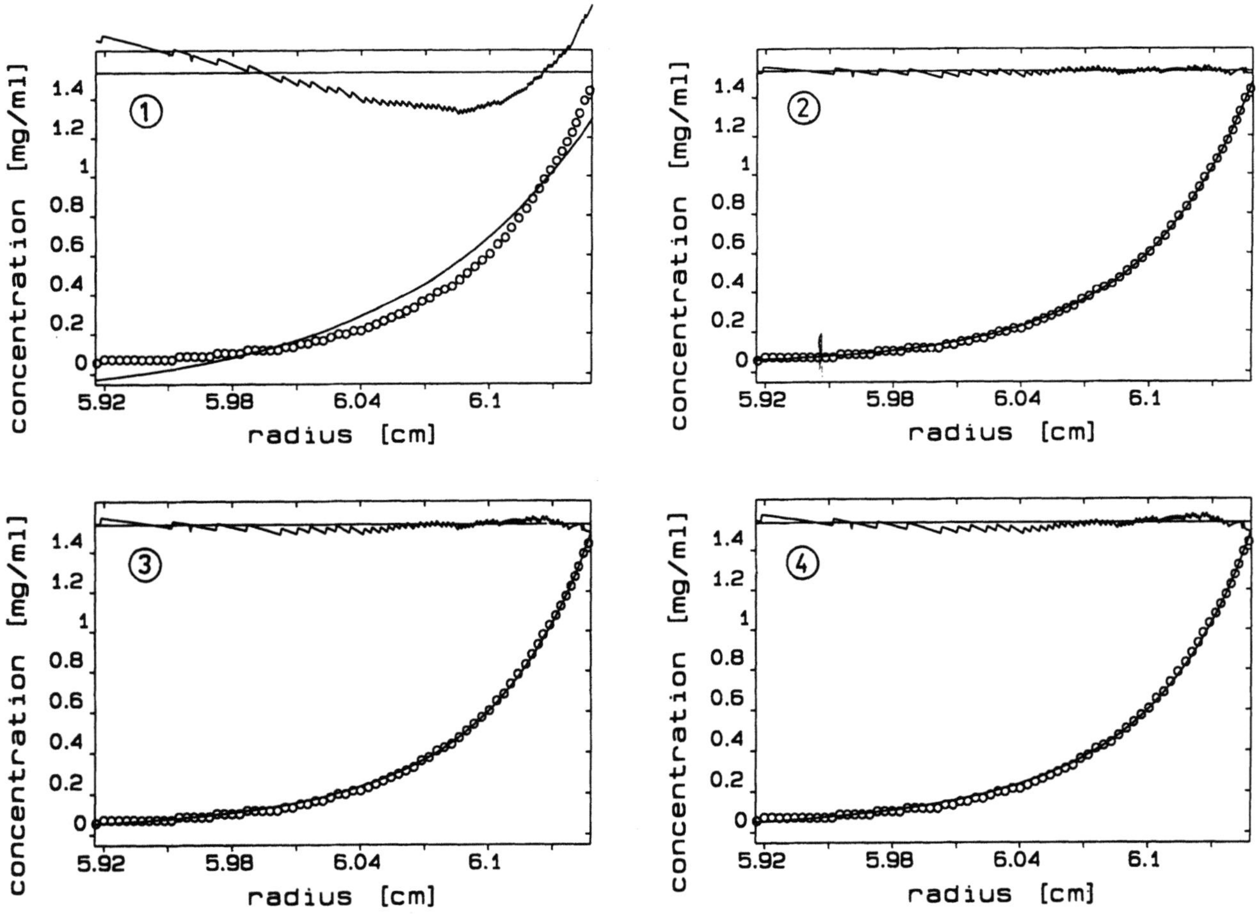

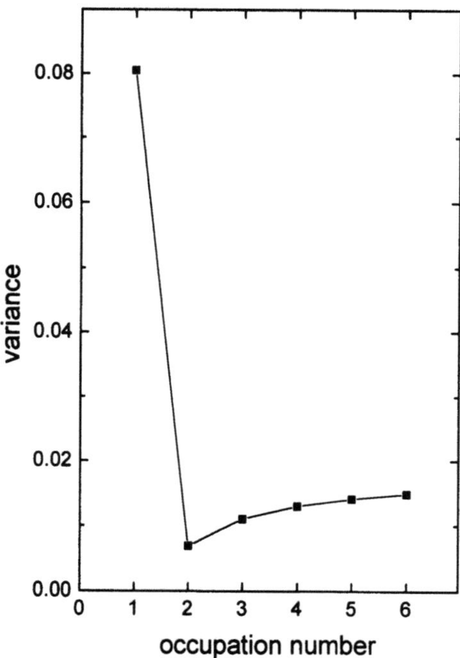

Fig. 5 Estimation of the stoichiometry for the binding of two HBsu dimers to one oligonucleotide duplex assuming different association numbers in s.d

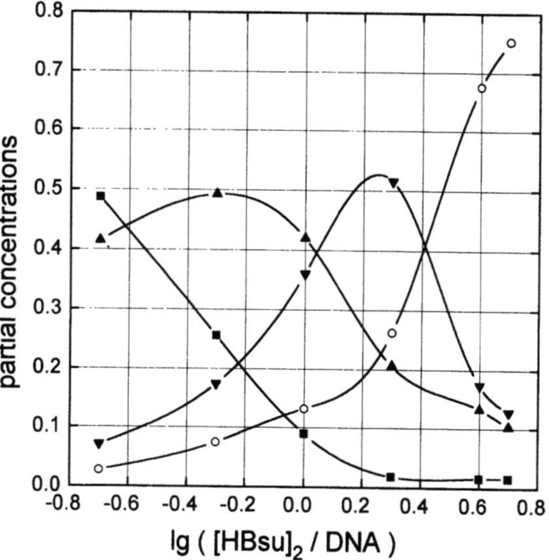

Fig. 6 Partial concentration of the oligonucleotide (■), the HBsu dimer (▲), and the 1:1 (○) and 2:1) (▼) complexes of the HBsu dimer and oligonucleotide monomer depending on different initial ratios. HBsu concentration (dimer):15 μM, DNA concentration (duplex):from a fivefold molar excess of protein over DNA to a fivefold excess of DNA over protein. Buffer conditions as given in Fig. 2. Speed:24 000 rpm, temperature 14.3 °C

per DNA decamer. To examine whether also complexes with higher stoichiometries (three, four or more protein dimers per oligonucleotide) are formed, especially at higher protein-DNA ratios, we have analyzed the concentration distribution according to the model function Eq. (8) assuming higher protein coordination numbers than 2. In all experiments the standard deviation for such a test shows the lowest value at $n = 2$, i.e., the 2:1 complex (HBsu dimer/oligonucleotide) is the most probable.

From the sedimentation equilibrium runs at different protein-DNA ratios partial concentration of free oligonucleotide, free protein and HBsu-DNA complexes can be obtained (Fig. 6). These results are based on the assumption of molecular masses of 6233 (DNA oligomer) and 19764 (HBsu dimer). In all experiments complexes between the reactants are formed. Whereas the concentration of 1:1 complex exceeds that of the 1:2 stoichiometry at higher oligonucleotide concentrations, at higher protein content in the mixtures we observe an opposite behaviour.

The weight-average values of the molecular mass, as well as the association constants calculated from the partial concentrations (given in Fig. 6) are presented in Fig. 7. The highest molecular mass obtained for the 2:1 stoichiometry reflects the strongest interaction between protein and oligonucleotide. Other DNA-protein ratios lead to higher concentrations of free reactants reducing the weight-average molecular mass. The association constant

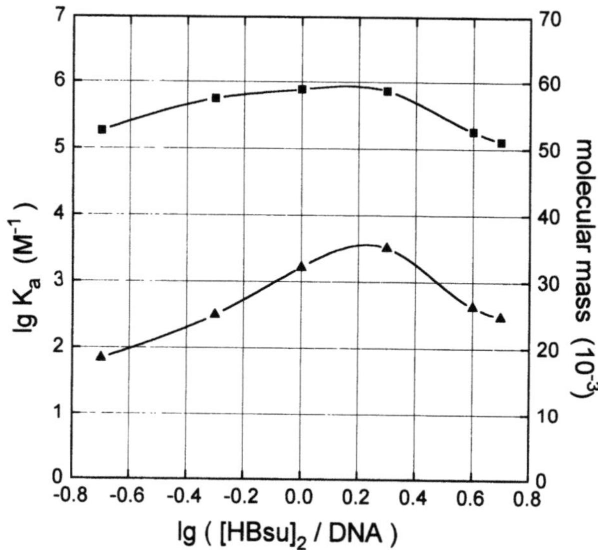

Fig. 7 Weight-average molecular masses (■) and association constants (▲) for the interaction of HBsu dimers with the oligonucleotide depending on different mixtures

for the first protein molecule binding on the nucleotide is calculated to be somewhat smaller than 10^6 M^{-1}. In the presence of an excess of either DNA oligomer or HBsu protein the association constants unexpectedly are somewhat smaller. This must be attributed to residual errors in data processing.

The association constant for the binding of the second protein molecule on the oligonucleotide amounts to about $K_2 = \frac{1}{4} K_1^2$. This means that both binding sites are occupied independently from each other in a statistic process without cooperativity.

Discussion

Studying the quaternary structure of HBsu by two techniques, protein concentration ranges from 80 to 430 μM in cross-linking experiments (not shown) and from near zero to 1 mM in analytical ultracentrifugation were covered. Assuming a number of 20 000 HBsu monomers per cell [30] and a cell volume of roughly 1 μm^3 an intracellular HBsu concentration of 10 to 100 μM can be estimated. The main quaternary structure of HBsu, determined from chemical cross-linking and from analytical ultracentrifugation under the chosen experimental conditions, is the dimer. This finding corresponds to the crystal structure of the HBsu homologue from *B. stearothermophilus*, HBst, showing a dimer with an architecture of closely interlocked monomers [14]. Interactions between subunits in a dimer are mostly hydrophobic and thus independent of ionic strength. Dissociation of the dimer requires denaturing conditions and leads to release of two denatured monomers in a two-state process [15–17]. For the complex composition under non-denaturing conditions as chosen in the ultracentrifugation experiments free monomer does not play a role. Formation of aggregates larger than the dimer is strongly dependent on a low ionic strength. At NaCl concentrations higher than 100 mM the dimer is nearly the only reaction product. As the physiological ionic strength of cells grown on conventional media exceeds this value, the dimer must be assumed to be the only quaternary structure species of HBsu with physiological relevance.

Addition of a short, double-stranded DNA oligomer prevents the formation of higher-order HBsu structures. This result contrasts the findings of Losso et al. [21] who observed a promoting effect of DNA on the formation of HU aggregates. The finding that binding of DNA prevents the formation of HBsu aggregates as does increased ionic strength corroborates the assumption that the unspecific binding of the protein to DNA is based on ionic interactions of basic sidechains located on the flexible "arms" of the structure [14] with the DNA backbone. Apparently, the same sidechains mediate the aggregation of HBsu dimers to higher-order structures at low ionic strength.

In the ultracentrifugation experiments the maximal number of HBsu molecules bound by the DNA decamer are two dimers. This corresponds to the results of Groch et al. [20] who determined a contact length of (3.5 ± 0.5) basepairs of DNA per HBsu dimer by means of fluorescence spectroscopy. For the closely homologous HBgl protein from *Bacillus globigii* contact lengths of 3.5 [31] and (6 ± 2) [32] basepairs were determined by different methods. A contact length that short does, in fact, not fit with the model of DNA binding proposed by Tanaka et al. [14] and White et al. [18] that requires a length of at least 8–10 basepairs of DNA. The co-crystallization of HBsu or one of its homologues with DNA is hoped to provide a clearer picture of the protein-DNA interactions in such a complex.

Acknowledgement The technical assistence of A. Knespel is gratefully acknowledged. Supported by the Deutsche Forschungsgemeinschaft under He 1318/6-2 and by the Fonds der Chemischen Industrie.

References

1. Drlica K, Rouvière-Yaniv J (1987) Microbiol Rev 51:301–319
2. Broyles SS, Pettijohn DE (1986) J Mol Biol 187:47–60
3. Le Hégarat F, Salti-Montesanto V, Hauck Y, Hirschbein L (1993) Biochim Biophys Acta 1172:101–107
4. Dixon NE, Kornberg A (1984) Proc Natl Acad Sci USA 81:424–428
5. Skarstad K, Baker TA, Kornberg A (1990) EMBO J 9:2341–2348
6. Goodman SD, Nash HA (1989) Nature 341:251–254
7. Goodman SD, Nicholson SC, Nash HA (1992) Proc Natl Acad Sci USA 89:11910–11914
8. Craigie R, Arndt-Jovin DJ, Mizuuchi K (1985) Proc Natl Acad Sci USA 82:7570–7574
9. Hodges-Garcia Y, Hagerman PJ, Pettijohn DE (1989) J Biol Chem 264:14621–14623
10. Bonnefoy E, Rouvière-Yaniv J (1992) EMBO J 11:4489–4496
11. Wada M, Kano Y, Ogawa T, Okazaki T, Imamoto F (1988) J Mol Biol 204:581–591
12. Huisman O, Faelen M, Girard D, Jaffé A, Toussaint A, Rouvière-Yaniv J (1989) J Bacteriol 171:3704–3712
13. Micka B, Marahiel MA (1992) Biochimie 74:641–650
14. Tanaka I, Appelt K, Dijk J, White SW, Wilson KS (1984) Nature 310:376–381
15. Welfle H, Misselwitz R, Welfle K, Groch N, Heinemann U (1992) Eur J Biochem 204:1049–1055
16. Welfle H, Misselwitz R, Welfle K, Groch N, Heinemann U (1993) J Biomol Struct Dyn 11:381–394
17. Welfle H, Misselwitz R, Welfle K, Schindelin H, Scholtz AS, Heinemann U (1993) Eur J Biochem 217:849–856
18. White SW, Appelt K, Wilson KS, Tanaka I (1989) Proteins:Struct Funct Genet 5:281–288
19. Groch N, Quaas R, Hahn U, Heinemann U (1988) Nucleosides Nucleotides 7:817–820
20. Groch N, Schindelin H, Scholtz AS,

Progr Colloid Polym Sci (1995) 99:74–81
© Steinkopff Verlag 1995

Hahn U, Heinemann U (1992) Eur J Biochem 207:677–685
21. Losso MA, Pawlik RT, Canonaco MA, Gualerzi CO (1986) Eur J Biochem 155:27–32
22. Gualerzi CO, Losso MA, Lammi M, Friedrich K, Pawlik RT, Canonaco MA, Gianfranceschi G, Pingoud A, Pon CL (1986) In:Gualerzi CO, Pon CL (eds) Bacterial Chromatin, Springer, Berlin, Heidelberg, pp 101–134
23. Behlke J, Ristau O, Knespel A (1994) Prog Colloid Polym Sci 94:40–45
24. Carpenter FH, Harrington KT (1972) J Biol Chem 247:5580–5586
25. Laemmli UK (1970) Nature 227:680–685
26. Yphantis DA (1964) Biochemistry 3:297–317
27. Wynne CG, Wormell PMJH (1963) Appl Optics 2:1233–1238
28. Wyman J, Gill SJ (1990) Binding and Linkage, University Science Books, Mill Valley
29. Behlke J, Ristau O, Gudkow AT (1984) Stud Biophys (Berlin) 99:45–52
30. Salti V, Le Hégarat F, Hirschbein L (1985) J Gen Microbiol 131:581–590
31. Imber R, Bächinger H, Bickle T (1982) Eur J Biochem 122:627–632
32. Imber R, Kimura M, Groch N, Heinemann U (1987) Eur J Biochem 165:547–552

Progr Colloid Polym Sci (1995) 99:82–86
© Steinkopff Verlag 1995

Hydrodynamic modelling of the solution conformation of 10 S myosin

O. Byron

Received: 10 May 1995
Accepted: 19 June 1995

Dr. O. Byron (✉)
NCMH, Department of Biochemistry
University of Leicester
University Road
Leicester LE1 7RH, United Kingdom

Abstract A great deal of work has been done to further the understanding of the solution conformation of the 6 S myosin monomer. Less extensively studied is the 10 S conformer in which part of the tail is observed to be attached to the heads. In this paper existing hydrodynamic data are further interpreted. A model for 10 S myosin is proposed which is more compact than previously reported structures.

Key words hydrodynamic bead modelling – 10 S conformation – myosin

Introduction

The solution conformations of myosin [1–5] and its subfragments [6, 7] have been widely studied and the data obtained have been analysed at a number of resolutions to further the understanding of the role of this molecule in the function of muscle. Specifically, whole myosin has been modelled hydrodynamically in two distinct ways. Its representation as a low resolution general triaxial ellipsoid [8, 9] results in an axial ratio of $(a/b, b/c) = (80, 1)$. But this strategy is insufficient to account for the mass distribution and inevitable flexibility of the molecule. The controversy surrounding myosin flexibility has been the subject of extensive work (see [10] and references cited therein) but it is almost certain that there are two points of flexibility within whole myosin: at the joint between the heads (S1 subfragments) and the tail (S2 and LMM) [10] and between the S2 and LMM sections of the tail itself [11, 12].

Myosin exists in a number of forms within the body. It is found in smooth muscle, in striated muscle and also in non-muscle cells. Monomer 6 S myosin self-associates to form myosin filaments. A 10 S conformer of myosin has also been observed both in smooth [13] and striated [4] muscle and in some non-muscle cells. In this inert conformer a region of the tail appears (under the electron microscope) to be attached to the neck region. The role of

these conformers is uncertain, their presence *in vivo* has yet to be confirmed. However the solution conformation has been studied with small-angle x-ray scattering [3, 5]. The apparent radius of gyration of cross-section was 17 Å for the 6 S and 32 Å for the 10 S myosin. Sphere models were constructed for both conformations and the wider angle scattering curves generated for these objects using Debye equations were compared with experimental data. The authors found it necessary to model 6 S myosin with an unexpectedly small angle between the S 1 subfragment heads. They also observed that their 10 S model with its openly-looped tail gave a comparatively poor fit to the scattering data.

In this present study sphere (or bead) models for 6 S and 10 S myosin are presented based upon the emulation of hydrodynamic data. In one model for the intact molecule the tail is represented, in common with Faruqi et al. (1991) by spheres of 20 Å diameter. But unlike the Debye sphere model reported by these authors, the heads of the hydrodynamic models reported here have been generated with an algorithm which facilitates the direct transformation of atomic coordinate data into bead coordinates [14]. With the release of the coordinates for the α-carbon backbone of the S 1 fragment [15] it is now possible to generate models confident of the limits imposed by steric hindrance in, for example, bringing the heads together in 6 S myosin.

Hydrodynamic bead modelling

The modelling of biological macromolecules with multi-sphere assemblies is well documented and will not be explored at length in this paper. For a good review of the theory upon which hydrodynamic modelling is based see [16]. The program TRV written by Garcia de la Torre [17] uses a modified Burgers–Oseen tensor to generate an array of reduced hydrodynamic parameters from the Cartesian coordinates and radii of the bead assembly. An updated version of this software is also available (HYDRO [18]) which, amongst other improvements over TRV, converts the reduced parameters to experimentally relevant data. The program used in this study is a version of TRV modified to do just this and to link directly with the model generation algorithm [19]. The hydrodynamic modelling was performed on a Silicon Graphics Instruments Challenge XL mainframe computer at the University of Leicester, UK.

The design of models

The remarkable discrepancies in hydrodynamic data obtained by different laboratories [10, 11] extend beyond the parameters of rotational relaxation times and rod length. A cursory survey of work published in the last four decades reveals a significant variation in molecular weight (M), sedimentation coefficient ($s^0_{20,w}$), diffusion coefficient (D), intrinsic viscosity ($[\eta]$) and radius of gyration (Rg) for intact 6 S myosin although the data for the S 1 subfragment are less variable [6]. It has been suggested that myosin isolated from the experimental animal might indeed be changing as the animal is inbred over time [20]. The modelling of the 6 S and 10 S conformers reported in this paper aims to reproduce the most recent data which are given in Table 1.

A further limitation to the level of accuracy fundamentally achievable in hydrodynamic modelling is the

Table 1 Comparison of experimentally determined and hydrodynamically modelled parameters for bead models of 6 S myosin

Parameter	Experimental value	Results from tropomyosin tail model	Results from synthetic tail model
M^0_w (kDa)	520[1]	–	–
$\bar{v}$ (ml/g)	0.728[2]	–	–
$s^0_{20,w}$ (S)	5.95 ± 0.06[3]	6.10	6.04
k_s (ml/g)	88.4 ± 8.0[3]	–	–
$[\eta]$ (ml/g)	234 ± 1[3]	314	317
k_η (ml/g)	92 ± 4[3]	–	–
Rg (Å)	468 ± 50[4]	510	510

[1] [26]; [2] [27]; [3] [1]; [4] [2].

uncertainty in the level of molecular hydration (δ). Although recently a reliable method for the determination of δ was reported [21] in this case an alternative derivation was possible for the intact molecule, requiring no further experimental work. The ratio of regression coefficients for the concentration dependence of viscosity and sedimentation coefficients (k_η and k_s respectively) is equal to the ratio of swollen and partial specific volumes ($\bar{v}_s$ and $\bar{v}$ respectively) [22].

$$\frac{k_\eta}{k_s} = \frac{\bar{v}_s}{\bar{v}} . \tag{1}$$

From the data in Table 1 this ratio is 1.054 ± 0.141. The hydration can then be calculated from the following relationship, as $\bar{v}$ for whole myosin is known to be 0.728 ml/g.

$$\frac{\bar{v}_s}{\bar{v}} = 1 + \frac{\delta}{v} . \tag{2}$$

As δ cannot be negative, the effective range of hydration is $(0 \leq \delta \leq 0.14)$g water/g protein. This is a low level of solvation (an average protein solvation is around 0.25–0.35 g water/g protein). To compensate for any underestimation in hydration the maximum value of $\delta = 0.14$ g/g was used in all modelling presented here. It was assumed that there would be no change in hydration upon change in conformation and that hydration is uniform.

The models were hydrated by uniform expansion [11]. The ratio of hydrated volume to dry volume given in (2) above for $\delta = 0.14$ g/g is 1.19. This ratio is the cube of the expansion factor (1.06). Thus in order to simulate the required hydration a 6% uniform expansion was applied so that all radii were increased by 6% and the coordinates were similarly modified so that the final model was essentially a magnified copy of the anhydrous starting assembly.

An edited version of the Brookhaven PDB [23, 24] file for the atomic coordinates of the α-carbon backbone of myosin subfragment S 1 [15] was submitted to the algorithm AtoB [14] for the generation of a bead assembly representation of the myosin head region. A resolution of 20 Å was adopted in order to reduce the number of constituent beads and accordingly the computational time required. This resolution notionally represents the maximum diameter a bead can have in the model. But in order to simultaneously satisfy the constrictions of $\bar{v}$, molecular weight and the atomic coordinates, some of the spheres in the model are larger than 20 Å in diameter and overlap with each other. In this instance TRV makes a good approximation as the mathematical basis for computing interaction tensors for overlapping beads of non-uniform size is as yet undeveloped.

Two models were constructed for 6 S myosin (Fig. 1)-one (T) generated solely from atomic resolution data, the

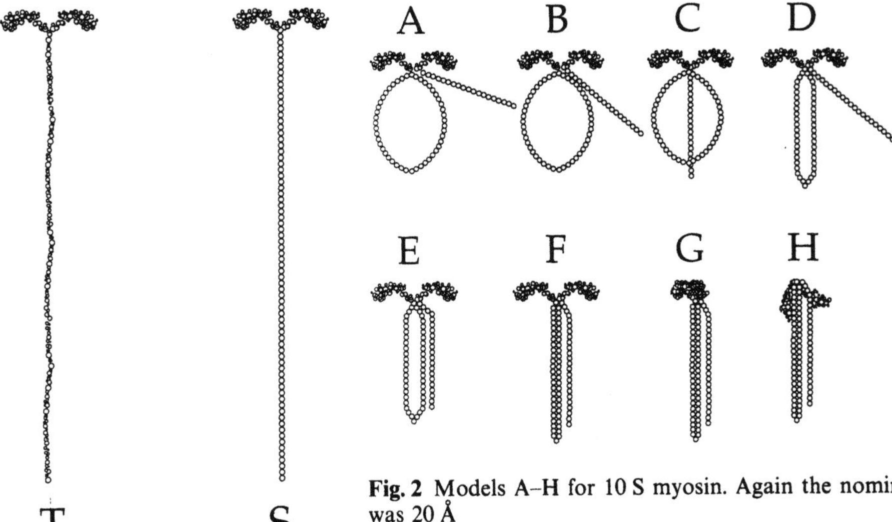

Fig. 1 Models T (with tail based on 3.5 lengths of tropomyosin) and S for 6 S myosin at 20 Å nominal resolution

Fig. 2 Models A–H for 10 S myosin. Again the nominal resolution was 20 Å

other (S) with a 'synthetic' tail. For model T the tail was modelled using the coordinates for 3.5 molecules of tropomyosin [25] laid end-to-end. The coiled-coil α-helical nature of the tail is preserved in the transformation to beads. The synthetic tail model was constructed as a precursor to the 10 S myosin model, as the manipulation of the tail was more straightforward for beads of a uniform size and symmetrical distribution. The tail was represented by 75 beads of 20 Å diameter. This is in general agreement with average dimensions as observed under the electron microscope and is in agreement with the model for 6 S (and 10 S) used by Faruqi et al. (1991).

Eight models were generated for 10 S myosin (Fig. 2), based on model S for 6 S myosin, ranging from configuration A, an extended structure, to H which is essentially the most compact form the conformer could adopt whilst maintaining an overall topology in agreement with that observed under the electron microscope [4]. Model B is very similar to the arrangement of beads used by Faruqi et al. (1991) in small angle x-ray scattering data curve simulations.

Results of modelling

6 S Myosin

There is little significant difference between models S and T for 6 S myosin which means that model S is as sound a starting model for the construction of models A–H for 10 S myosin as model T. Whilst the sedimentation coefficients for S & T are in good agreement with that measured, both the intrinsic viscosities and radii of gyration indicate a need for further refinement of the starting models. It is

probable that the myosin tail is not totally extended in solution, and it is also possible that the introduction of a kink or bend into the tail does reduce the intrinsic viscosity and radius of gyration without the reduction of the sedimentation coefficient below the bounds of the experimental value. This observation is in accordance with conclusions drawn by Garcia de la Torre (1994) who proposes that if a considerable degree of flexibility is built into the myosin rod via a dimensionless rigidity constant (Q) R_g and $[\eta]$ will be smaller. This is certainly observed by the same author for myosin rod alone.

The fact that these models can satisfy the replication of one hydrodynamic parameter but not others acts to highlight the importance of modelling to many experimentally determined data. Given that all three parameters were reproducible simply by the introduction of a nominal kink in the tail (not shown) it was acceptable to use model S as a starting basis for the 10 S models.

10 S Myosin

The results from hydrodynamic modelling of the 10 S assemblies are presented in Fig. 3. Strikingly, models A–E are totally excluded from the final set of possible conformations as none of these yields $s^0_{20,w} = 10$ S. Model F is also practically excluded as $s^0_{20,w}$ for this configuration lies in the region of 10 S only at unfeasibly low hydrations $(\delta < 0.02$ g water/g protein). The two remaining models (G and H) differ in the extent to which the myosin heads are folded down against the looped tail region. In H the heads are almost touching the tail whilst they are somewhat more orthogonal to the tail in model G.

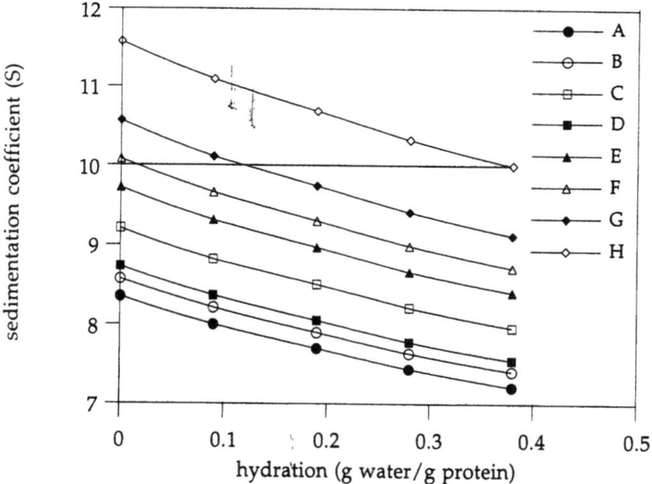

Fig. 3 Sedimentation coefficient generated by TRV as a function of hydration for models A–H of 10 S myosin. The horizontal line indicates the region of experimental data to which the model should conform

Models G and H are both in approximate agreement with the small-angle x-ray data obtained by Faruqi et al. (1991). For both 6 S and 10 S myosin the authors measured the radius of gyration of the cross-section (R_{xs}). This value can be related to the radius of gyration (R_g) through the overall length of the molecule (L) by Eq. (3)

$$L = \{ 12(R_g^2 - R_{xs}^2) \}^{1/2} . \tag{3}$$

For 6 S myosin $R_g = 46$ nm [10] and $R_{xs} = 1.7$ nm [5] this yields a length of $L = 160$ nm which is not unreasonable given the overall dimensions of 6 S myosin (the rod alone has a length of approximately 150 nm). In the compact, rod-like conformations of models G and H it is possible to repeat this calculation for 10 S myosin in order to obtain an estimate for R_g. The overall length of model G is approximately 60 nm. This value for L combined with $R_{xs} = 3.2$ nm for 10 S myosin [5] yields $R_g = 17.6$ nm. For model G the radius of gyration generated by TRV is 19 nm. An imperfect agreement is not surprising given that model G is not a truly rod-shaped structure and Eq. (3) represents an approximation to this geometry. The same treatment of model H yields $R_g = 15.6$ nm which is actually in better agreement with $R_g = 14.9$ nm as generated by

TRV. This is an expected consequence of the more rod-like nature of model H.

The factor which discriminates between models G and H is the intersection of the dependence of $s_{20,w}^0$ upon hydration with the $s_{20,w}^0 = 10$ S line (Fig. 3). A hydration of approximately 0.13 g water/g protein is required for model G whilst model H only intersects with the experimental data at $\delta = 0.38$ g water/g protein. It is unlikely that δ will vary greatly upon conformational change, and δ for 6 S myosin is 0.14 g water/g protein. Model G therefore offers the best representation of the solution conformation of 10 S myosin.

Discussion

It is possible that the highly compact conformation represented by model H for 10 S myosin would give a more satisfactory fit to the wide-angle x-ray scattering data for 10 S myosin [5]. The next step with this work is to construct similar models with beads of uniform size so that small-angle x-ray scattering curve simulation can be performed for these more compact models and the resultant curves compared to experimental data. The intrinsic viscosity of 10 S myosin does not appear to have been measured as yet. It is probable that this is prohibited by the relative scarcity of the conformer. But this would also facilitate more stringent testing of this model.

Finally, given the wealth of data available from hydrodynamic studies of the S 1 subfragment the head regions should be studied further as separate entities on the basis of the atomic resolution data. A higher bead resolution should be used in order to preserve the structure of the actin binding and active sites. This level of detail is beyond the scope of the current paper, the aim of which was to propose a likely solution conformation of the intact 10 S myosin conformer: the solution suprastructure of the individual head regions is unlikely to be directly related to their structure in this form of the molecule.

Acknowledgements I am extremely grateful to Prof Jose Garcia de la Torre for the code for TRV and for his advice in its effective use. I also thank Dr. Arthur Rowe for his continuing general support and his advice, together with that of Dr Clive Bagshaw, on this particular paper. This contribution was presented to the IX Symposium on Analytical Ultracentrifugation in Berlin (March 2nd & 3rd, 1995).

References

1. Emes CH, Rowe AJ (1978) Biochim Biophys Act 537:125–144
2. Hoeltzer A, Lowey S (1958) Biochem 81 (March 20):1370–1377
3. Faruqi AR, Cross R, Kendrick-Jones J (1990) J Musc Res Cell Mot 11 (6):544
4. Ankrett RJ, Rowe AJ, Cross RA, Kendrick-Jones J, Bagshaw CR (1991) J Mol Biol 217:323–335
5. Faruqi AR, Cross RA, Kendrick-Jones J (1991) J Cell Sci (Suppl) 14:23–26.

6. Garrigos M, Morel JE, Garcia de la Torre J (1983) Biochem 22:4961–4969

7. Sugimoto Y, Tokunaga M, Takezawa Y, Ikebe M, Wakabayashi K (1995) Biophys J 68 (April):29s–34s

8. Harding SE (1987) Biophys J 51:673–680

9. Harding SE (1989) In: Harding SE, Rowe AJ, Editors Dynamic Properties of Biomolecular Assemblies. Royal Society of Chemistry: Cambridge, UK

10. Garcia de la Torre J (1994) Eur Biophys J 23:307–322

11. Garcia de la Torre J, Bloomfield VA (1980) Biochem 19:5118–5123

12. Solvez JA, Iniesta A, Garcia de la Torre J (1988) Int J Biol Macromol 10:39–43

13. Onishi H, Wakabayashi T (1982) J Biochem (Tokyo) 92:871–879

14. Byron O (1995) in preparation

15. Rayment I, Rypniewski WR, Schmidt-Base K, Smith R, Tomchick DR, Benning MM, Winkelmann DA, Wesenberg G, Holden HM (1993) Science 261(2 July):50–58

16. Garcia de la Torre J, Bloomfield VA (1981) Quart Rev Biphys 14(1):81–139

17. Garcia de la Torre J (1989) In: Harding SE, Rowe AJ, Editors Dynamic Properties of Biomolecular Assemblies. Royal Society of Chemistry: Cambridge, UK

18. Garcia de la Torre J, Navarro S, Lopez Martinez MC, Diaz FG, Lopez Cascales JJ (1994) Biophys J 67 (August):530–531

19. Byron O (1992) Solution Studies on the Conformation and Assembly of the Monoclonal Antibody B72.3. PhD. University of Nottingham

20. Rowe AJ Personal Communication (1995)

21. Eisenberg H (1994) Biophys Chem 53 (1–2):57–68

22. Rowe AJ (1977) Biopolymers 16:2595–2611

23. Abola EE, Bernstein FC, Bryant SH, Koetzle TF, Weng J (1987) In: Allen FH, Bergerhoff G, Sievers R, Editors Crystallographic Databases – Information Content Software Systems Scientific Applications. Data Commission of the International Union of Crystallography: Bonn, Cambridge, Chester. 107–132

24. Bernstein FC, Koetzle TF, Williams GJB, Meyer EF Jr., Brice MD, Rodgers JR, Kennard O, Shimanouchi T, Tasumi M (1977) J Mol Biol 112:535–542

25. Phillips GN, Cohen C (1986) J Mol Biol 192:128–131

26. Bagshaw CR (1993) Muscle Contraction. 2nd ed. Chapman & Hall: London, UK

27. Parrish RG, Mommaerts WFHM (1954) J Biol Chem 209:901

Progr Colloid Polym Sci (1995) 99:87–93
© Steinkopff Verlag 1995

J. Behlke
P. Dube
M. van Heel
M. Wieske
K. Hayeß
R. Benndorf
G. Lutsch

Supramolecular structure of the small heat shock protein Hsp25

Received: 25 April 1995
Accepted: 22 May 1995

J. Behlke
Humboldt University Berlin
c/o Max Delbrück Center for Molecular
Medicine
Robert-Rössle-Str. 10
13122 Berlin, Germany

Prof. Dr. J. Behlke (✉) · M. Wieske
K. Hayeß · R. Benndorf
G. Lutsch
Max Delbrück Center for Molecular
Medicine
Robert-Rössle-Straße 10
13122 Berlin, Germany

P. Dube · M. van Heel
Fritz Haber Institute of the Max Planck
Society
Faradayweg 4-6
14195 Berlin, Germany

Abstract As determined by hydrodynamic measurements the small heat shock protein Hsp25 isolated from Ehrlich ascites carcinoma cells occurs in the form of monomers with a molecular mass of 23 kDa and a diameter of about 3.8 nm as well as in the form of highly associated complexes with molecular masses of about 180 kDa up to 740 kDa. Electron microscopic investigation of Hsp25 complexes shows ring-like particles with a diameter of about 16 nm and a probably eightfold symmetry. Considering different arrangements of the monomers and using the program HYDRO [Garcia de la Torre et al. Biophysical J. 67:530–531 (1994)] theoretical sedimentation and diffusion coefficients were calculated which are in consensus with the experimental data when assuming ring-like structures. From these data a model for the 3D structure of Hsp25 complexes is derived which consists of four stacked rings each containing eight Hsp25 molecules.

Key words Heat shock protein – supramolecular structure – sedimentation coefficients – diffusion coefficients – electron microscopy – image analysis

Introduction

The increased synthesis of stress or heat shock proteins in response to unphysiological conditions is an evolutionary conserved process of cells to protect important functions. But, also in normal cell metabolism stress proteins assist folding and degradation processes [1]. In comparison with other stress proteins only little information is available about the function of the small heat shock protein (Hsp25) for which a molecular mass of about 23 kDa was derived from the amino acid sequence [2]. At the cellular level Hsp25 is involved in the acquisition of thermotolerance [3–5]. Furthermore, monomeric Hsp25 is observed as an inhibitor of actin polymerization [6–8]. A chaperoning effect has been demonstrated for the recombinant mammalian Hsp25 [9] as well as for the structurally related αB-crystallin [10]. Both proteins are able to form supramolecular complexes of either compact or ring-like structure [11–19]. To get more insight in the possible function of Hsp25 we have analyzed the high molecular mass complexes from Ehrlich ascites carcinoma cells using hydrodynamic and electron microscopic methods. From these data a model of the three-dimensional structure of multimeric Hsp25 complexes is derived.

Material and methods

Preparation of Hsp25

Hsp25 was isolated from Ehrlich ascites carcinoma cells expressing constitutively high amounts of the protein by ammonium sulfate precipitation, column chromatography and ultracentrifugation [8]. Hsp25 particles were separated from the low molecular mass Hsp25 (monomers) by centrifugation through a cushion of 30% sucrose followed by Sepharose 6B chromatography [8].

Hydrodynamics

Sedimentation (s) and diffusion coefficients (D) of Hsp25 were determined by means of the analytical ultracentrifuge Spinco E (Beckman) equipped with UV optics and a photoelectric scanner using a synthetic boundary cell as described earlier [13]. From these data and the partial specific volume $\bar{v} = 0.7326$ ml/g, obtained from the amino acid composition of the protein [2] and the v_i increments of the amino acids, the molar mass (M) was determined by the Svedberg equation

$$M = \frac{s \cdot R \cdot T}{D(1 - \rho\bar{v})} \tag{1}$$

with R the gas constant, T the absolute temperature and ρ the solvent density. Because Hsp25 complexes contain long chain fatty acids [20], $\bar{v} = 0.7447$ ml/g has to be used for these species.

Proposals about the shape of Hsp25 molecules or the deviation from sphere were derived from data of the frictional ratio according to Eq. (2)

$$f/f_0 = 10^{-8}\left(\frac{1 - \rho\bar{v}}{D^2 \cdot s \cdot \bar{v}}\right)^{1/3} \tag{2}$$

Assuming sphere-like molecules the radius (r) of Hsp25 monomers was calculated by formula (3)

$$r = \left(\frac{3M \cdot \bar{v}}{4\pi \cdot N_A}\right)^{1/3} \tag{3}$$

with N_A the Avogadro number.

Sedimentation coefficients of Hsp25 associates (s_n) were estimated from the sedimentation coefficient of the monomer (s_1) and the association number (n) by Eq. (4).

$$s_n = s_1 \cdot n^{2/3} \tag{4}$$

When the associates are members of a polymer homologous series formula (5) can be used

$$s = k_s \cdot M^{a_s} \tag{5}$$

in which the values of the constants k_s and a_s depend upon the nature of the solvent, the conformation of the polymer molecule in it and the extent to which the latter is permeated by this solvent.

For the calculation of model-dependent hydrodynamic properties the computer program HYDRO [21] was used. It is based on the geometry of beads which have been considered for single molecules and requires their radii and the Cartesian coordinates of the centers of beads. Further data as molecular mass, buoyancy factor, solvent viscosity and temperature are necessary to yield such physical parameters as sedimentation and diffusion coefficients. Using different arrangements of the beads the possible model was obtained by fitting the calculated to the experimentally determined hydrodynamic data.

Electron microscopy

Hsp25 samples were prepared for electron microscopy using a double carbon film technique and 1% uranyl formate as negative stain. Micrographs were taken on a Philips EM 400T operating at 80 kV or a Philips CM 12 at 120 kV acceleration voltage. The nominal magnification of 60 000x and a defocus range of -3000 to -5000 Å was used. Selected micrographs were digitised using a datacopy 610 F CCD densitometer with a sampling size of 0.38 nm at the specimen scale. All image processing was performed using the IMAGIC-5 software (Image Science GmbH, Berlin) on VAX-3100 work stations and IBM-PC computers. From 25 electron micrographs, about 3000 molecular images were selected interactively. The molecular images were band-pass filtered (spatial frequencies $< 1/85$ Å and $> 1/15$ Å were suppressed) and normalized. All molecular images were then centered by an iterative translational alignment using a rotationally averaged reference image which was created from the total average of the data set. At the end of each cycle of translational alignment, the reference image was created a new from the total average of the aligned data set. Multivariate Statistical Analysis (MSA) eigenvector-eigenvalue data compression [22] and classification techniques [23, 24] were used to separate various molecular views of Hsp25.

Results

Hydrodynamics

By centrifugation on a sucrose cushion (see material and methods) supernatant material (fraction 1) was separated from heavy material containing Hsp25 complexes. The latter seems to be not uniform but at least of three different sizes. These fractions called fraction 2, 3 and 4 were partially separated by column chromatography on Sepharose 6B.

As expected the supernatant of the separation experiments (fraction 1) contains low molecular mass material. It is characterized by a sedimentation coefficient $s_{20,w} = 2.5S$ and a diffusion coefficient $D_{20,w} = 10.1 * 10^{-7}$ cm^2/s

(Fig. 1). Using the Svedberg equation a molecular mass of 23 kDa was calculated. This result is in agreement with that value obtained from the amino acid composition indicating the supernatant contains monomeric Hsp25. The frictional ratio $f/f_0 = 1.126$ derived from Eq. (2) demonstrates Hsp25 monomers are sphere-like molecules. According to formula (3) we can assume these molecules possess a radius of about 1.9 nm.

For the Hsp25 complexes partially fractionated by Sepharose 6B chromatography we obtained the following results. Fraction 2 is characterized by a sedimentation coefficient of 8.5S and a diffusion coefficient of $D_{20,w} = 4.5*10^{-7}$ cm^2/s (Fig. 2). According to its sedimentation

and diffusion coefficients the molecular mass of fraction 2 amounts to about 180 kDa indicating nearly the eightfold value of Hsp25 momoners. Fraction 3 is a minor one with a sedimentation coefficient of 14.0S (Fig. 3). Because of the low amount of this material the diffusion coefficient could not be determined. When comparing the sedimentation coefficients of fractions 2 and 3 by Eq. (4) the material of fraction 3 seems to be a dimer of fraction 2. Fraction 4 is characterized by $s_{20,w} = 21.8S$ and $D_{20,w} = 2.8*10^{-7}$ cm^2/s (Fig. 4). Its molecular mass of about 740 kDa exceeds the value of Hsp25 monomers 32 times.

According to Eq. (5) the Hsp25 complexes seem to be members of a polymer homologous series with

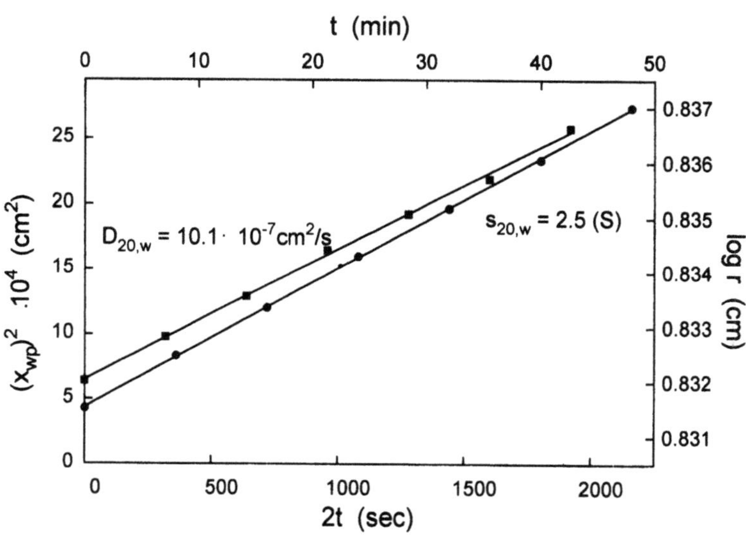

Fig. 1 Plots for the determination of sedimentation (●) and diffusion coefficients (■) of the supernatant Hsp25 solution (fraction 1). Protein concentration: 0.4 mg/ml in 2 mM Tris-HCl, pH 7.5, containing 20 mM NaCl, 1 mM MgCl$_2$, 0.2 mM Na$_2$ATP, 0.5 mM 2-mercaptoethanol. For recording the moving boundary the wave length 280 nm was used. Rotor speed: 6000 rpm for diffusion and 40000 rpm for sedimentation. Temperature 20 °C. The following axis are paired: the upper abscissae with the right ordinate (sedimentation) and the lower abscissae with the left ordinate (diffusion). X_{wp} means the amount of boundary broadening at 0.16 or 0.84 of the plateau, respectively

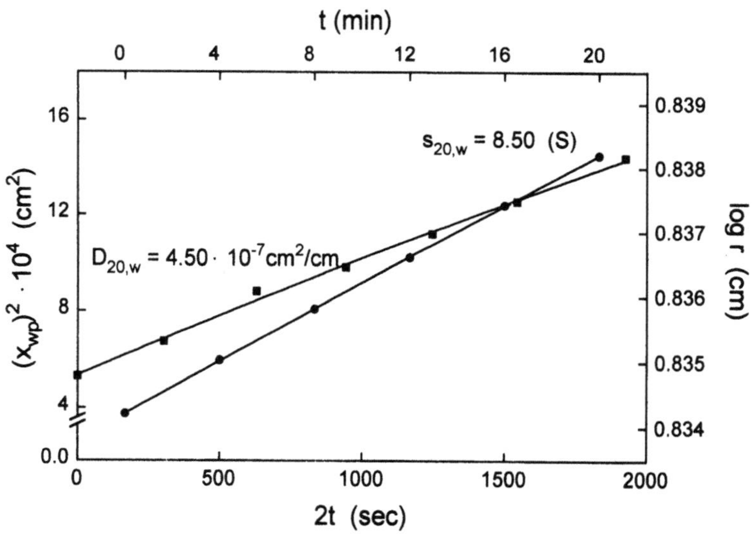

Fig. 2 Plots for the determination of sedimentation (●) and diffusion coefficients (■) of Hsp25 complexes (fraction 2). Conditions as given in Fig. 1 with the exception of sedimentation velocity speed: 20000 rpm

$k_s = (2.56 \pm 0.05) * 10^{-3}$ $S*Mol/g$ and $a_s = 0.67$ meaning globular partciles [25]. However, in comparison with the Hsp25 monomers their sedimentation rate is too small by 15–20%. Lower particle density and/ or a higher surface/volume ratio could be the reasons for that behaviour. In order to clarify this question electron microscopic investigations of Hsp25 complexes were performed.

Electron microscopy

The different fractions of Hsp25 obtained after centrifugation and column chromatography were analyzed by negative staining. In accordance with the hydrodynamic data the supernatant obtained after high speed centrifugation (fraction 1) contains only low molecular mass material (Fig. 5A), high molecular mass material is virtually absent. The different fractions obtained after column chromatography do not differ significantly in their electron microscopic appearance. They contain ring-like particles with a diameter of about 16 nm and a cavity of about 8 nm (Fig. 5B). The particles seem to be penetrated to different extent by negative stain. To get more insight into the fine structure of these particles image analysis has been performed (see material and methods). The resulting class averages representing the different molecular views of Hsp25 are shown as a gallery in Fig. 6A. These views show an approximate circular profile of the molecule indicating a top view orientation of a somewhat cylindrical structure of the Hsp25 complexes. Some of the class averages (see central row in Fig. 6A) appear more compact than the others. The eigenimages generated during MSA show a four-fold, two-fold and to some extent eight-fold symmetric character of the Hsp25 complexes (see Fig. 6B); the first eigenimage being the total sum of the aligned data set and the subsequent three eigenimages showing a slight tilt of the molecules with respect to the supporting film. Although the overall pointgroup symmetry of the multimeric Hsp25 particles could not be unambiguously determined from the present data set, electron microscopy clearly demonstrates that Hsp25 monomers associate to form higher multimeric assemblies not in a dense packing but rather in a ring-like or cylindrical manner.

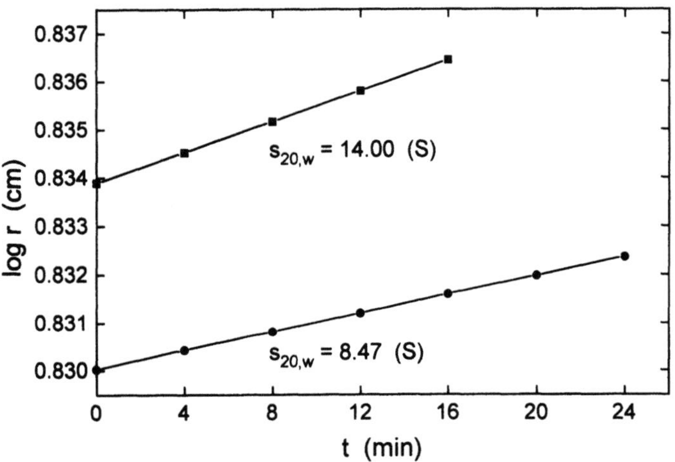

Fig. 3 Determination of sedimentation coefficients of fraction 2 (●) and fraction 3 (■) obtained from the moving boundary in one high velocity experiment at 20 000 rpm. The separation of both fractions was possible after about 50 minutes. Conditions as given in Fig. 1

Modelling of Hsp25 structure

On the basis of the described hydrodynamic and electron microscopic data we tried to approach the structure of multimeric Hsp25 particles by modelling. Various models

Fig. 4 Plots of the determination of sedimentation (●) and diffusion coefficients (■) of Hsp25 complexes (fraction 4). Conditions as given in Fig. 1 with the exception of sedimentation velocity speed: 20 000 rpm

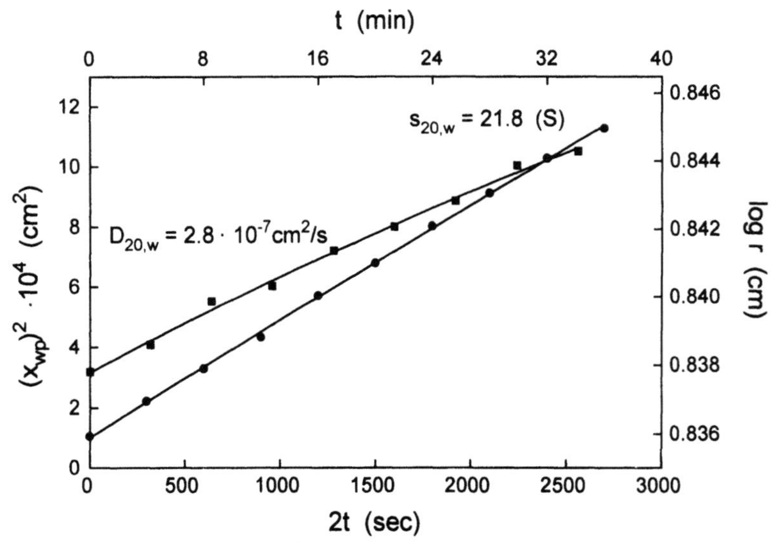

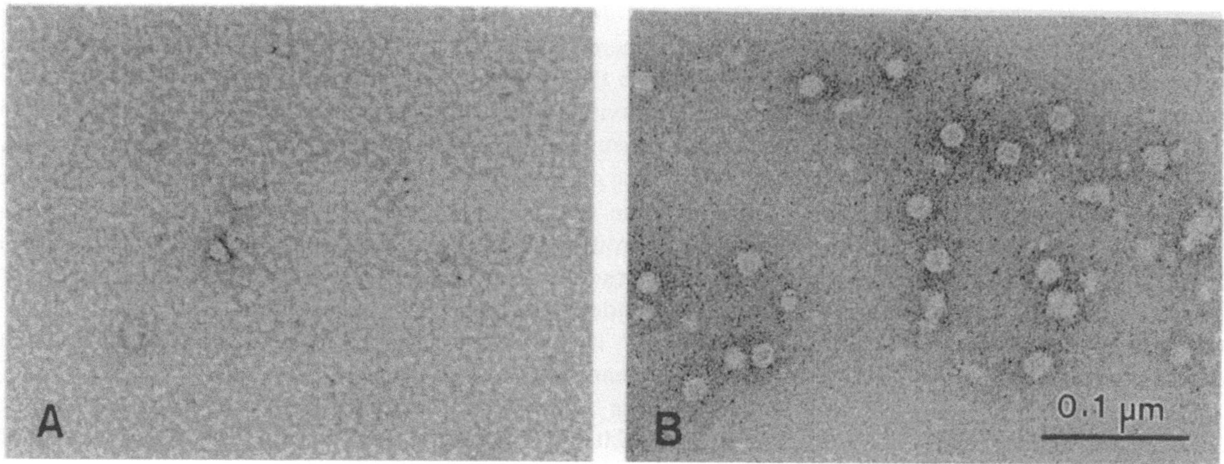

Fig. 5 Electron micrographs of negatively stained Hsp25. A. Hsp25 monomers. B. Hsp25 complexes. Magnification 210000x

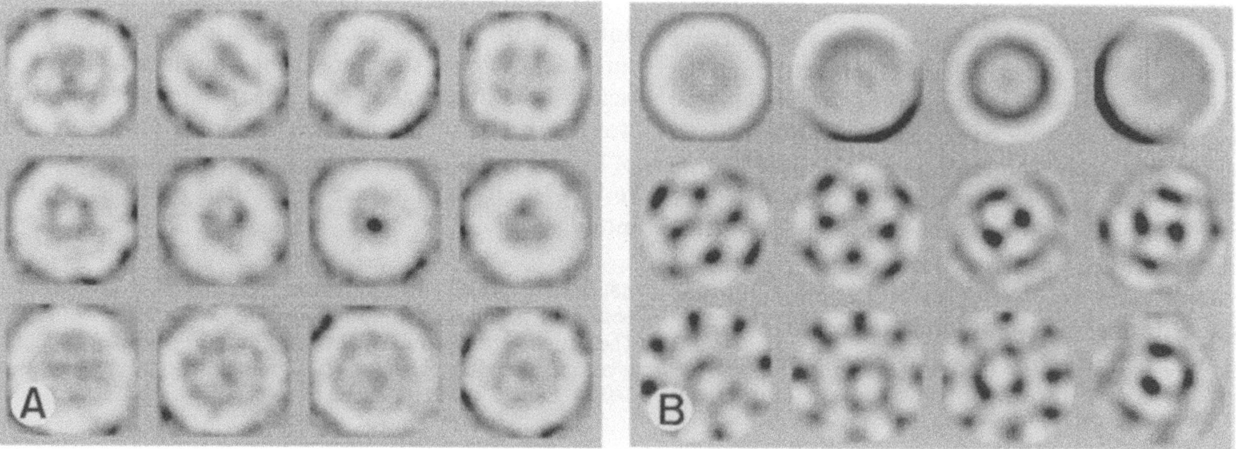

Fig. 6 The result of image processing of negatively stained Hsp25 complexes. A. A gallery showing characteristic views of Hsp25 complexes. B. The first 12 eigenimages generated during image processing. The symmetry properties of the data set are indicated in the middle and bottom row of this mount

Fig. 7 Models of the three-dimensional structure of Hsp25 complexes together with the experimentally determined and calculated (in brackets) parameters. The dimension [F] means 10^{-7} cm^2/s
$\varnothing$ = diameter, h = height

$s_{20,W}$	[S]	2.5		8.5	(8.5)	14.0	(14.20)	21.8	(21.96)
$D_{20,W}$	[F]	10.1		4.5	(4.43)	n.d.	(3.70)	2.8	(2.86)
Mr	[kD]	23		183	(184)	366	(368)	732	(736)
$\varnothing$	[nm]	(4)		14-17	(14)	14-17	(14)	14-17	(14)
h	[nm]	(4)		n.d.	(4)	8-10	(8)	18-20	(16)

with different arrangements of the monomers (beads with $r = 1.9$ nm) were constructed. Using the program HYDRO [21] and as input data the molecular mass of Hsp25 complexes, number, size and the cartesian coordinates of the monomers (beads) and the buoyancy term (here = 0.2553) we obtain among other parameters also the theoretical sedimentation and translational diffusion coefficient for each model. The calculated values were compared with the experimentally determined ones. A good agreement was obtained when assuming stacked rings each containing eight molecules for the three-dimensional structure of Hsp25 complexes (Fig. 7).

Discussion

By application of hydrodynamic and electron microscopic methods a model for the three-dimensional structure of multimeric complexes of the mammalian small heat shock protein Hsp25 has been derived. The model consists of four stacked rings each of them containing eight Hsp25 molecules. Furthermore, hydrodynamic data support the existence also of single and double rings. This structure has some similarity with that of multimeric complexes of other heat shock proteins as well as proteasomes. Members of the 60-kD class of heat shock proteins, i.e. GroEL from eubacteria, TF55 and thermosome from archaebacteria, Hsp60 from mitochondria, rubisco binding protein from chloroplasts, TCPI complexes or TRiC from the cytosol of eukaryotic cells were described previously to be composed of two stacked rings of seven to nine molecules of Hsp60 (see [26] and literature cited therein). A similar cylindrical structure was found for the proteasome, a ubiquitous multicatalytic proteinase (see [27] and literature cited therein). In the case of Hsp60 the corresponding substrate, i.e. the protein to be folded, was found to be located in the cavity of these multimeric particles [28, 29]. Similarly, it is assumed that protein degradation occurs in the cavity of the proteasome. Therefore, it can be speculated that also in the case of multimeric Hsp25 complexes the supposed protein folding may occur in the interior of the cylindrical structure. Further electron microscopic experiments are under way to check this possibility.

In contrast to Hsp25 isolated from Ehrlich ascites tumour cells as described in this paper ("native" Hsp25), the recombinant form of Hsp25 and the structurally related αB-crystallin were shown to possess a compact globular structure [13, 16, 18]. Reasons for the differences between rcombinant and "native" Hsp25 may be posttranslational modifications which cannot be realized in the bacterial cell. This raises the question in which way protein folding occurs on these structures which were described to be active as chaperones [9, 10].

In summary, on the basis of correlative hydrodynamic and electron microscopic investigations a model for the three-dimensional structure of high molecular mass particles of the mammalian small heat shock protein Hsp25 was deduced which has some similarity with the structure of Hsp60 complexes. The functional significance of these results, i.e. chaperoning activity of these particles, has to be established.

Acknowledgements This contribution was presented at the IX Symposium on Analytical Ultracentrifugation March 2/3 1995 in Berlin. The authors are grateful to Dr Garcia de la Torre for providing the program HYDRO. These studies were supported by grants Be 1404/1-1 and Lu 499/3-1 of the Deutsche Forschungsgemeinschaft.

References

1. Gething MJ, Sambrook JF (1992) Nature 355:33–45
2. Gaestel M, Gross B, Benndorf R, Strauss M, Schunck WH, Kraft R, Otto A, Bohm H, Stahl J, Drabsch H, Bielka H (1989) Eur J Biochem 179:209–213
3. Chretien P, Landry (1988) J Cell Physiol 137:157–166
4. Knauf U, Bielka H, Gaestel M (1992) FEBS Lett 309:297–302
5. Lavoie JN, Gingras-Breton G, Tanguay RM, Landry J (1993) J Biol Chem 268:3420–3429
6. Miron T, Wilchek M, Geiger B (1988) Eur J Biochem 178:543–553
7. Miron T, Vancompernolle K, Vandekerckhove J, Wilcheck M, Geiger B (1991) J Cell Biol 114:255–261
8. Benndorf R, Hayeß K, Ryazantsev S, Wieske M, Behlke J, Lutsch G (1994) J Biol Chem 269:20780–20785
9. Jakob U, Gaestel M, Engel K, Buchner J (1993) J Biol Chem 268:1517–1520
10. Horwitz J (1993) Proc Natl Acad Sci USA 89:1049–1053
11. Arrigo AP, Suhan JP, Welch JP (1988) Molec Cell Biol 8:5059–5071
12. Longoni S, Lattonen S, Bullock G, Chiesi M (1990) Molec Cell Biochem 97:121–128
13. Behlke J, Lutsch G, Gaestel M, Bielka H (1991) FEBS Lett 288:119–122
14. Collier NC, Heuser J, Levy MA, Schlesinger MJ (1988) J Cell Biol 106:1131–1139
15. Spector A, Li LK, Augsteyn RC, Schneider A, Freund T (1971) Biochem J 124:337–343
16. Siezen R, Bindels JG, Hoenders J (1978) Eur J Biochem 91:387–396
17. Koretz FJ, Augusteyn RC (1988) Curr Eye Res 7:25–30
18. Augusteyn RC, Koretz JF, Schurtenberger P (1989) Biochim Biophys Acta 999:293–299
19. Merck KB, Groenen PJTA, Voorter CEM, de Haard-Hoeckman WA, Horwitz J, Bloemendahl H, de Jong WW (1993) J Biol Chem 268:1046–1052
20. Oesterreich S, Benndorf R, Reichmann G, Bielka H (1991) NATO ASI Series 56:489–493

21. Garcia de la Torre J, Navarro S, Lopez Martinez MC, Diaz FG, Lopez Cascales JJ (1994) Biophys J 67:530–531
22. Van Heel M, Frank J (1981) Ultramicroscopy 6:113–130
23. Van Heel M (1984) Ultramicroscopy 13:165–184
24. Van Heel M (1989) Optik 82:114–126
25. Cox CD (1969) Arch Biochem Biophys 129:106–123
26. Waldmann T, Nimmesgern E, Nitsch M, Peters J, Pfeifer G, Müller S, Kellermann J, Engel A, Hartl F-U, Baumeister W (1995) Eur J Biochem 227:848–856
27. Zwickl P, Pfeifer G, Lottspeich F, Kopp F, Dahlmann B, Baumeister W (1990) J Struct Biol 103:197–203
28. Langer T, Pfeifer G, Martin J, Baumeister, W, Hartl, F-U (1992) EMBO J 11:4757–4765
29. Braig K, Simon M, Furuya F, Hainfeld JF, Horwich AL (1993) Proc Natl Acad Sci USA 90:3978–3982

Progr Colloid Polym Sci (1995) 99:94–100
© Steinkopff Verlag 1995

Physical-chemical characterization of the different individual cortical alfa-crystallin fractions from bovine lenses

T. Aerts
Q.H. Wang
S. Tatarkova
J. Clauwaert

Received: 23 March 1995
Accepted: 22 May 1995

T. Aerts · Q.H. Wang · S. Tatarkova
Dr. J. Clauwaert (⊠)
Biophysics Research Group
Department of Biochemistry
University of Antwerp
Universiteitsplein 1
2610 Antwerp, Belgium

Abstract The soluble bovine eye lens cytoplasm has been separated in its α-, β- and γ-crystallin fractions, using size exclusion chromatography. The individual α-crystallin fractions have been characterized using moving boundary and equilibrium sedimentation and photon correlation spectroscopy. The α-crystallin fractions represent a very broad distribution in molar mass from 500 000 to 1 100 000 g/mole, with a maximum at 650 000 g/mole. This population contains also a broad distribution of hydrodynamic structures, ranging from compact to more asymmetric structures; the main population has a frictional coefficient f/f_0 ratio of (1.45 ± 0.05). These structural properties perfectly fit the micellar model.

Key words Eye-lens – alfa-crystallin – hydrodynamic structure

Introduction

The eye lens of mammalians is a biconvex, avascular, colourless and almost completely transparent structure, located in the anterior part of the eye behind the pupil-iris diaphragm. The major role of the cytoplasm of the vertebrate eye lens fibre cells is to form a high refractive transparent medium so that the lens can contribute to focus the images on the retina. This high refractive medium (n ranging from 1.37 to 1.44) is obtained by a high concentration of soluble proteins. A rough idea about this protein concentration can be obtained from the relation

$$n = n_0 + \frac{\partial n}{\partial c} \cdot \Delta c \ . \tag{1}$$

If we take $n_0 = 1.33$ and $\partial n/\partial c = 0.200 \text{ ml/g}$, we obtain a protein concentration Δc ranging from 20 to 55 g/100 ml. The gradient of refractive index perfectly corrects the spherical aberration due to the convex surfaces so that the lens can be considered as an almost "perfect lens".

The lens crystallins are the main contributors to this high protein concentration. On a physical, biochemical and immunological basis, three main classes of crystallins can be distinguished in the eye lens of mammalians. In order of decreasing molar mass, are the α-crystallins with a molar mass of about 6.10^5 g/mole, the β_H- and the β_L-crystallins with a molar mass of 2.10^5 and 5.10^4 and the γ-crystallins with a molar mass of about 2.10^4 g/mole. Their concentration in the lens is 45, 20, 20 and 15% respectively. In spite of its high protein content, the eye lens is virtually completely transparent under normal healthy conditions. A theoretical explanation for this apparent contradiction was given by Benedek in the early 1970s [1]. He showed that a limited degree of order in the lens cytoplasm could account for the observed transparency. This was proven experimentally to be correct by Delaye and Tardieu more than a decade later [2]. α-crystallin is the largest protein and it is present in the highest concentration in the cytoplasm so it contributes for more than 90% to the light scattering. It is an oligomeric protein which mainly contains 4 peptides αA_1, αA_2, αB_1, αB_2 where the A peptides have an isoelectric point below pH 7 (acidic) and the B peptides have an isoelectric point above pH 7 (basic). αA_2, the major α-crystallin peptide, and αB_2 are the only primary gene products. αA_1 and αB_1 arise from these peptides by a specific postsynthetic phosphorylation [3, 4]. In addition to these "intact" peptides, α-crystallin contains degraded peptides; these degraded peptides arise on maturing and or aging by specific cleavages of the A or B peptides. For example $\alpha A_{2,1-169}$ represents a peptide identical to αA_2 but only containing the first 169 amino acids.

The tertiary and quaternary structure is still a matter of controversy. The characterization of the native α-crystallin suggests a relation between peptide composition and quaternary structure. α-crystallin, isolated from the cytoplasm of newly synthesized fibre cells mainly contain the 4 undegraded A and B peptides: the young α-crystallin proteins also form a quite homogeneous population with a mean molar mass around 650 000 g/mole. The α-crystallins, isolated from older cells, contain a broad pattern of peptides: the 4 undegraded peptides and a whole set of degraded and modified peptides, originating from the 4 peptides by quite specific degradation and chemical modification such as phosphorylation and deamidation. These α-crystallin solutions contain a quite broad distribution of aggregates [5]. It is still an open question if this broad population of protein molecules is formed by a continuous collection of proteins each differing by one peptide, or if there are some discrete subclasses. Spectroscopic and light scattering studies have suggested a multi-layer tetrahedral model [6] which could give rise to a continuous set of sizes by filling up the different layers continuously or to a discrete set of molecules by giving the proteins with filled layers an extra stability. Some years ago, a micellar model was suggested for the quaternary structure of α-crystallin [7]: this includes that hydrophobic forces are responsible for the aggregation of the peptides and that polar interactions with solvent molecules keep the aggregates soluble and limit their size. Augusteyn and Koretz also claim arguments for a two-dimensional arrangement of the peptides instead of a more globular arrangement as suggested by hydrodynamic and light scattering methods [6, 8] but the experimental evidence for a two-dimensional arrangement is rather indirect [9]. A three-layer structure model has been proposed for the native α-crystallin [10]: this model combines the multilayer and the micelle model. The inner layer (core) forms a micellar structure containing 12 αA subunits: the polar groups are distributed in different parts of the molecule and the apolar region forms the micelle core. Six more subunits can easily be added to form a second layer: the apolar surfaces are oriented toward the hydrophobic inner core. The third layer just adds more subunits and can accommodate 24 αA or αB in a cuboctahedron type symmetrical structure. This model easily explains the changes of the molecular mass according to the solvent conditions: the third layer peptides are loosely bound to the inner two layers by weak polar interactions and these peptides can be easily removed.

Transient electric-birefringence measurements have proven the existence of a fraction of α-crystallin in a more extended quaternary structure [11].

Chaotropic agents such as urea or guanidine hydrochloride reduce the size of the α-crystallin and the addition of high concentrations of these agents to α-crystallin solutions results in a mixture of denatured αA and αB peptides. The finding that intermediate concentrations of urea and guanidine hydrochloride resulted in a step-wise dissociation from larger oligomers to 20 000 g/mole proteins, supported the above mentioned multilayer model [12]. For a long time it has not been possible to renature the denatured peptides to the original oligomers [13] but it has been shown that by carefully controlling the conditions such as protein concentration, pH and/or ionic strength and/or temperature of the solvent, a denatured sample can be renatured to every preset mean molar mass population [14]: an increase in repulsive interactions, obtained by a lower ionic strength or a pH further removed from the isoelectric pH, decreases the size of the renatured proteins; conditions which favour hydrophobic interactions, such as a higher temperature, increase the size of renatured α-crystallins.

The interest for α-crystallin recently reemerged due to the evidence of its chaperone activity [15].

We have examined individual populations of α-crystallins, selected on their elution position after gelfiltration, in order to find out if the physical-chemical properties of these individual populations vary as a function of their elution position.

Materials and methods

Preparation of α-crystallin

The lenses of 6-month ($\pm$ 2 weeks) old calves were freshly obtained at a local slaughterhouse within 3 h after slaughtering and were subsequently stored at 4 °C. The lens capsule was removed and the lenses were mixed with a sixth-fold quantity of buffer (containing 10 mM Hepes, 120 mM KCl, 25 mM NaCl, 0.02% NaN_3, pH 7.0) and gently stirred at 4 °C for 20 min. In this way only the outer cortical cells were dissolved. This suspension was centrifuged at 12 000 g for 30 min to remove the insoluble material.

About 20 ml of this cortical protein solution, dissolved in the above-mentioned buffer (containing about 2000 $A^{1 cm}_{280 nm}$ units), was loaded on a Bio-Gel A-5m column (ϕ 5 cm × 85 cm, Pharmacia) at 4 °C and the eluent was collected in 15 ml fractions. The top fractions of the low molecular mass α-crystallin elution zone were collected.

Photon correlation spectroscopy

Photon correlation spectroscopy has been used for the determination of the diffusion coefficient of the α-crystallin samples and the equivalent hydrodynamic radius. Light scattered by the solutions was detected with an ITT FW 130 photomultiplier and the photocurrent output of the photomultiplier was analyzed using a Brookhaven BI-8000 AT correlator. The setup was installed in a thermostated room and the temperature was monitored directly in the scattering cell.

The quality of our setup was routinely checked by measurements at scattering angles of 50, 90 and 130 degrees.

With a homodyne correlation setup the measured intensity correlation function (or second order correlation function) of a diluted homogeneous solution, containing spherical particles which are small as compared to the wavelength of the light, becomes

$$g^2(t) = A + B \cdot \exp(-2 \cdot D \cdot k^2 \cdot i \cdot \tau) \quad (2)$$

which is usually normalized to

$$g^2(t) = 1 + a \cdot \exp(-2 \cdot D \cdot k^2 \cdot i \cdot \tau) \quad (3)$$

where a: an experimental constant which depends on the correlation volume and the quality of the optical set up; k: the scattering vector $k = 4 \cdot \pi \cdot n \cdot \sin(\theta/2)/\lambda$ D: the translation diffusion coefficient; τ: the sample time and i: the channel number.

When a continuous set of particle sizes is present in the scattering sample, it is assumed that the first-order correlation function (or electric field correlation function) $g^1(k, t)$ can be written as an integral over the relaxation times $\Gamma = D \cdot k^2$

$$g^1(k, t) = \int F(\Gamma) \cdot \exp(-\Gamma t) \cdot d\Gamma. \quad (4)$$

The interesting quantity in Eq. (4) is the distribution function $F(\Gamma)$ which gives the probability that a certain particle size class, with relaxation time Γ, is present in the scattering volume.

We have routinely used the CONTIN method for the extraction of the distribution $F(\Gamma)$ [16]. This method of analysis gives the number distribution, the weight distribution or the z distribution function, depending on the setting of the parameters. We have routinely choosen the setup for the weight distribution function, as this function is directly comparable to the weight distribution function $g(s)$, obtained from boundary sedimentation [17].

Ultracentrifugation: boundary sedimentation and equilibrium sedimentation

The Beckman Optima XL-A analytical ultracentrifuge was employed to perform sedimentation velocity experiments. Solute distributions at $20.0°$ C were recorded via their absorption at 280 nm. Consecutive scans were recorded at regular intervals, utilizing the "autoscan" facility. Sedimentation coefficient $(S_{20,w})$ values were determined in the standard way by plotting the $\ln(r_{\text{first moment of the boundary}})$ versus time or from the time derivative of the sedimentation velocity concentration profile [17].

For the sedimentation equilibrium runs, the run conditions (angular velocity ω and duration of run) were calculated from the preset molecular parameters (sedimentation coefficient, molar mass, 3 mm solution column), using the method proposed by Yphantis [18]. After reaching the equilibrium and having taken the equilibrium absorbance profiles, the angular velocity ω was increased to high speed (40 000 rpm) for another 24 h so that all the proteinous material was sedimented. The remaining absorbance profiles were considered as the best estimate of the residual blanco absorbance and were subtracted from the sample absorbance profiles to obtain the c_r values as a function of r.

The Beckman software was used to analyze the standard equilibrium equation

$$c_r = c_0 \cdot \exp[(M \cdot (1 - v \cdot \varrho) \cdot \omega^2/2 \cdot R \cdot T) \cdot (r^2 - r_0^2)] \quad (5)$$

The solutions have been considered to be ideal.

Results and discussion

Nine individual fractions have been selected in the α-crystallin elution zone afer gel-filtration of the soluble eye lens cytoplasm on a Bio-Gel A-5m column (see Fig. 1). They have been studied with photon correlation spectroscopy, boundary sedimentation and equilibrium sedimentation.

The CONTIN program has been used for the analysis of the experimental correlation functions (16). It is based on a least squares fit of the first-order autocorrelation function to a sum of exponentially spaced set of decaying exponentials. Non-negativity of the amplitudes of the distribution function is built into the minimalization procedure as **a priori** constraint. As photon correlation spectroscopy is actually a low resolution method, the resulting distribution function $F(\Gamma)$ can overestimate the width of the distribution, but the relative trend between the different fractions is reliable. Although photon correlation spectroscopy is related to the Z-average

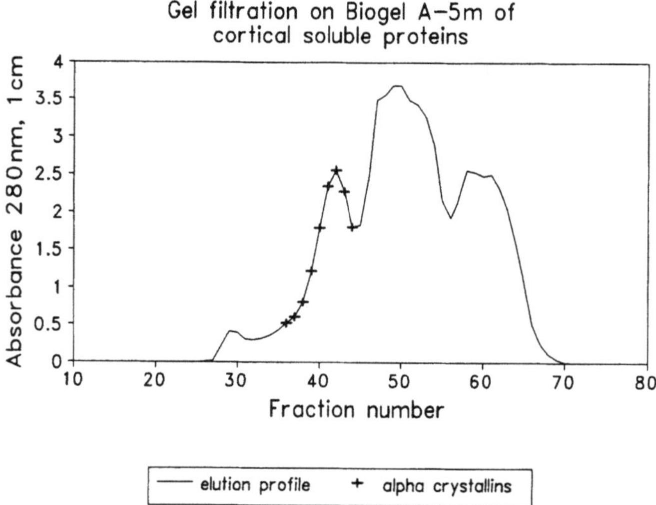

Fig. 1 Gel-filtration elution profile of the soluble cortical lens proteins; the α-crystallin fractions 36 to 44 have been further used

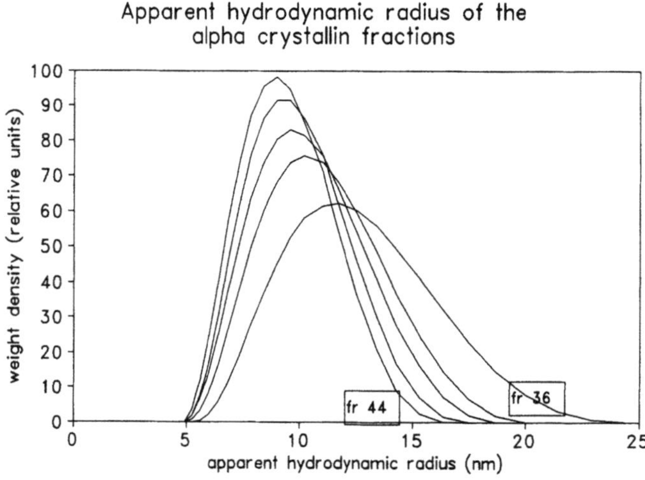

Fig. 2 Distribution function $F(R)$ of the apparent hydrodynamic radius of some α-crystallin fractions, namely 36, 38, 40, 42 and 44, as obtained from the Contin analysis of the experimental intensity correlation functions of the light scattered by the different fractions

diffusion coefficient, it is possible to calculate the weight density distribution function if a model is introduced for the scattering particles (in our case spherical particles). Figure 2 shows the weight density distribution function of the equivalent hydrodynamic radii of the particles present in the different fractions. It is clear that the larger fraction numbers contain a population of smaller particles, as expected from a separation based on size exclusion, but the distribution function is much more narrow and has a maximum at an equivalent radius of (8.0 ± 0.2) nm. The samples with smaller elution number have larger sizes; the weight distribution function has now a maximum at an equivalent radius of (11.5 ± 0.5) nm but the distribution function is much more asymmetric. A larger equivalent hydrodynamic radius can be related to a particle with a similar molar mass but having a more asymmetric structure. These results are in agreement with the studies of van Haeringen and coworkers [11]: they did transient electric-birefringence measurements and used ultraviolet linear-dichroism spectroscopy to prove the presence of more extended, ellipsoidal or cylindrically shaped particles in the α-crystallin fractions, eluting in the front (smaller fraction number) of the α-crystallin elution zone.

The boundary sedimentation profiles can also yield a distribution function $g(s)$ on using the analysis method, as proposed by Stafford [17]. The analysis, in its present form, does not correct for the diffusion broadening, so it also gives a overestimation of the broadness of the real distribution function (see Fig. 3). Both types of distribution functions, $F(R)$ from photon correlation spectroscopy and $g(s)$ from boundary sedimentation profiles, give equivalent and consistent results. Both

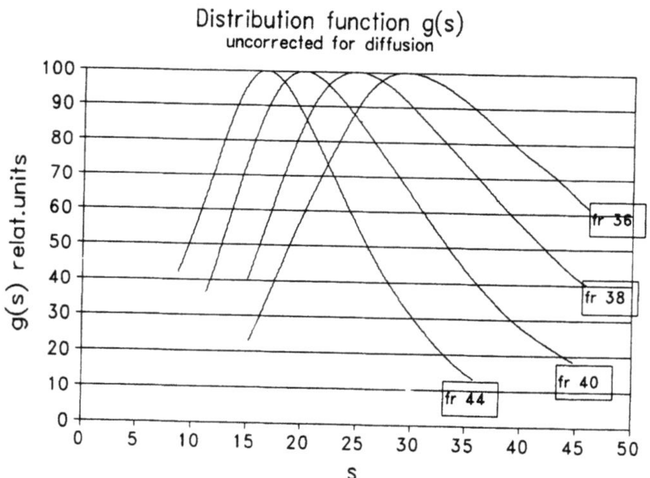

Fig. 3 Apparent distribution function $g(s)$ of the sedimentation coefficient of some α-crystallin fractions, as obtained from the boundary sedimentation of these fractions and using the Stafford analysis method [17]

distribution functions, in their present form, are weight distribution functions. We have used both distribution functions to calculate R_w and s_w using the expressions

$$R_w = \frac{\int F(R) \cdot R \cdot dR}{\int F(R) \cdot dR} \tag{6}$$

$$s_w = \frac{\int g(s) \cdot s \cdot ds}{\int g(s) \cdot ds} \ . \tag{7}$$

We can use R_w and s_w to calculate the molar mass M_w

using the Stokes and Svedberg relations

$$D_w = \frac{k_B \cdot T}{6 \cdot \pi \cdot \eta \cdot R_w} \tag{8}$$

$$M_w = \frac{s_w \cdot k_B \cdot N_A \cdot T}{D_w \cdot (1 - v \cdot \rho)}, \tag{9}$$

where k_B: Boltzmann's constant N_A: Avogadro number η: viscosity of solvent ρ: density of solvent v: partial specific volume T: temperature °Kelvin.

It is now possible to compare the molar mass values, obtained from the above calculations, with the molar mass values directly obtained from equilibrium measurements (see Fig. 4). As can be seen from this figure, there is good agreement between both sets of values. As can be expected from the way the fractions have been selected, the larger fraction numbers, eluting later from the gel-filtration column, have a smaller molar mass than the fractions eluting earlier (smaller fraction number). It is also worthwhile to remark that the fractions, eluting around the top fractions, namely fractions 40 to 44, have a molar mass ranging from 500 000 to 700 000 g/mole.

If mixing up the 4, 6 or 8 top fractions, as usually is done on preparing α-crystallin for biochemical or physical chemical studies, similar results have been obtained. If mixing up the fractions 40 to 44, the following results have been obtained: a s_w value of (18.4 ± 0.3) S and a R_w value of (8.6 ± 0.4) nm results in a mean molar mass M_w of $(675\,000 \pm 40\,000)$ from the Svedberg relation. This value is consistent with the value of $(640\,000 \pm 30\,000)$ g/mole, we obtained directly from equilibrium sedimentation. But this value is appreciably lower than the value of 800 000 g/mole which is usually accepted for α-crystallin [5, 6, 8, 19]: the latter high value results from a combination of the sedimentation and diffusion coefficient in the Svedberg relation or from light or x-ray scattering measurements. The high molar mass, obtained from the Svedberg relation, is probably the result of a too low diffusion coefficient, obtained from photon correlation spectroscopy. Photon correlation spectroscopy and the cumulant analysis [20] yield the D_z average. From the broad distribution functions in Figs. 2 and 3, we can easily understand that the D_z overestimates the mean size and molar mass as compared to the number or weight average values. For the quantitative interpretation of light scattering measurements, the concentration c of the protein has to be determined. The concentration is usually determined from the absorbance at 280 nm. Values of $A_{1cm}^{1\%}$ between 8.0 anc 8.45 have been used [5, 6 and 8]. We have calculated the $A_{280\,nm}^{0.1\%}$ from the absorbance at 205 nm, 207 nm and 209 nm [21] and obtained a value of (0.657 ± 0.013). The introduction of this $A_{280\,nm}^{0.1\%}$ value in the expression for light and x-ray scattering [5, 6 and 8]

also results in molar mass values in the range of 650 000 g/mole.

Augusteyn and coworkers have studied the different fractions, resulting from gelfiltration of bovine lens nuclear extracts [5] or complete lenses [22]. Only the latter data can be compared. As they calculate the molar mass from quasi-elastic light scattering measurements and from the D_z, their values are appreciably larger than our values in Fig. 4, but they follow the same trend. van Haeringen and coworkers [11, 23] have shown that the α-crystallin fractions, isolated using by size-exclusion chromatography and eluting before the peak fraction, are multimodal: most of the proteins are spherical but a reasonable proportion are ellipsoidal and/or cylindrically shaped. This completely agrees with our distribution functions of the apparent hydrodynamic radii for the different fractions (see Fig. 2). The larger fraction numbers are symmetrically distributed around their maximal values. The lower fraction numbers, have an asymmetric distribution function towards large apparent hydrodynamic radii: the latter asymmetric particles are seen by transient-electric-birefringence and UV linear-dichroism measurements; so the data of van Haeringen and coworkers mainly refer to these particles.

As we have now all hydrodynamic data, we can calculate the hydrodynamic friction coefficients f from the hydrodynamic radius and the minimal friction coefficient f_0 from the molar mass M and the partial specific volume v, accepting a hard dry sphere.

Figure 5 gives the experimental $\ln(D)$ versus $\ln(s)$ for the nine selected α-crystallin fraction 36 to 44. On the same graph we gave represented curves of $\ln(D)$, $\ln(s)$ values of

Fig. 4 Molar mass of some α-crystallin fractions obtained from equilibrium sedimentation and or a combination of the experimental s and D values in the Svedberg relation

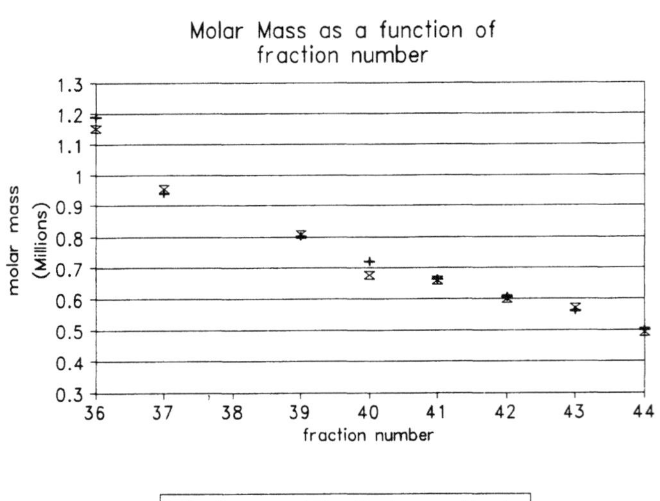

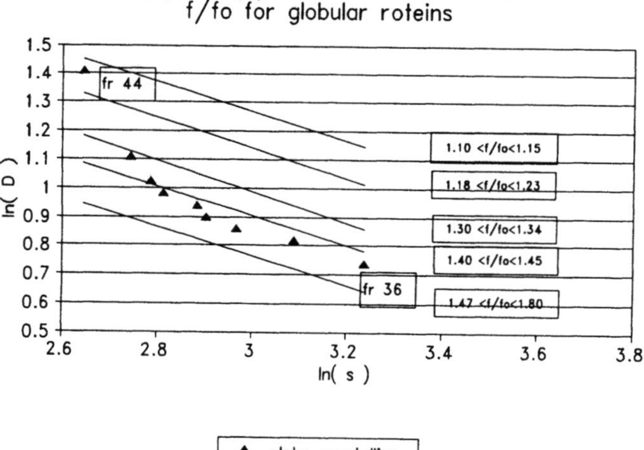

Fig. 5 Graph of the $\ln(D)$ versus $\ln(s)$ of the α-crystallin fractions; this graph contains also theoretical curves for the $\ln(D)$ versus $\ln(s)$ of several classes of globular proteins, classified according to their ratio of the friction coefficient f verus the minimal friction coefficient f_0

proteins within a certain range of f/f_0

$$f = 6 \cdot \pi \cdot \eta \cdot R \tag{10}$$

$$f_0 = 6 \cdot \pi \cdot \eta \cdot \left(\frac{3 \cdot M \cdot v}{4 \cdot \pi \cdot N_A} \right)^{\frac{1}{3}} \tag{11}$$

values; e.g. the line $1.40 < f/f_0 < 1.45$ represents the best line for the $\ln(D)$ values versus $\ln(s)$ values of proteins, with a f/f_0 in the range 1.40 to 1.45. These values have been taken from the Handbook of Biochemistry, Section C, Table I, full data, globular proteins [24].

From this graph, we can conclude that most of the α-crystallin fractions have a large f/f_0 value, ranging from 1.46 to 1.30. There is a clear tendency of a decrease of this parameter for the larger fraction numbers. A large f/f_0 is related to an asymmetric particle or a highly hydrated particle or a combination of both factors. Theoretical models of α-crystallin suggest that the subunits occupy less than 35% of the volume occupied by α-crystallin [25]. This high hydrodynamic hydration contributes a factor 1.4 to the ratio f/f_0. This allows us to conclude that the larger fraction numbers 40 to 44 are highly hydrated but spherical aggregates. The lower fraction numbers 36 to 39 combine this large hydration with asymmetric arrangement of the subunits into ellipsoidal or cylindrical shapes.

Conclusions

Our experimental data on the different α-crystallin fractions allow us to conclude that the α-crystallin protein population contains a very broad distribution of molar masses, ranging from 500 000 to 1 200 000 g/mole, with a maximum in the distribution around 650 000 g/mole. These particles all have a rather high f/f_0, which is mainly related to the high hydrodynamic hydration. Any realistic model on α-crystallin has to take into account this broad size distribution and this large f/f_0 value [26, 27]. But these physical-chemical properties of α-crystallin, namely its broad distribution in size and shape, perfectly fit the micelle model [28].

References

1. Benedek GB (1971) Appl Optics 10: 459–475
2. Delaye M, Tardieu A (1983) Nature 302:415–417
3. Spector A, Chiesa R, Sredy J, Garner W (1985) Proc Natl Acad Soc 82: 4712–4716
4. Voorter CEM, de Haard-Hoekman WA, Roersma ES, Meyer HE, Bloemendal H, De Jong W (1989) FEBS Letters 259:50–52
5. Schurterberger P, Augusteyn RC (1991) Biopolymers 31:1229–1240
6. Tardieu A, Laporte D, Licinio P, Krop B, Delaye M (1986) J Mol Biol 192:711–724
7. Augusteyn RC, Koretz JF (1987) FEBS Lett 222:1–5
8. Andries C, Backhovens H, Clauwaert J, De Block J, De Voeght F, Dhont C (1982) Exp Eye Res 34:239–255
9. Radlick JW, Koretz JF (1992) Biochim Biophys Acta 1120:193–200
10. Walsh MT, Sen AC, Chakrabarti B (1991) J Biol Chem 266:20079–20084
11. van Haeringen B, Eden D, van den Bogaerde MR, van Grondelle R, Bloemendal M (1992) Eur J Biochem 210:211–216
12. Siezen RJ, Bindels JG (1982) Exp Eye Res 34:969–983
13. Thomson JA, Augusteyn RC (1984) J Biol Chem 259:4339–4345
14. Clauwaert J, Ellerton HD, Koretz JF, Thomson K, Augusteyn RC (1989) Curr Eye Res 8:397–403
15. Horwitz J (1992) Proc Natl Acad Sci USA 89:10449–10453
16. Provencher SW (1982) Comput Phys Commun 27:213–227; 229–242
17. Stafford WF (1992) Anal Biochem 203:295–301
18. Yphantis DA (1964) Biochemistry 3: 297–317
19. Groenen PTJA, Merck KB, de Jong WW, Bloemendal H (1994) Eur J Biochem 225:1–19
20. Koppel D (1972) J Chem Phys 57: 4814–4820
21. van Iersel J, Jen JF, Duine JA (1985) Anal Biochem 151:196–204
22. Augusteyn RC, Parkhill EM, Stevens A (1992) Exp Eye Res 54:219–228
23. van Haeringen B, van den Bogaerde MR, Eden D, van Grondelle R, Bloemendal M (1993) Eur J Biochem 217:143–150

24. Sober HA (ed) Handbook of Biochemistry, 2nd edition, The Chemical Rubber Co, Cleveland, pp c10–c12
25. Groth-Vaselli B, Kumosinski TF, Farnsworth PN (1995) Computer-generated model of the quaternary structure of alpha crystallin in the lens. Exp Eye Res, in press
26. Wistow G (1993) Exp Eye Res 56: 729–732
27. Carver JA, Aquilina JA, Truscott RJW (1994) Exp Eye Res 59:231–234
28. Tanford Ch (1973) The hydrophobic effect: formation of micelles and biological membranes. John Wiley & Sons, New York, pp 60–70

Progr Colloid Polym Sci (1995) 99:101–108
© Steinkopff Verlag 1995

G. Pavlov
S. Frenkel

Sedimentation parameter of linear polymers

Received: 3 March 1995
Accepted: 26 June 1995

Dr. G. Pavlov (✉)
Institute of Physics
University
Ulyanovskaya 1
Petergof
St. Petersburg 198904, Russia

S. Frenkel
Institute of Macromolecular Compounds
Russian Academy of Sciences
Bol'shoi 31
St. Petersburg 199004, Russia

Abstract Relationship between the sedimentation coefficient S_0 and its concentration coefficient K_S obtained in experiments on velocity sedimentation are discussed. The values of S_0, K_S, and independently determined molecular masses reported by different researchers for different linear unchanged polymers are considered. It was established that the scaling index $K \sim S_0^v$ unambiguously related to scaling index $S_0 \sim M^b$. The generalization of experimental data, which is based on Svedberg's equation for S_0 and on the expression $K_S = B(h^2)^{3/2} M^{-1}$ made it possible to introduce the dimensionless sedimentation parameter β_S. The average experimental values of β_S were calculated for different polymer types and, to the first approximation, do not depend on molecular mass. Therefore, it is possible to write the generalized Wales-van Holde equation $M_{KS} = (N_A/\beta_S)^{3/2}[S]^{3/2}K_S^{1/2}$.

The adequacy of evaluation of M is illustrated taking as an example the determination of equilibrium rigidity of a polymer chain (flexible-chain polymers under θ-conditions and flexible-chain polymers in a thermodynamically good solvents) and of the unit length mass of the polymer chain (rigid-chain polymers). A rigorous theory of the concentration dependence of the sedimentation coefficient should take into account quantitatively the relationship between polymer chain draining and the solvent counterflow in the sedimentation of macromolecules.

Key words Velocity sedimentation – concentration dependence – sedimentation parameter – molecular characteristics

Introduction

Analytical ultracentrifugation still allows a wide range of investigations of soluble macromolecules and macromolecular systems [1–5]. Among the corresponding methods velocity sedimentation is especially valuable: it is the only method giving directly the mass spectra which are obtained from diagrams by means of a nearly linear transformation (in contrast to sedimentation equilibrium which involves Laplacian transformation, thus leading to uncertainties connected with first type Fredholm's equation and Goedel's theorem [6]). Although, just as 65 years ago [1], velocity sedimentation is less theoretically founded than different versions of sedimentation equilibrium, which still fails to establish molecular mass distribution (MMD) without a reverse transformation. During this reverse transformation based on a sequence of average molecular masses the MMD can be distorted, especially if it is multimodal. The sedimentation velocity approaches give directly the general shape (numbers of modes, or maxima) of MMD. Therefore, an experimental (methodological) approximation under mass spectrometric conditions (i.e., velocity sedimentation) makes it possible to estimate not only the mass spectra but also many other topical macromolecular and conformational characteristics and provides possibility of converting this method-which some researchers treat as obsolete-into an absolute method. Considerable improvements of theory and methodology are doubtless necessary, and the present work is just a part of the corresponding program. Essential progress has recently been attained in the use of absorption optics (Optima XL-A) for the recording the sedimentation boundary. This makes it possible to pass to the field of polymer concentrations in which concentration effects are negligible [7, 8]. However, in this case the concentration dependence coefficient is "last," and just the combination of this coefficient with the sedimentation constant provides a route for the "absolutization" of velocity ultracentrifugation.

Different kinds of concentration effects connected with sedimentation velocity are well known [9–11]. Their combination can be expressed by a single concentration dependence coefficient depending, in turn, on the effective hydrodynamic volume of macromolecules or more complex particles in the same manner as the intrinsic viscosity. However, this viscosity even obtained by extrapolation to infinite dilution still retains some features of forced rotation induced by the method of measurement. In the case of velocity ultracentrifugation the concentration dependence parameter is affected only by translational flow. (The rotational Brownian motion appearing perhaps only as an implicit parameter both in the sedimentation and concentration dependence coefficients.)

A short summary of previous theoretical studies of concentration dependence of sedimentation coefficients [9–20] can be expressed by the following equation:

$$S^{-1} = S_0^{-1}(1 + \Lambda\phi + \Lambda_1\phi^2 + \cdots)$$
$$= S_0^{-1}(1 + (K_S + \bar{v})c + \cdots), \qquad (1)$$

although interpreted in different ways. Here, S, S_0 and $\bar{v}$ have the usual meaning, ϕ is the volume fraction of the polymer in solution, c is the m/v concentration, K_S is Gralen's coefficients, Λ and Λ_1 are dimensionless coefficients.

For rigid spheres the problem has been considered in [9, 12–15] starting with papers by Kermack et al. [12] and Burgers [9]. The recent multiple scattering theory calculations of concentration dependence of the cooperative friction coefficients for suspensions of interacting spheres [14] give estimates of Λ (431/64) almost coinciding with Burgers' value (440/64).

The theories of the concentration dependence of S for macromolecules take into account their chain nature and consider, first, the influence of the thermodynamic quality of the solvent on hydrodynamic interactions in the chain in the Gaussian range of molecular masses [10, 16–20].

Hence, the theoretical results [17] can be presented in the following form:

$$K_S = \gamma(\varepsilon)\,\Phi(\varepsilon)\,(h^2)^{3/2}\,M^{-1} \qquad (2)$$

$\gamma(\varepsilon) = K_S/[\eta]$ is the dimensionless function of the parameter ε characterizing the thermodynamic quality of the solvent ($h^2 \sim M^{1+\varepsilon}$ [21]), $\Phi(\varepsilon)$ is the viscometric Flory parameter, h is the end-to-end distance of the chain and M is the molecular mass.

In a general form K_S can be expressed as

$$K_S = B\,(h^2)^{3/2}\,M^{-1} \qquad (3)$$

$B = B(L/A;\ d/A;\ \varepsilon)$ is a dimensionless parameter depending both on hydrodynamic interactions in the sedimentating system and the thermodynamic quality of the solvent (L is the contour length of the macromolecular chain, d is the effective diameter, and A is the length of the Kuhn segment). In most cases the possibility of the dependence of K_S on hydrostatic pressure in the region of the sedimentation boundary must also be taken into account.

We will use Eq. (3) in subsequent considerations.

Discussion

The scaling relation between K_S and S_0

We suppose that the Mark–Kuhn–Houwink relation for S_0 holds:

$$S_0 = KM^b \qquad (4)$$

The Svedberg equation [1, 2] is used

$$S_0 = \frac{(1 - \bar{v}\rho_0)M}{N_A\,P_0\,\eta_0\,(\overline{h^2})^{1/2}}\,, \qquad (5)$$

where P_0 is the Flory parameter for the translational friction coefficient and other symbols have their usual meaning. Hence,

$$K_S = B\left(\frac{1 - \bar{v}\rho_0}{N_A\,P_0\,\eta_0}\right)\frac{M^2}{S_0^3} = \left(\frac{1 - \bar{v}\rho_0}{N_A\,P_0\,\eta_0}\right)\frac{B}{K^{2/b_s}}\,S_0^{(2-3b_s)/b_s} \qquad (6)$$

Therefore, $K_S \sim S_0^v$, where $v = (2 - 3b_s)/b_s$.

These relationships of K_S, S_0, v, and b_S, as shown by a detailed analysis [22, 23], are fulfilled for many types of polymers. For flexible-chain polymers in θ-conditions $b = 0.5$ and $K_S \sim S_0$, which is a well-established fact. The excluded volume effects in athermal solvents lead to inequalities $0.4 < b_S < 0.5$ and $(2 - 3b_S)/b_S > 1$. Concentration dependence is especially pronounced in rigid-chain polymers for some of which $K_S \sim S_0^4$ [23].

Another ultimate case of a weak dependence of K_S on S_0 is observed for globular proteins (and in general for globular macromolecules where $d[\eta]/dM = 0$, i.e. $[\eta] = K \times M^a$ with $a = 0$, bearing in mind that $b_S = 1 - b$, b being the exponent in the expression for the translational friction coefficient: $f \sim M^b$, and $b = (1 + a)/3$ [24]). In [25] data are given for 10 globular proteins corresponding to $S_0 = 4.6 \times 10^{-3} M^{0.63 \pm 0.04}$ $(r = 0.9942)$; $K_S = 4.0 \times S_0^{0.25 \pm 0.07}$ $(r = 0.7879)$, where r is the coefficient of linear correlation and S_0 and K_S are expressed in Svedberg units and in cm^3/g, respectively (Fig. 1). Therefore, as predicted by theory [26], the scaling relation (6) is fulfilled in this case as well.

Therefore, the stronger the dependence of the sedimentation coefficient on molecular mass, the weaker is its concentration dependence (and vice versa), as follows directly from the preceding expressions.

Sedimentation parameter

Equations (3) and (5) give

$$M = \left(\frac{N_A P_0}{B^{1/3}}\right)^{3/2} [S]^{3/2} K_S^{1/2} = \left(\frac{N_A}{\beta_S}\right)^{3/2} [S]^{3/2} K_S^{1/2} \quad (7)$$

$[S] = S_0 \eta_0/(1 - v\rho)$ is the intrinsic sedimentation coefficient.

The product $B^{1/3} P_0^{-1} \equiv \beta_S$ is defined by us as the sedimentation parameter. Equation (7) is a generalization of Wales–van Holde formula [27].

Fig. 1 Dependences of $\log S_0$ on $\log M$ and $\log K_S$ for globular proteins [25]

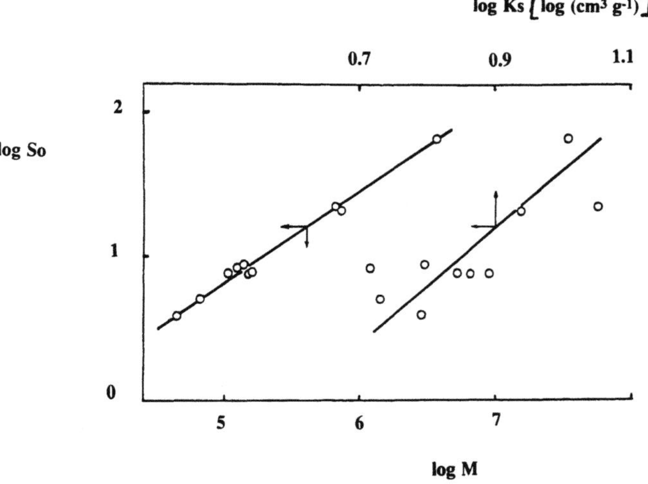

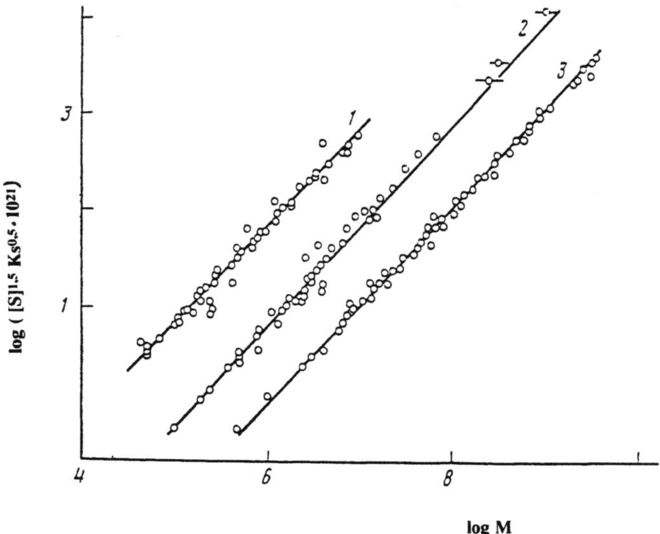

Fig. 2 Double logarithmic plot of ($[S]^{1.5} K_S$ 0.5) on M_W (or M_{SD}) for 1) flexible-chain polymers under the θ-conditions [33–45], 2) rigid-chain polymers (refs. see [23]), 3) flexible-chain polymers in thermodynamically good solvents [33, 36, 40, 41, 46, 52]. Lines 2 and 3 are displaced with respect to line 1 along the M axis by 1 and 2 decades, respectively

Figure 2 represents a comparison of sedimentation data with independently evaluated molecular masses (either M_W from light scattering or M_{SD}) for polymers belonging to different classes. The data are plots of $\log([S]^3 K_S)$ vs $\log M$. The slopes of three straight lines corresponding to different groups of polymers are equal to unity. One line corresponds to flexible chain polymers in θ-conditions (a 250-fold change of MM); the second line corresponds to rigid-chain polymers (10^4-fold M range); the third one to flexible chain polymers in theoretically good athermal solvents (also a 10^4-fold M range). These data relate to fractions or samples of unchanged linear macromolecules with a unimodal MMD.

Figure 2 shows a distinct correlation between $([S]^3 K_S)^{1/2}$ and independently measured M values and makes it possible to estimate the value of the sedimentation parameter averaged for different polymer-solvent groups. The average values are given in Table 1.

Munk et al. [28] have recently published a detailed study of hydrodynamic properties and masses of micelles formed by polystyrene-poly(methacrylicacid) block copolymers (PS-PMA) of SA and ASA types. In this case a distinct correlation of independently measured values of M and $([S]^3 K_S)^{1/2}$ is also observed (Fig. 3).

Table 1 also gives the values calculated for cellulose, its derivatives, and other polysaccharides [29]; for globular proteins [25] and PS-PMA micelles [28].

The comparison of experimental data for β_S with theoretical values presented in Table 2 is possible. Table 2 also

Table 1 Averaged experimental values of the sedimentation parameter β_S

Polymer-Solvent	$\beta_S \times 10^{-7}$ mol$^{-1/3}$	$\Delta\beta_S \times 10^{-7}$ mol$^{-1/3}$
Flexible-chain polymers in Θ-conditions	1.0	0.02
Flexible-chain polymers in thermodynamically good solvents	1.25	0.02
Rigid-chain polymers	1.0	0.02
Cellulose and its derivatives	1.0	0.03
Globular proteins	1.17	0.03
Micelles	1.43	0.09

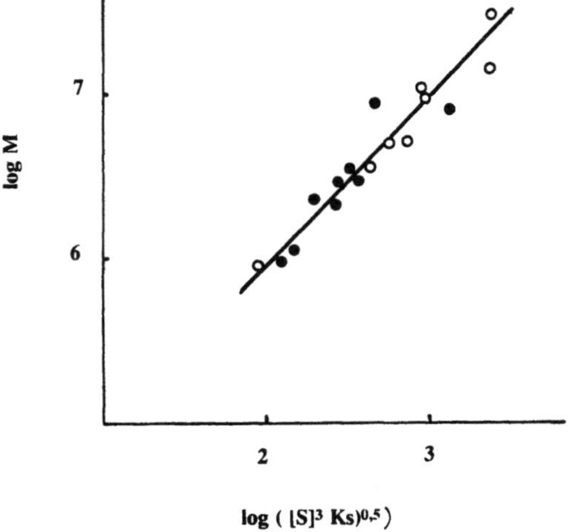

Fig. 3 Dependence of $\log M_W$ (or M_{SD}) on $\log([S]^{1.5}K_S^{0.5})$ for PS-PMA micelles [28] (open circles: SA micelles; filled circles: ASA micelles)

gives the theoretical values of concentration dependence parameters Λ and B. For calculating the theoretical values, one must use the ultimate theoretical values of hydrodynamic parameters P_0 and, in some theories Φ_0, which themselves differ depending on the adopted hydrodynamic models and mathematical approximations. The values of P_0 and Φ_0 obtained after a preliminary averaging of the hydrodynamic Oseen's tensor are $P_0 = 5.11$ and $\Phi_0 = 2.87 \times 10^{23}$ [30]. If the preliminary averaging of Oseen's tensor is omitted and one proceeds with computer simulation extrapolating data to infinite M, the corresponding values become $P_0 = 6.00$, $\Phi_0 = 2.50 \times 10^{23}$ [31]. The use of the renorm group theory [32] leads to $P_0 = 6.20$; $\Phi_0 = 2.36 \times 10^{23}$.

Both theory and experiment lead to higher β_S values in thermodynamically good solvents as compared to θ-conditions and also to higher β_S values for inpenetrable spheres.

Higher experimental β_S values and correspondingly higher values of K_S can probably be due to a stronger counterflow of the solvent during the sedimentation of these systems.

At the same time, when molecular or, in the case of micelles, supermolecular homology is maintained (for macromolecules, this corresponds to persistance of mass per unit length M_L and/or of the effective diameter d) the values of β_S in the homologious series remain virtually constant.

Therefore, the change in the value of $([S]^3K_S)^{1/2}$ characterizes the change in the molecular mass of the system under investigation, and the molecular mass itself can be

Table 2 Limiting theoretical values of the parameters of concentration dependence S and sedimentation parameter β_S

Λ	$B \times 10^{-23}$/mol^{-1}	$\beta_S \times 10^{-7}$/mol$^{-1/3}$	Remarks	Ref.
6.535–7.1	3.29–3.465	1.35–1.39	Rigid sphere	9, 12–14
	$N_A (3\pi^5 2^{-13})^{1/2}$	1.15	Sedimentation of solvated spherical molecules	11
	$N_A (3\pi^5 2^{-15})^{1/2}$	0.91	Sedimentation of solvated molecules	11
7.01; 7.16	3.53; 3.61	1.38; 1.39	Thermodynamically good athermal solvents	16
	1.85 ($\varepsilon = 0.2$)	1.33		17
2.23; 2.30	1.12; 1.16	0.94; 0.95		16
1.0	0.50	0.72	θ-conditions	18, 30
		1.29*; 1.05**; 1.00***		10
	0.62	1.10*; 0.90**; 0.85***		19
	0.75	1.17*; 0.95**; 0.91***		17, 20

* $P_0 = 5.11$; $\Phi_0 = 2.87 \times 10^{23}$
** $P_0 = 6.0$; $\Phi_0 = 2.50 \times 10^{23}$
*** $P_0 = 6.2$; $\Phi_0 = 2.36 \times 10^{23}$

estimated using the generalized Wales–van Holde equation (Eq. (7)) applying the corresponding β_S values.

In the next section we show the adequacy of these estimates.

Comparison of estimates of conformational parameters of macromolecules, obtained using either independently evaluated molecular masses or only data of velocity sedimentation

Let us consider the experimental data presented in ref. [33–54] where in addition to S_0 and K_S, M_w values of samples or fractions were also measured by light scattering.

The estimates of the parameters (A, M_L, etc.) correspond either to the combination of S_0 and M_w or only of S_0, K_S and β_S.

Fig. 4A Dependence of $[S]M_0^{-1}$ on $(M/M_0)^{1/2}$ for PS (open circles) and PαMS (filled circles) in cyclohexane (θ-conditions) [33–45]. **B** Dependence of $[S]M_0^{-1}$ on $([S]^{3/2}K_S^{1/2}M_0^{-1})^{1/2}$ for systems shown in Fig. 4a

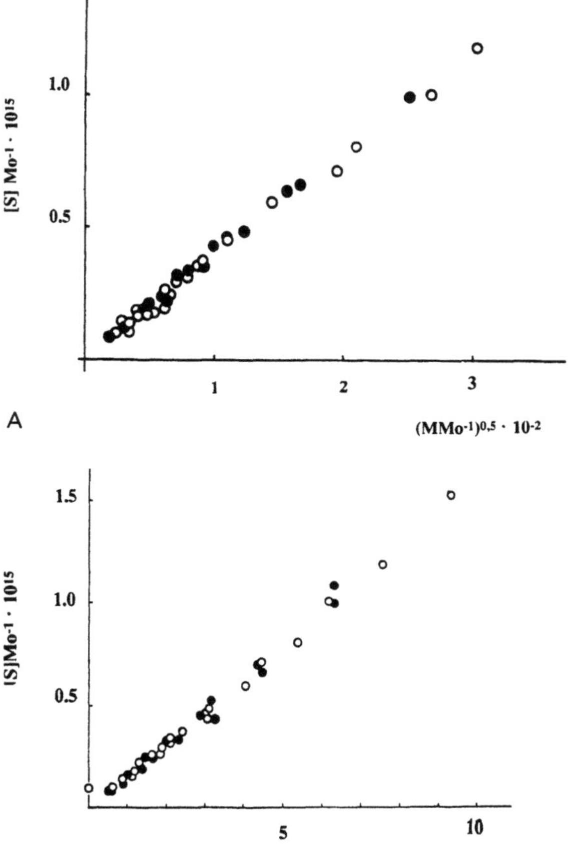

Polystyrene and poly-α-methylstyrene in cyclohexane (θ-conditions) [33–45]

In θ-conditions (absence of excluded volume effects) the dependence of the translational friction coefficient on molecular mass follows either the theory of Hearst–Stockmayer [55] or the Yamakawa–Fujii theory [56]. In cases when $L/A > 2.3$ the following relation is fulfilled:

$$[S]M_0^{-1} = (P_0 N_A)^{-1} [(\lambda A)^{-0.5} (MM_0^{-1})^{0.5}$$
$$+ (P_0/3\,\pi\lambda)(\ln A/d - \varphi(0))] \tag{8}$$

M_0 is the molecular mass of the repeat unit (for polystyrene $M_{01} = 104$ and for poly-α-methylstyrene $M_{02} = 118$), and λ is its projection on the direction of the main chain ($\lambda_1 = \lambda_2 = 2.52 \times 10^{-8}$ cm, $\varphi(0) = 1.43$ [55] or 1.056 [56]. Figure 4a shows the dependence of $[S]M_0^{-1}$ on $MM_0^{-1})^{0.5}$; from its slope one obtains $A = (27.0 \pm 0.6) \times 10^{-8}$ cm ($r = 0.9963$). Figure 4b shows the plot using only S_0 and K_S on the assumption that $\beta_S = 1.0 \times 10^7$ mol$^{1/3}$; here again $A = (27.0 \pm 0.6) \times 10^{-8}$ cm ($r = 0.9956$).

PS and P-α-MS in toluene (a thermodynamically good solvent) [33, 36, 40, 41, 46,–52]

Here, the thermodynamic polymer-solvent interactions are strong and instead of (8) in accord with the theory [57], one must use (as it was shown for flexible polymers [58]) for the whole range of M the expression

$$[S]M_0^{-1} = (P_0 N_A)^{-1} \left[\frac{3}{(1 - \varepsilon)\,(3 - \varepsilon)} \right.$$
$$\times (A^{(1 - \varepsilon/2)}\lambda^{(1 + \varepsilon/2)})^{-1}(MM_0^{-1})^{(1 - \varepsilon/2)}$$
$$\left. + (P_0/3\,\pi\lambda)(\ln A/d - \varphi(\varepsilon)) \right] \tag{9}$$

This equation has the same symbols as above, and $\varphi(\varepsilon)$ is tabulated in [57] ($\varphi(0) = 1.43$).

Figure 5a shows the dependence of $[S]M_0^{-1}$ on $MM_0^{-1})^{(1 - \varepsilon)/2}$. The parameter ε was calculated from the relation $\varepsilon = 1 - 2b_S$ [21], and b_S was evaluated from a Mark–Kuhn–Houwink-type plot $[S]M_0^{-1} \sim Z^{0.404 + 0.005}$ ($r = 0.9964$), Z being the degree of polymerization. Therefore, $\varepsilon = 0.192 \pm 0.012$ and $A = (25.0 \pm 0.8) \times 10^{-8}$ cm ($r = 0.9975$).

Figure 5b shows the same dependence using only sedimentation data (S_0 and K_S). The value of b_S was estimated independently from the scaling relation (6) $K_S M_0 \sim ([S]M_0^{-1})^{1.74 \pm 0.038}$ ($r = 0.9903$). Hence, $b_S = 0.422 \pm 0.003$; $\varepsilon = 0.156 \pm 0.006$; and by using the mean experimental value $\beta_S = 1.25 \times 10^7$ mol$^{1/3}$ obtained for flexible chain polymers in athermal solvents (Table 1), one obtains $A = (28.0 \pm 0.35) \times 10^{-8}$ cm ($r = 0.9994$).

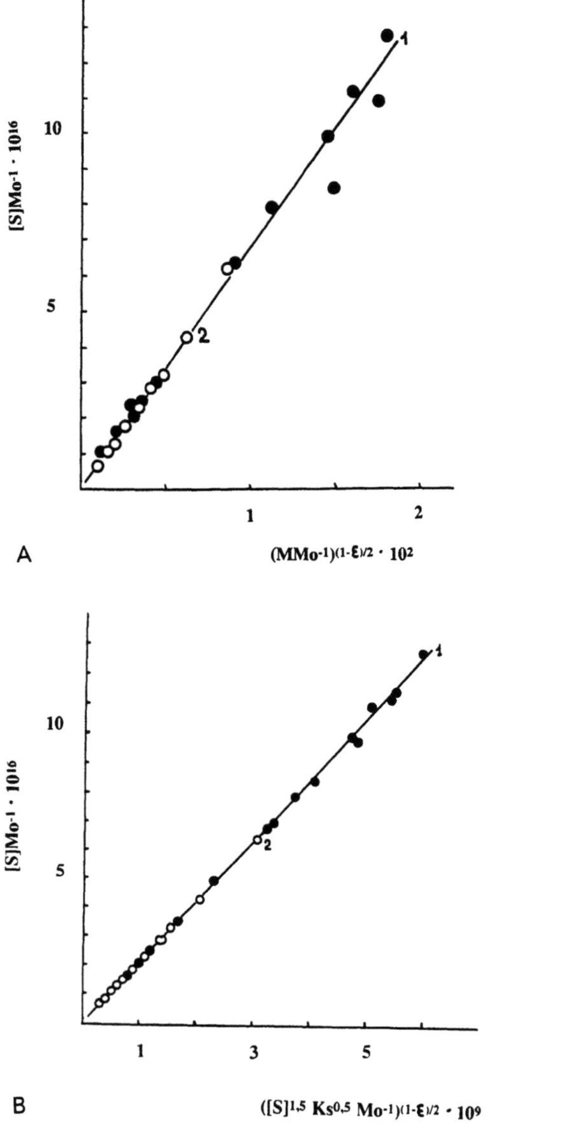

Fig. 5A Dependence of $[S]M_0^{-1}$ on $(M/M_0)^{1-\varepsilon/2}$ for PS (filled circles) and PαMS (open circles) in toluene [33, 36, 40, 41, 46–52]. **B** Dependence of $[S]M_0^{-1}$ of $([S]^{3/2}K_S^{1/2}M_0^{-1})^{1-\varepsilon/2}$ for systems shown in Fig. 5a

Rigid-chain polymers

Up to the present the highest equilibrium rigidity was observed for a polysaccharide Schizophyllum commune whose macromolecules are three-strand helices. This polysaccharide is being studied in detail by Japanese researchers. Its molecular and hydrodynamic characteristics have been presented in [53].

The hydrodynamic properties of rigid-chain polymers in the range of small contour lengths ($L/A < 2.3$; $L \gg d$) can be described using a model of a weakly bending rod or a cylinder. In accord with theory [30, 56] for this

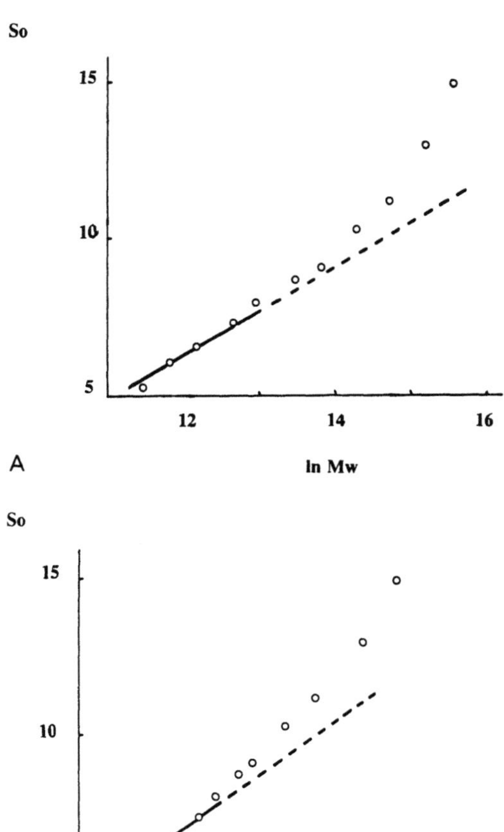

Fig. 6A Dependence of $[S]$ on $\ln M$ for Schizophyllum in water [53], **B** Dependence of $[S]$ on $\ln(S_0^3 K_S)$ for the same system

model we have:

$$[S] = (M_L/3\pi N_A)\,[\ln M - \ln(M_L d) + 0.3863]. \tag{10}$$

The initial slope of the corresponding plot of $[S]$ vs $\ln M$ allows the evaluation of $M_L = M/L = M_0/\lambda$. The data obtained in [53] are presented in Fig. 6. If M_w is used one obtains an estimate of $M_L = (2.3 \pm 0.16) \times 10^{10}$ cm^{-1}, and the use of sedimentation data alone (S_0 and K_S) give $M_L = (2.47 \pm 0.03) \times 10^{10}$ cm^{-1}.

The corresponding slopes were calculated for the first four points. It should be mentioned that the knowledge of the precise value of β_S is not necessary for the estimation of M_L based on S_0 and K_S, the assumption of its constancy being sufficient.

A similar comparison can be made using data obtained for another rigid chain polymer, polyhexylisocyanate [54].

Hence, the estimates of equilibrium rigidity of chains and of M_L obtained either with the use of independently measured M_w and S_0 or using only velocity sedimentation data (S_0 and K_S) are quantitatively comparable (they virtually coincide), which supports the adequacy of estimates of M based on the sedimentation parameter β_S.

Conclusions

The combination of two parameters obtained in one series of experiments (S_0 and K_S) for the study of polymerhomologues is an additional method for obtaining information on some equilibrium properties of macromolecules. This possibility is based on the concept of the persistence of the sedimentation parameter β_S $(=B^{1/3}P_0^{-1})$ in a polymer homologous series. A general theory of concentration effects during sedimentation (i.e., cooperative translational motion of macromolecules) must solve the problem about the way in which the dimensionless parameter B (Eq. 3) depends on the relative contour length and effective diameter of macromolecules, or, in other words, the problem of quantitative relationship between the draining of macromolecules and the solvent counterflow during sedimentation.

Presently, two remarks should be made: 1) the product $B^{1/3}P_0^{-1}$ is similar not only in appearance, but also in the physical sense to the Flory–Mandelkern invariant $\Phi_0^{1/3}P_0^{-1}$ since both K_S and $[\eta]$ depend on the effective

hydrodynamic volume. However, in the case of sedimentation (or cooperative translation of diffusion) this volume is affected by the Brownian rotational motion only, whereas it is also affected by the impellent rotation unavoidable during viscosity measurements. Even in this case the macromolecules with respect to $[\eta]$ are "more drainable" by the solvent than with respect to the cooperative translational friction (sedimentational or translational diffusion coefficients) [59]. Therefore, the use of sedimentation velocity measurements alone may be preferable to that of a combination of $[\eta]$ and S_0. 2) even until this main problem is completely solved, the proposed approach, in combination with "graphic fractionation", makes it possible to obtain simultaneously both the distribution of sedimentation coefficients and the hydrodynamic and conformational data on all macromolecules covered by this distribution [60]. If they are homologous, the next step to the determination of MMD is achieved without difficulty, but if for some reason the assumption of homology is false, this procedure will give the MMD as well and will show when the homology fails.

Acknowledgments The authors thank the Russian Foundation for Fundamental Research (grant #93-03-5791), the International Science Foundation and the Organizing Committee of the IXth Symposium on analytical ultracentrifugation for support of this study.

References

1. Svedberg T, Pedersen K (1940) The Ultracentrifuge, Clarendon Press, Oxford
2. Tanford Ch (1963) Physical Chemistry of macromolecules, J Wiley, New York
3. Fujita H (1975) Foundation of Utracentrifugal Analysis, J Wiley, New York
4. Cantor Ch, Schimmel P (1980) Biophysical Chemistry, Part II, Freeman, San Francisco
5. Munk P (1991) In "Modern Methods of Polymer Characterization" (Barth H, May J, Eds), J Wiley, New York, P 271
6. Belenkii B, Vilenchik L (1983) Modern Liquid Chromatography of Macromolecules, Elsevier, Amsterdam
7. Lechner M, Machtle W (1992) Makromol Chem Rapid Commun 13:555
8. Kehrhahn J-H, Lechner M, Machtle W (1993) Polymer 34:2447
9. Burgers J (1945) Proc Kon Ned Acad Wet 45:125
10. Freed K (1976) J Chem Phys 64:1976
11. Rowe A (1977) Biopolymers 16:2595
12. Kermack W, M'Kendrick A, Pouder E (1929) Proc Roy Soc Edinburgh 49:170
13. Batchelor G (1972) J Fluid Mech 52:245
14. Muthukumar M, Freed K (1983) J Chem Phys 78:511
15. Ogston A (1961) J Chem Phys 65:57
16. Puin C, Fixman M (1964) J Chem Phys 45:937
17. DeMeuse M, Muthukumar M (1985) Macromolecules 18:1173
18. Ymai S (1970) J Chem Phys 52:4212
19. Perico A, Freed K (1983) J Chem Phys 78:2058
20. Muthukumar M (1983) J Chem Phys 78:2764
21. Ptitsyn O, Eizner Yu (1959) Zh Tekh Fiz 29:1105
22. Pavlov G, Frenkel S (1982) Vysokomol Soedin B24:178
23. Pavlov G, Frenkel S (1988) Acta Polymerica 39:107
24. Tsvetkov V, Eskin V, Frenkel S´(1970) Structure of macromolecules in solution, Butterworths, London
25. Creeth J, Knight C (1965) Biochim Biophys Acta 102:549
26. Frenkel S (1965) Introduction to Statistical Theory of Polymerization, Nauka, Moscow
27. Wales M, van Holde K (1954) J Polym Sci 14:81
28. Qin A, Tian M, Ramireddy C, Webber S, Munk P, Tuzar Z (1994) Macromolecules 27:120
29. Pavlov G (1989) Wood Chemistry (Riga) 4:3
30. Yamakawa H (1971) Modern Theory of Polymer Solution, Harper and Row, New York
31. Zimm B (1980) Macromolecules 13:592
32. Oono Y, Kohmoto M (1983) J Chem Phys 78:520
33. Billick I (1962) J Chem Phys 66:1941
34. McIntyre D, Wims A, Williams L, Mandelkern L (1962) J Chem Phys 66:1932
35. Homma T, Kawahara K, Fujita H (1963) Makromol Chemie 67:132
36. Noda J, Saito S, Fujimoto T, Nagasawa M (1967) J Chem Phys 71:4048
37. Abe M, Sakato K, Kageyama T (1968) Bull Chem Soc Japan 41:2330
38. Petrus V, Danihel J, Bohdanecky M (1971) Europ Polym J 7:143
39. Kotera A, Saito T, Hamada T (1972) Polymer J 3:421
40. Noda J, Mizutani K, Kato T Macromolecules 10:618
41. Appelt B, Meyerhoff G (1980) Macromolecules 13:657

42. Mulderije J (1980) Macromolecules 13:1707
43. Peeters F, Smits H (1981) Bull Soc Chim Belg 90:111
44. Lavrenko P, Boikov A, Andreeva L (1981) Vysokomol Soedin A23:1937
45. Vidakovic P, Allain C, Rondolez F (1982) Macromolecules 15:1571
46. Schulz G, Cantow H, Meyerhoff G (1953) J Polym Sci 10:79
47. Meyerhoff G (1955) Z Phys Chem 4:355
48. Lutje H, Meyerhoff G (1963) Makromol Chemie 18:180
49. Jerome R, Desreux V (1970) Europ Polym J 6:411
50. Hadjichristidis N, Devaleriola M, Desreux V (1972) Europ Polym J 8:1193
51. Skazka V, Yamshikov V, Tarasova G (1973) Leningrad Univ Vestnik, Ser Fiz Khim 16:59
52. Tricot M, Blens J, Riga Y, Desreux V (1974) Makromol Chemie 175:913
53. Yanaki T, Norisuye T, Fujita H (1980) Macromolecules 13:1462
54. Murakami H, Norisuye T, Fujita H (1980) Macromolecules 13:345
55. Hearst J, Stockmayer W (1962) J Chem Phys 37:1425
56. Yamakawa H, Fujii M (1973) Macromolecules 6:407
57. Gray G, Bloomfield V, Hearst J (1967) J Chem Phys 46:1493
58. Pavlov G, Panarin E, Korneeva E, Kurochkin C, Baikov V, Ushakova V (1990) Makromol Chem 191:2889
59. Debye P, Bueche A (1948) J Chem Phys 16:573
60. Pavlov G, Tarabukina E, Frenkel S (1995) Polymer 36:2043

Progr Colloid Polym Sci (1995) 99:109–113
© Steinkopff Verlag 1995

Velocity sedimentation of water-soluble methyl cellulose

G. Pavlov
N. Michailova
E. Tarabukina
E. Korneeva

Received: 3 March 1995
Accepted: 28 June 1995

Dr. G. Pavlov (✉) · N. Michailova
Institute of Physics
University Ulianovskaya str. 1
Petergof
198904 St. Petersburg, Russia

E. Tarabukina · E. Korneeva
Institute of Macromolecular compounds
RAS, 199004 St. Petersburg, Russia

Abstract Velocity sedimentation and other hydrodynamic properties were studied for water soluble methyl cellulose (MC) samples (with the substitution degree 1.7) in molecular weight range $19 \leq M \cdot 10^{-3} \leq 408$. The correlations between hydrodynamic characteristics and molecular weights are obtained. The values of persistence length and hydrodynamic diameter of MC chains were estimated. The comparison of investigations of concentration dependence of sedimentation coefficient S_0 for real samples and graphic fractions was carried out. It was shown that use of S_0 and concentration coefficient k_s leads to adequate estimates of equilibrium rigidity of MC chains.

Key words Velocity sedimentation– hydrodynamic properties – methyl cellulose – molecular characteristics

Introduction

Methyl cellulose (MC) is the first term of the series of O-alkyl cellulose derivatives. In a certain range of the degrees of substitution MC is a water soluble polymer [1–2]. It is widely used in industry [3]. Moreover, the study of MC is of interest for comparing its equilibrium properties with those of water–soluble polysaccharides differing in the way of insertion of glucopyranose rings into the main chain and the position of OH-groups in the monosaccharide ring. Variations in the structure of linear polysaccharides lead to considerable changes in the equilibrium properties of their molecules [4–6].

However, the information about molecular characteristics of water soluble MC is not plentiful and is contradictory. Thus, the exponents in the Kuhn–Mark–Houwink–Sakurada [K-M-H-S] equation are reported by different workers for MC with similar degrees of substitution (DS) in water range from 0.55 to 0.86 [7–10].

We began a systematic study of hydrodynamic characteristics of water soluble MC. This paper presents prelimi-nary results obtained in the investigation of some MC samples in water. These investigations were carried out by the methods of molecular hydrodynamics.

Experimental

Materials and methods

Water soluble MC was obtained by the alkylation of alkaline cellulose with methyl chloride [1]. Quantitative analysis has shown that the contents of $OCH_3 -$ groups in the samples virtually coincide and average value is $(28.4 \pm 1.2)\%$ which corresponds to DS of (1.68 ± 0.08).

In order to extend the range, one of the samples (sample 5, Table 1) was subjected to ultrasonic degrada-tion. The initial solution at a concentration $c = 0.5 \cdot 10^{-2}$ g/cm^3 was subjected to ultrasonic treatment. After different periods of time parts of the solution were collected, filtered, and lyophilically dried. It was assumed that DS remains invariable. In this way five samples were obtained (8–12, Table 1).

Velocity sedimentation

Velocity sedimentation was studied on a MOM3180 ultracentrifuge (Hungary) at a rotor speed of 40 000 rpm in a double-sector cell with the formation of an artificial boundary (Fig. 1). For six samples the concentration dependence of S was studied (Fig. 2) and approximated by the linear equation

$$S^{-1} = S_0^{-1}(1 + k_s c + \cdots) . \tag{1}$$

A correlation between the values of S_0 and k_s determined by this method was established. It obeys the following equation:

$$k_S = 85.6 S_0^{1.87 \pm 0.16} . \tag{2}$$

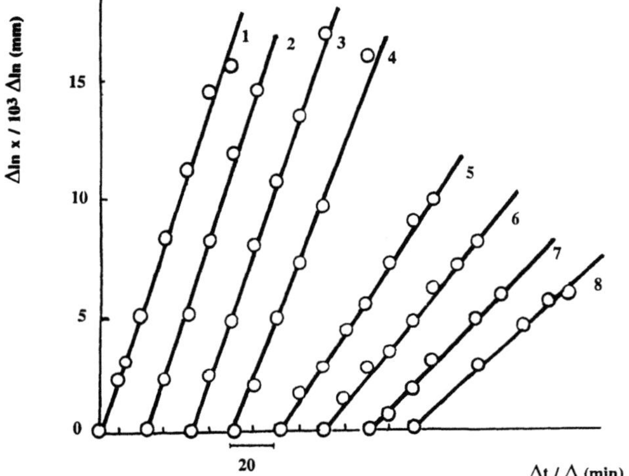

Fig. 1 Dependencies of Δlnx on sedimentation time Δt, where x is the position of the sedimentation peak maximum for solutions of sample 1 at concentrations $c \cdot 10^2$ g/cm^3 = 0.041 (1); 0.055 (2); 0.083 (3); 0.140 (4) and sample 12 at concentrations $c \cdot 10^2$ g/cm^3 = 0.130 (5); 0.216 (6); 0.282 (7); 0.404 (8)

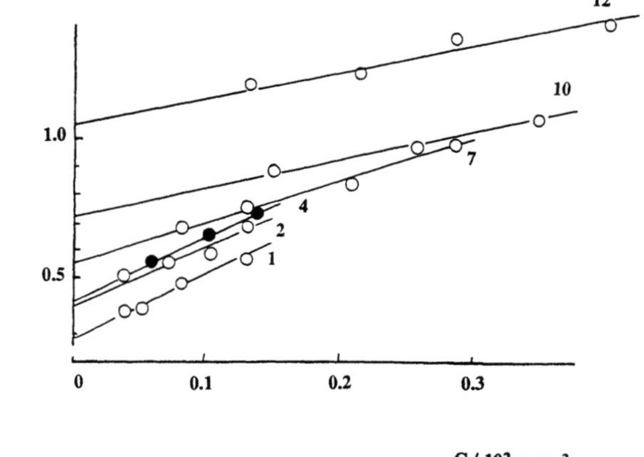

Fig. 2 Concentration dependencies of the sedimentation coefficient S for methyl cellulose. The number of indicates the sample numbers on Table 1

This equation was used to exclude concentration effects in all other cases. Single experiments were carried out at $c < 0.09 \cdot 10^{-2}$ g/cm^3. The optical system of boundary recording was a polarizing interferometer in the investigations of both sedimentation and translational diffusion [11].

The buoyancy factor $(1 - \bar{v}\rho_0) = 0.286 \pm 0.003$ of the MC-water system was determined with a pycnometer.

Translational diffusion

Isothermal translational diffusion was studied by the classical method of forming an interface between the solution and the solvent. The dispersion of the diffusion interface (σ^2) was calculated by the method of maximum ordinate and area [11]. The translational diffusion coefficient D was determined from the time dependence of dispersion

Table 1 Hydrodynamic and molecular characteristics of methyl cellulose in water

N	$[\eta]$ cm^3/g	k'	S_0 10^{13} c	k_S cm^3/g	D_0 10^7 cm^2/c	$\Delta n/\Delta c$ cm^3/g	β_S 10^{-7}	M 10^{-3}
1	830	1.12	3.44	740	0.73	0.106	1.06	408
2	760	0.90	2.62	620	0.90	0.105	1.05	252
3	500	1.00	2.57	–	1.17	0.105	–	190
4	450	1.35	2.50	540	1.33	0.118	1.28	163
5	310	0.35	2.07	–	1.75	0.102	–	102
6	270	0.64	1.90	–	–	–	–	87
7	260	0.74	1.80	270	1.75	0.111	1.09	89
8	238	0.65	1.71	–	1.63	0.111	–	91
9	242	0.64	1.62	–	2.44	0.117	–	58
10	138	0.94	1.40	140	3.12	0.115	1.19	39
11	110	0.80	1.36	–	3.80	0.116	–	31
12	67	0.82	0.94	78	4.34	0.105	1.07	19

$\sigma^2 = \sigma_0^2 + 2Dt$ where σ_0^2 is the initial dispersion characterizing the quality of interface formation. Diffusion experiments were carried out at an average concentration $c = 0.036 \cdot 10^{-2}$ g/cm^3. It was assumed that in this range of c concentration effects may be neglected. The refractive index increment $\Delta n/\Delta c$ (Table 1) was determined from the area spanned by the interference curve, and its average value was (0.110 ± 0.002) cm^3/g at $\lambda = 550$ nm.

Viscometry

Intrinsic viscosity $[\eta]$ and Huggins parameter k' were calculated according to Huggins equation from relative viscosities measured in an Ostwald viscometer with a flow time of water $\tau_0 = 85.6$ s.

Experimental data were obtained at 25° (Table 1).

Discussion

The values of molecular weights were obtained according to Svedberg [11] on the basis of S_0 and D_0:

$$M = R[S]/[D] , \qquad (3)$$

where R is the universal gas constant; $[S] = S_0\eta_0/(1 - \bar{v}\rho_0)$ is the characteristic sedimentation coefficient;

$[D] = D_0\eta_0/T$ is the characteristic diffusion coefficient; η_0 is the solvent viscosity, and T is the temperature.

The consideration of correlations between M and hydrodynamic characteristics $[\eta]$, D_0, and S_0 (Fig. 3) leads to the establishment of known K-M-H-S equations ($Pr_i = K_i M_i^b$, where $Pr_i = S_0$, D_0 or $[\eta]$) the parameters of which were calculated by the least-squares method (Table 2). In the same table the parameters of the corresponding correlations between pairs of hydrodynamic characteristics are given. It is clear that (taking into account experimental errors) a correlation is observed between $b_1 = b_5/b_4$, $b_2 = b_6/b_4$, $b_5 = -\frac{1}{3}(b_4 + 1)$, $b_3 = (2 - 3b_6)/b_6$ [11–13]. This is characteristic of polymer homologues.

The latter relationship between b_3 and b_6 scaling values has been established previously [14]. This relationship usually valid under the condition of homology in a series of samples (fractions) of polymers (also as the other relationships between scaling indexes) including polysaccharides [13–17]. The deviation from this relationship can be caused (among the other reasons) by the fact that homology in the series of polymer molecules under investigation cannot exist.

The difference between 0.5 and the b_4 (b_η), b_5 (b_D) values for cellulose derivatives, just like for cellulose itself, may be related to drainage effects of molecules, whereas volume effects may be neglected to the first approximation. In this case hydrodynamic data can be interpreted on the basis of the following MW dependences:

$$[S]N_A P_0 = (M^2 \Phi_0/[\eta])^{1/3}$$
$$= (M_L/A)^{1/2}M^{1/2} + (P_0 M_L/3\pi) [\ln(A/d) - \varphi(0)] , \qquad (4)$$

where P_0 and Φ_0 are Flory's hydrodynamic parameter, $M_L = M/L$, L is the contour length of the macromolecule, A is the length of the Kuhn segment, and d is the hydrodynamic chain diameter. The dependence $[S] = f(M^{1/2})$ (Fig. 4a) is an analytical expression following from the theories of Hearst–Stockmaier ($\varphi(0) = 1.431$) and Yamakawa–Fujii (($\varphi(0) = 1.056$) for the case $L/A > 2.3$ [18, 19].

Fig. 3 Kuhn–Mark–Houwink–Sakurada dependencies: (1) $[\eta]$-M; (2) $S_0 \cdot 10^{14} - M$; (3) $D_0 \cdot 10^9 - M$

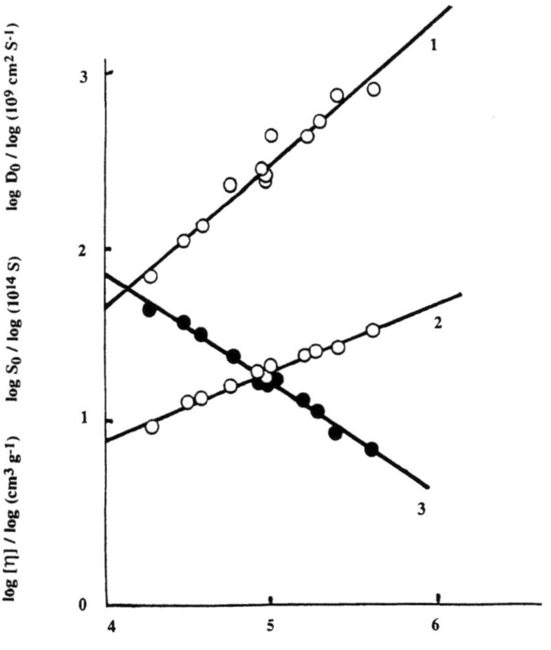

Table 2 Parameters in Mark–Kuhn–Houwink–Sakurada equations for methyl cellulose in water

i	Correlation values	b_i	$\pm \Delta b_i$	K_i	r_i^*
1	D_0-$[\eta]$	-0.73	0.039	$1.07 \cdot 10^{-5}$	0.9871
2	S_0-$[\eta]$	0.46	0.030	$1.43 \cdot 10^{-14}$	0.9793
3	k_s-S_0	1.87	0.16	85.6	0.9902
4	$[\eta]$-M	0.83	0.043	$2.08 \cdot 10^{-2}$	0.9879
5	D_0-M	-0.61	0.021	$1.91 \cdot 10^{-4}$	0.9945
6	S_0-M	0.39	0.020	$2.21 \cdot 10^{-15}$	0.9872

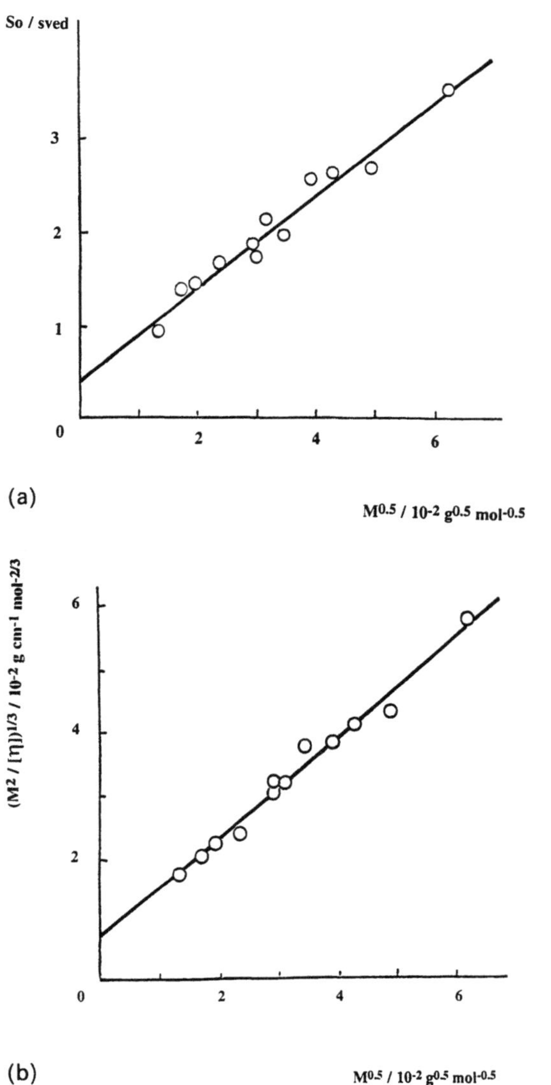

Fig. 4 Dependencies of (a) S_0 and (b) $(M^2/[\eta])^{1/3}$ on $M^{1/2}$ according to Eq. (4)

The dependence $(M^2/[\eta])^{1/3} = f(M^{1/2})$ (Fig. 4b) was first proposed in ref. [20] and was repeatedly used to interpret viscometric data obtained for rigid-chain polymers [21] including cellulose and its derivatives [4, 22]. This dependence has also been discussed in detail in ref. [23].

The values of A and d depend on the limiting values of hydrodynamic parameters P_0 and Φ_0. The conventional values of P_0 and Φ_0, calculated with preliminary averaging of Oseen's hydrodynamic tensor are $P_0 = 5.11$ and $\Phi_0 = 2.87 \cdot 10^{23}$ mol^{-1} respectively [21]. If other limiting values of P_0 and Φ_0 are used, the values of A and d will be displaced. The character of these changes is illustrated in Table 3 which also gives the values of P_0 and Φ_0 obtained by computer simulation [24] and by the renorm group theory [25].

In the study of the velocity sedimentation of one MC sample [26], we have previously used the method of graphic fractionation [27] and have determined the values of S_0 and k_s for graphic fractions (experiments were carried out in a single sector cell at the same rotor speed). By using the concept of the sedimentation parameter β_s [13, 15, 17] it is possible to obtain from these data the values of conformational parameters for polymers. Table 1 gives the values of β_S calculated for real MC samples from the equation

$$\beta_s = N_A (R^{-2}[D]^2[S]k_s)^{1/3} . \tag{5}$$

The average value of β_S is $\beta_S = (1.12 \pm 0.08) \cdot 10^7$ mol$^{1/3}$. This value is characteristic of cellulose and its derivatives and only slightly differs from the average experimental values obtained for this polymer class: $\beta_S = 1.0 \cdot 10^7$ mol$^{1/3}$ [15, 16]. Using β_S value it is possible to determine the MW of samples from the values of S_0 and k_s.

Figure 5 shows a dependence analogous to Eq. (4) in which $(S^3 k_s)^{1/2}$ was used instead of M [13]. Figure 5 contains experimental data obtained only by studying velocity sedimentation. In plotting this figure both the values of S_0 and k_s obtained for real samples and fractions (Table 1 and data in ref. [8] reduced to 25 °C) and those obtained for graphic fractions are well. Satisfactory correlation is observed between these two systems of data which fit a linear dependence according to Eq. (4). The slope of this dependence is 0.227 ± 0.007, the intercept is 0.445 ± 0.056, and $r = 0.9911$. The use of the average value of $\beta_S = 1.1 \cdot 10^7$ mol$^{1/3}$ obtained in this work leads to the following values: $A = (170 \pm 10) \cdot 10^{-8}$ cm, $d = (6.8 \pm 1.4) \cdot 10^{-8}$ cm. The use of the value of $\beta_S = 1.0 \cdot 10^7$ mol$^{1/3}$ gives $A = (195 \pm 10) \cdot 10^{-8}$ cm, $d = (7.8 \pm 1.6) \cdot 10^{-8}$ cm. These values are in satisfactory agreement with those listed in Table 3 and illustrate the possibility of evaluating correctly the molecular weights and equilibrium rigidity of chains by using only the data on velocity sedimentation.

The values of A for MC are close to those for other water soluble cellulose derivatives and for cellulose itself

Table 3 Theoretical values of the parameters P_0, Φ_0, and the values of A and d of the chains simulating MC molecules

P_0	$A_f \cdot 10^8$ cm	$\varphi(0)$	$d \cdot 10^3$ cm	$\Phi_0 \cdot 10^{-23}$	$A_\eta \cdot 10^8$ cm	ref
5.11	180 ± 20	1.431	5.1 ± 1.7	2.20	157 ± 12	[18, 21]
5.11	180 ± 20	1.056	7.4 ± 2.5	2.87	130 ± 10	[15, 21]
6.00	130 ± 14	–	–	2.50	144 ± 11	[24]
6.20	120 ± 14	–	–	2.36	149 ± 11	[25]

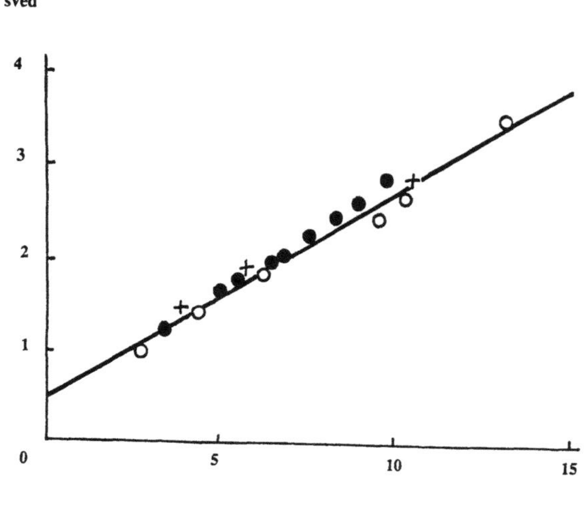

Fig. 5 Dependence of S_0 on $(S_0^3 k_s)^{0.25}$ according to the data: for real samples: (○)-Table 1; (+)-[8] and for graphic fractions: (●)-[26]. S_0 are expressed in Svedberg units

[4, 28, 29]. When the equilibrium properties of MC and other polysaccharides are compared, it is necessary to take into account the effect of different paths of monosaccharides insertion into a linear chain ($1 \to 4$, $1 \to 3$, and $1 \to 6$ bonds), different types of glucoside bond (β or α) and different position of OH groups in different types of sugars [30]. (In principle, for polysaccharide derivatives the degree of substitution and the character of distribution of OH groups substituents along the chain should also be taken into account). The system of intrachain hydrogen bonds, which determines to a considerable extent the equilibrium properties of polysaccharide chains, depends on the above factors. In nature, mostly those polysaccharides exist in which these factors change simultaneously. Hence, it can be only tentatively supposed which of them plays the predominant role in each case. Among water soluble neutral single-strand polysaccharides experimentally studied up to the present, water soluble cellulose derivatives including MC probably exhibit the highest equilibrium rigidity.

Acknowledgements We are grateful for the financial support of the Russian Foundation of Fundamental Science Studies (grant N 93-03-5791). Thanks are due to Dr. G.N. Smirnova for providing the MC samples.

References

1. Rogovin Z (1972) Cellulose chemistry, Khimia, Moscow
2. Petropavlovskij G (1983) Hydrophylic partially substituted cellulose ethers, Nauka, Leningrad
3. Onda Y (1994) In: Preprints of the Kyoto Conference on Cellulosics, p 49
4. Pavlov G, Shildiaeva N (1988) Wood Chemistry N4:10
5. Pavlov G, Korneeva E, Michailova N, Ananyeva E (1992) Carbohydrate Polymers 19:243
6. Pavlov G, Korneeva E, Yevlampieva N (1994) Int J Biol Macromol 16:318
7. Singer K, Tavel P (1938) Helv Chim Acta 21:535
8. Uda K, Meyerhoff G (1961) Makrom Chem 67:168
9. Neely W (1963) J Polym Sci A-1:311
10. Vink H (1966) Makromol Chem 94:1
11. Tsvetkov V, Eskin V, Frenkel S (1970) Structure of Macromolecules in Solution, Butterworth, London
12. Budtov V (1992) Physical Chemistry of Polymer Solution, Khimia, St. Petersburg
13. Pavlov G, Frenkel S (1988) Acta Polymerica 39:107
14. Pavlov G, Frenkel S (1982) Vysokomol Soedin 24B:178
15. Pavlov G (1989) Wood Chemistry N4:3
16. Pavlov G (1994) In: Preprints of the Kyoto Conference on Cellulosics, p 68
17. Pavlov G, Frenkel S Progress Colloid and Polymer Sci (in press)
18. Hearst J, Stockmayer W (1962) J Chem Phys 37:1425
19. Yamakawa H, Fujii M (1973) Macromolecules 6:407
20. Bushin S, Tsvetkov V, Lysenko E, Emelianov V (1981) Vysokomol Soedin 23A:2494
21. Tsvetkov V (1989) Rigid-chain Polymers, Consultants Bureau, New York, London
22. Pavlov G, Kozlov A, Martchenko G, Tsvetkov V (1982) Vysokomol Soedin 24B:284
23. Bohdanecky M (1983) Macromolecules 16:1483
24. Zimm B (1980) Macromolecules 13:592
25. Oono Y (1985) Adv Chem Phys 61:301
26. Tarabukina E (1992) Ph. D. Thesis, Institute of Macromolecular Compounds, St. Petersburg
27. Pavlov G, Tarabukina E, Frenkel S (1995) Polymer 36:2043
28. Shtennikova I, Lavrenko P, Korneeva E, Kolbina G, Strelina I (1995) Vysokomol Soedin (in press)
29. Yalpani M (1988) Polysaccharides, Elsevier, Amsterdam
30. Dashevskiy V (1987) Conformational analysis of macromolecules, Nauka, Moscow

Progr Colloid Polym Sci (1995) 99:114–119
© Steinkopff Verlag 1995

H.G. Müller
F. Herrmann

Simultaneous determination of particle and density distributions of dispersions by analytical ultracentrifugation

Received: 10 March 1995
Accepted: 17 May 1995

Dr. H.G. Müller (✉) · Herrmann
Bayer AG
Central Research Division
ZF-TPP 3, Geb. E-41
51368 Leverkusen, Germany

Abstract Mixtures of dispersions and grafted latices are of great importance in industry. These products are usually chemically heterogeneous and have not only a distribution in particle size but also in particle density. The analysis of such products can be carried out by analytical ultracentifugation. This is done by determining sedimentation coefficient distributions of the sample in two isorefractive dispersing media which have different densities.

In this way it is possible to determine the particle size distribution and particle density distribution simultaneously, i.e., for each particle size of the size distribution the correlated particle density can be determined. This is demonstrated by a number of samples. This method represents a major advancement in the study of the structure of more complicated dispersions. So the composition of latex mixtures can be analysed and valuable insight into the mechanism of grafting reactions is available.

Key words Analytical ultracentrifuge – particle size distribution – particle density distribution – sedimentation analysis – isorefractive media

Determination of particle size distributions of dispersions by analytical ultracentrifugation is a well known and widely used method (1–4) in the diameter range between 0.001–10 μm. The high resolution of this method is shown in Fig. 1, which represents the result of a mixture of nine monodisperse latices measured in one cell during one run of the ultracentrifuge.

The method is based on measurements of the sedimentation velocity v (or the sedimentation coefficient $s = v/w^2 r$) which depends on particle diameter d, particle density ρ_p, the density of the dispersing medium ρ_0, the centrifugal acceleration $\omega^2 \cdot r$, and its viscosity η, (Stokes' law), following Eq. (1):

$$v = d^2(\rho_p - \rho_0) \cdot \frac{\omega^2 \cdot r}{18 \cdot \eta} \tag{1}$$

The experimental set up has recently been improved compared with that in (1). So the preparative Beckman ultracentrifuge L 5–75 has been replaced by a Beckman Optima XL (Palo Alto, CA, USA) which has been modified by implementation of a simple optical path and an eight-hole rotor of Heraeus-Christ Corp. Osterode, Germany, cp. Fig. 2.

The sedimentation velocity (or better the velocity distribution) is measured by a sort of simplified absorption optics: The intensity of the transmitted light (transmission), HE-Ne laser, wavelength 633 nm, from which the absorption is computed is measured only in the middle of the cell as a function of time cp. Fig. 3.

This arrangement allows to calculate v as a quotient of the fixed path length $r-r_m$, r being the distance between the axis of rotation and the radius where the laser beam transmitts the cell, r_m is the meniscus, and the running time t (in the case of sedimentation).

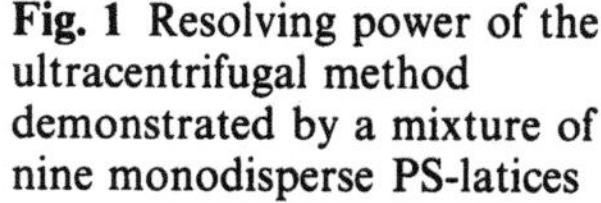

Fig. 1 Resolving power of the ultracentrifugal method demonstrated by a mixture of nine monodisperse PS-latices

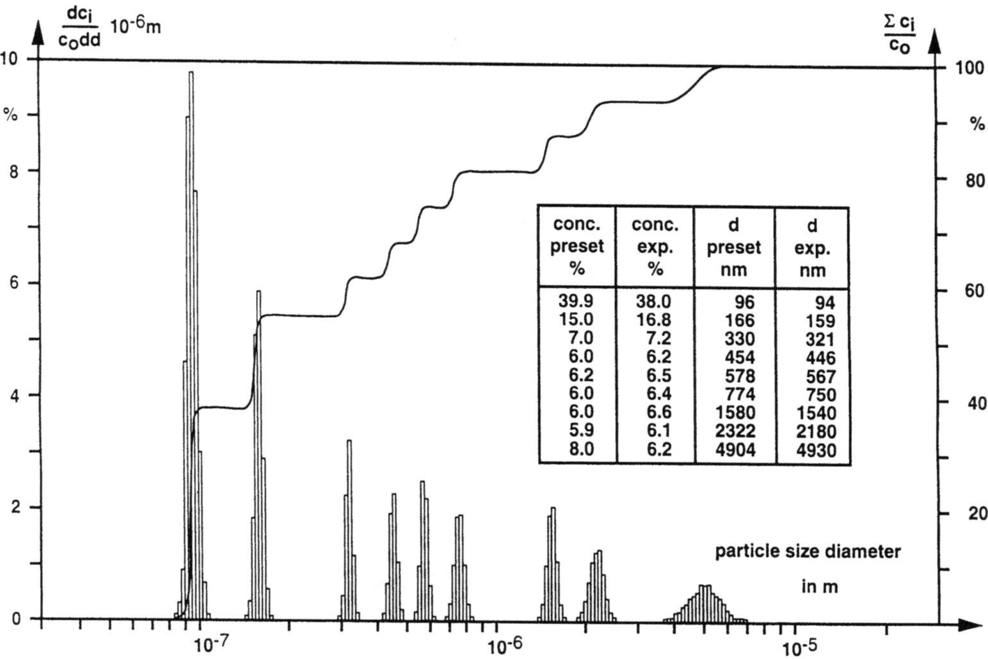

conc. preset %	conc. exp. %	d preset nm	d exp. nm
39.9	38.0	96	94
15.0	16.8	166	159
7.0	7.2	330	321
6.0	6.2	454	446
6.2	6.5	578	567
6.0	6.4	774	750
6.0	6.6	1580	1540
5.9	6.1	2322	2180
8.0	6.2	4904	4930

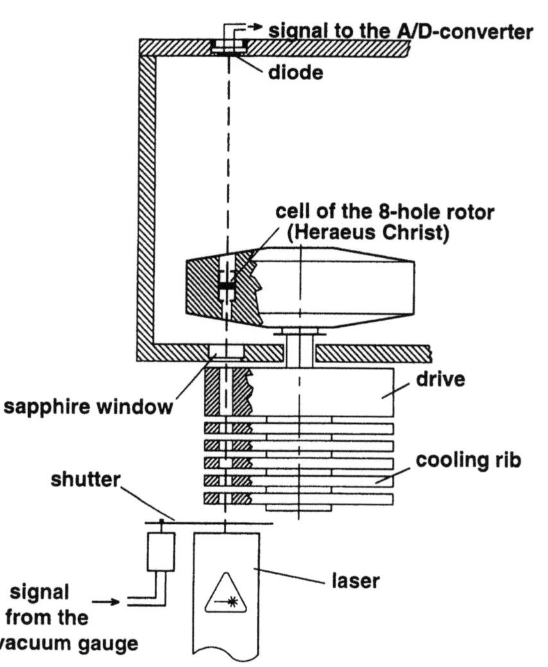

Fig. 2 Experimental set up of the modified Beckman Optima XL with the implemented optical path

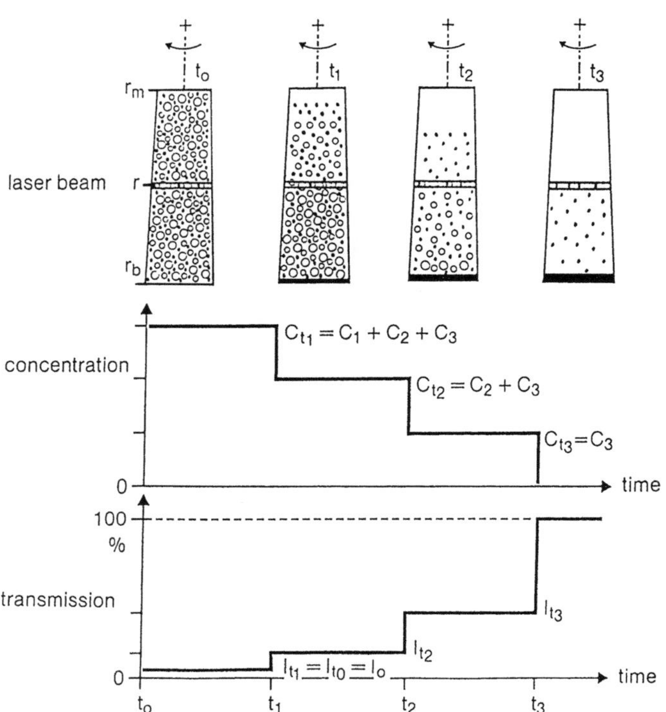

Fig. 3 Change of concentration and transmission in the middle of the cell during the measurement of a three-modal particle size distribution

In Fig. 3 the sedimentation of a three-modal particle size distribution as a function of time is depicted schematically as well for the concentration and transmission at that position of the cell. For the sake of clarity, we draw horizontal lines instead of inclined, which means we ignore the dilution-effect in the used sector-shaped cells (see refs. [2, 3]).

As long as a dispersion with particles of the same chemical composition (i.e., of identical particle density ρ_p) is analyzed the evaluation of particle diameter can be carried out easily. However, mixtures of dispersions and

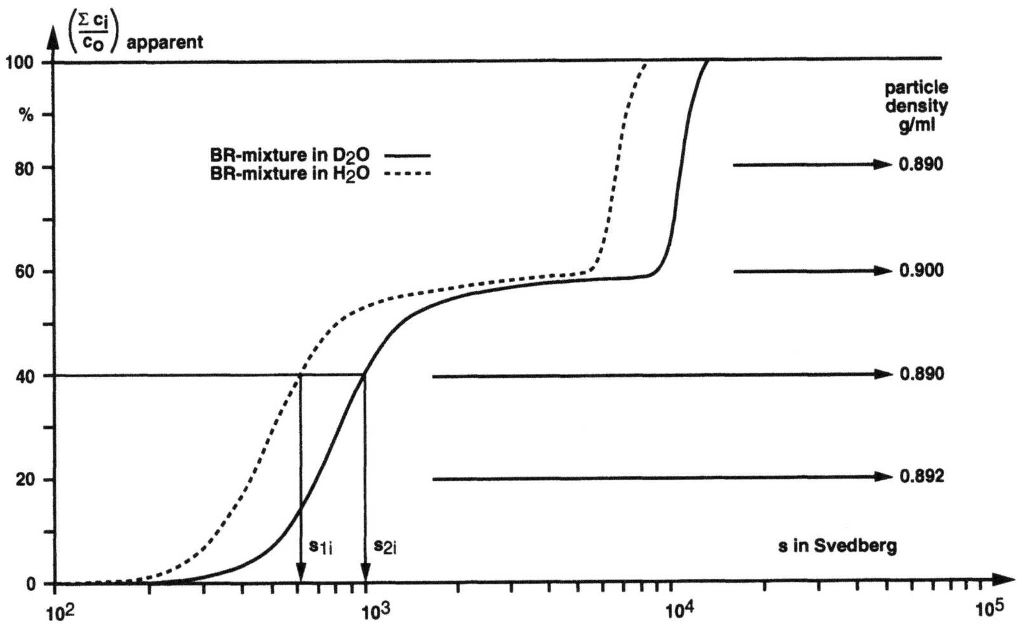

Fig. 4 Distribution of sedimentation coefficients of a mixture of small and large polybutadiene particles and resulting values of density

grafted latices are also of high industrial importance. In this case particles are chemically heterogeneous and most often have different particle densities. In many systems the particle density is unknown and Eq. (1) cannot be resolved, because d and ρ_p are both unknown.

This difficulty can be overcome (according to ref. [7]) by measuring the sedimentation velocity in a second dispersing medium with a different density ρ_{02}. Then instead of one equation with two unknowns, we have two equations. These can be solved resulting in the simultaneous determination of particle size and particle density for each value of the distribution. The solutions are

$$\rho_p = \frac{s_1 \cdot \eta_{01} \cdot \rho_{02} - s_2 \cdot \eta_{02} \cdot \rho_{01}}{s_1 \cdot \eta_{01} - s_2 \cdot \eta_{02}}, \tag{2}$$

the corresponding equation for the particle diameter is

$$d = \sqrt{\frac{18(s_2 \cdot \eta_{02} - s_1 \cdot \eta_{01})}{\rho_{01} - \rho_{02}}}, \tag{3}$$

where s_1 is the sedimentation coefficient of the particles in the first dispersing medium, η_{01} the viscosity of the dispersing medium 1, ρ_{01} the density of the same medium. s_2, η_{02} and ρ_{02} are the corresponding values in (or of) the second medium.

As an example for this method a mixture of fine and coarse polybutadiene particles has been investigated at the generally used temperature of 25 °C. First the distributions of sedimentation coefficients were calculated from the sedimentation velocity distributions in two dispersing media, H_2O and D_2O, comp. Fig. 4.

Then, for the same value of apparent concentration (value of the ordinate axis) the corresponding sedimentation coefficients s_{1i} and s_{2i} were read and from these the particle density and particle diameter of the corresponding particle fraction i were calculated from Eq. (2) and (3). The particle density was determined to be in the range of 0.89–0.90 g/ml for the whole distribution which is in good agreement with the generally accepted value of 0.894 g/ml. In Fig. 4 only four values of density have been plotted for the sake of clarity.

For each horizontal line in Fig. 4 that means for each particle fraction i we get a coupled pair of s_{1i}/s_{2i} values, which yield the corresponding two values ρ_{pi} and d_i. It should be added here that this method implies that in each of the two dispersing media either sedimentation or flotation takes place: It must be avoided to have sedimentating and flotating particles in one cell at the same time. This can be made sure by a foregoing preparative ultracentrifugation in both dispersing media.

As a second example a mixture of SBR-latex (polystyrene butadiene copolymer) and PS (polystyrene)-latex has been examined.

In this case it is necessary to correlate s values belonging to the same species of particles. That is why this method requires dispersing media with different densities but with the same refractive index.

So in this case solutions with the same amount of mesoerythrose in water and in heavy water have been used as isorefractive media, other types of isorefractive media being mixtures of water and heavy water with methanol.

Figure 5 shows the s-distributions of this mixture in these two dispersing media with 25 wt% mesoerythrose in

Fig. 5 Distribution of sedimentation coefficients of a mixture of PS and SBR-latex and resulting values of density

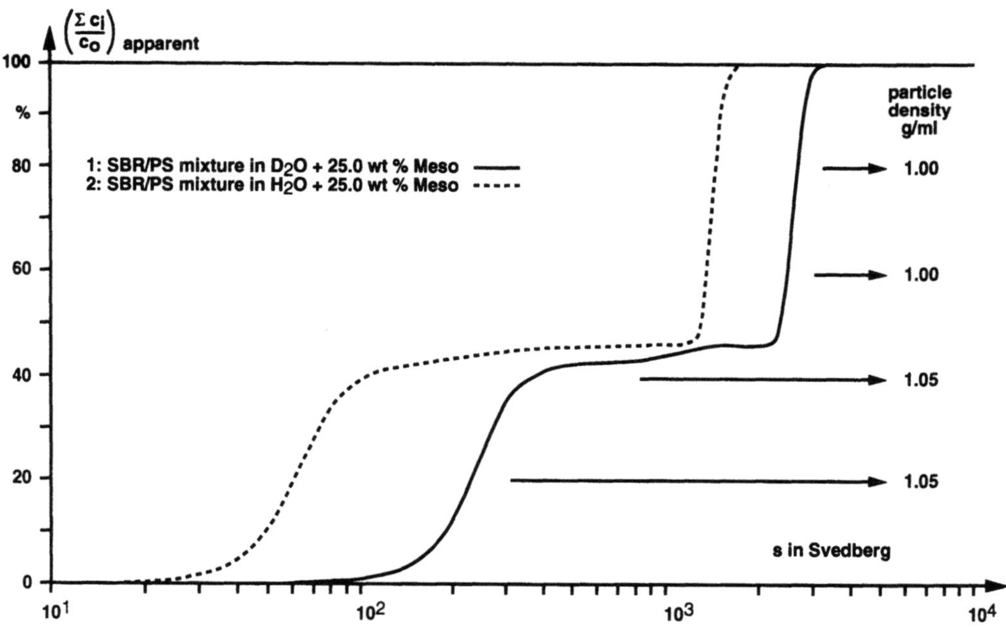

Fig. 6 Integral and differential particle size distribution evaluated from Fig. 5 by application of the Mie light-scattering theory

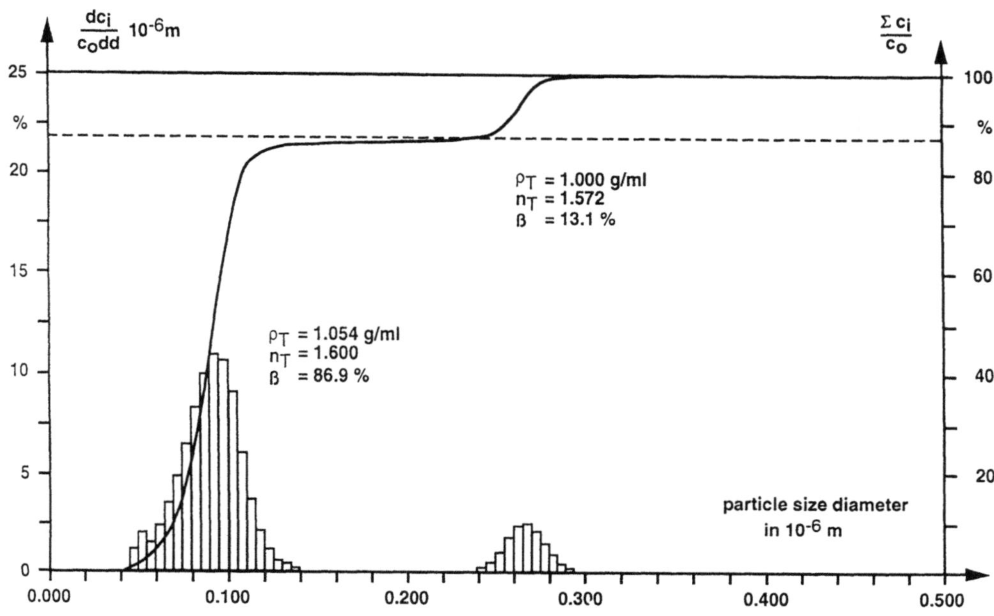

this special case, resulting in media densities of 1.072 g/ml and 1.163 g/ml and the calculated particle densities of 1.05 and 1.00 g/ml from the corresponding horizontal lines which are the known densities of polystyrene and this special type of SBR.

So the two components of this mixture have been identified and with help of the light scattering theory of Mie [2, 5] the mass distribution of this latex mixture can be thus evaluated from these s-distributions (Fig. 6)

The required two different particle refractive indices for the two different kinds of particles indicated in Fig. 6 we got for the case of polystyrene from literature. For the case of SBR we calculated its composition from the experimentally determined density and from that the refractive index by the equation of Biot-Arago.

A comparison with the original components of this mixture shows that the mass distribution has also been determined correctly (present mixing ration in Fig. 6 is

H.G. Müller and F. Herrmann
Simultaneous determination of particle and density distributions of dispersions

indicated by the broken line). In this way analytical ultracentrifugation helps to analyze mixtures of latices. The density of the components often tells directly of which type of components the mixture consists, e.g., a density of 1.05 g/ml is typical for polystyrene latex.

A third example is a mixture of fine and coarse particles of polybutadiene which has been grafted with a mixture of monomeric styrene and acrylonitrile (SAN), Fig. 7.

For this bimodal s-distribution particle densities of 1.00 and 0.96 g/ml were found. Simultaneously the particle size distribution in Fig. 8 has been calculated.

It shows that this grafted latex at first sight consists of about 70% fine particles with a density of about 1.00 g/ml and about 30% big particles with a density of 0.96 g/ml. On closer inspection it reveals that the fine particles consist of about 49% smaller particles with a density of about 1.005 g/ml and about 17% coarser particles with a density of 0.991 g/ml. As the density of the grafting material SAN is above 1.00 g/ml, it can easily be deduced that the fine particles are more heavily grafted than the bigger ones, because grafting in this special case is proportional to the specific surface area of the different particle fractions. The

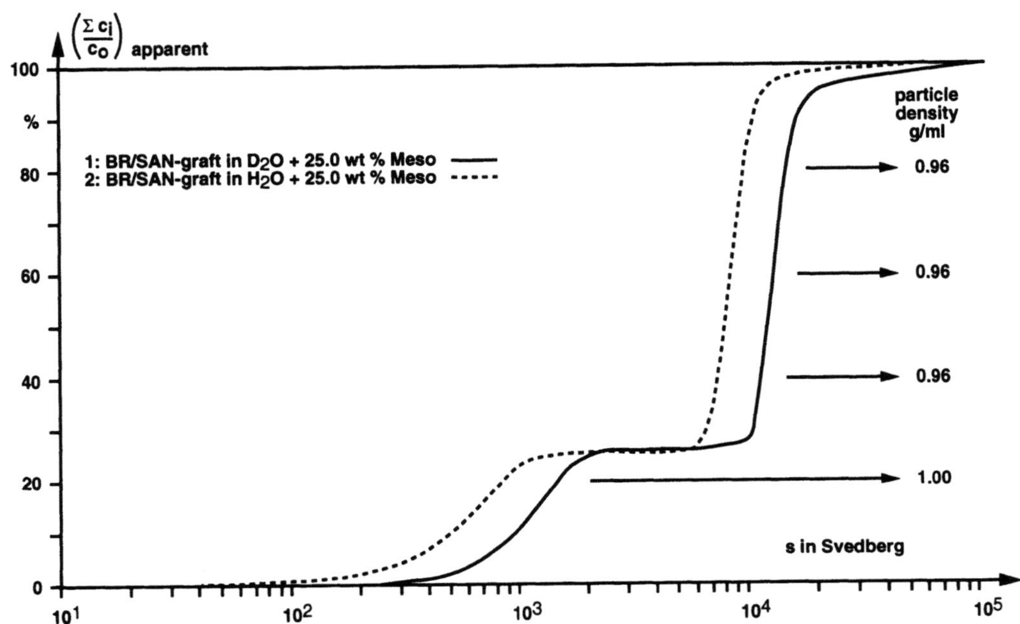

Fig. 7 Distribution of sedimentation coefficients of a mixture of SAN-grafted polybutadiene particles and resulting values of density

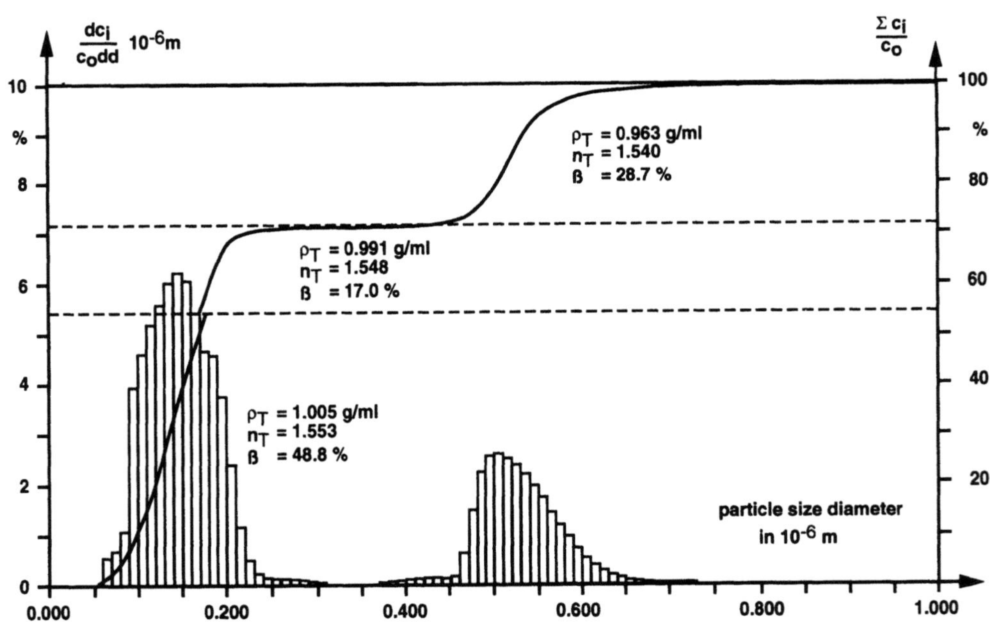

Fig. 8 Integral and differential particle size distribution evaluated from Fig. 7 by application of the Mie light-scattering theory

same result was found in ref. [7]. With known density of the SAN the thickness of the grafted SAN shell thickness can be computed as a function of particle diameter.

$$th = \frac{d_3 - d_1}{2}, \tag{4}$$

where th = thickness of the grafted shell, d_3 = determined diameter of the grafted particles and d_1 = diameter of the ungrafted particles.

d_1 can be determined by

$$d_1 = \sqrt[3]{\frac{d_3^3 \cdot \rho_3 \cdot \bar{x}_1}{\rho_1}}, \tag{5}$$

where ρ_3 = measured density of the grafted particles, ρ_1 = known density of the ungrafted particles and $\bar{x}_1$ = mass fraction of the core material in the grafted particles.

$\bar{x}_1$ is given by

$$\bar{x}_1 = \left(\frac{1}{\rho_3} - \frac{1}{\rho_2}\right)\left(\frac{\rho_1 \rho_2}{\rho_2 - \rho_1}\right), \tag{6}$$

where ρ_2 = known density of the grafting polymer.

In this way the thickness of the grafted shell can separately be determined for the small and the coarse particles giving a valuable insight into the mechanism of the grafting reaction.

Results obtained in this way are in agreement with previous results [6] from a much more time-consuming method.

The method published here has already been described in [7], but new dispersing media make it possible to analyze a much broader range of particle densities and applications.

References

1. Müller HG (1989) Colloid Polymer Sci 267:1113–1116
2. Scholtan W, Lange H (1972) Kolloid-Z u Z Polymere 250:782–796
3. Mächtle W (1988) Die Angew Makromol Chem 162:35–52
4. Mächtle W (1992) Makromol Chem, Macromol Symp 61:131–142
5. Mie G (1908) Ann Phys 25:377
6. Kuhn R, Müller HG, Bayer G, Krämer-Lucas H et al. (1993) Colloid Polymer Sci 271:133–142
7. Mächtle W (1984) Makromol Chem 185:1025–1039

Progr Colloid Polym Sci (1995) 99:120–124
© Steinkopff Verlag 1995

Molar mass distribution of polymers from sedimentation velocity in an analytical ultracentrifuge

M.D. Lechner
W. Mächtle

Received: 27 March 1995
Accepted: 6 June 1995

Prof. Dr. M.D. Lechner (✉)
Physical Chemistry
University of Osnabrück
Barbarastraße 7
49069 Osnabrück, Germany

Dr. W. Mächtle
Kunststofflaboratorium
Polymerphysik
BASF Aktiengesellschaft
67056 Ludwigshafen, Germany

Abstract An alternate procedure, the run time integral method, is suggested for precise determinations of molar mass distributions of polymers. The method is based on the fact that modern analytical ultracentrifuges directly measure the variable angular velocity ω as a function of time. This allows an exact calculation of the run time integral $\int \omega^2 \, dt$ and replaces the physically uncorrect zero time correction t_0 by $\int \omega^2 \, dt$. The procedure is demonstrated by narrow and broad distributed polymers.

Key words Molar mass distribution – sedimentation velocity – analytical ultracentrifuge – run time integral method

Introduction

Analytical ultracentrifugation (AUC) and size exclusion chromatography (SEC) are the most powerful methods for the determination of molar mass averages and molar mass distributions. SEC is a fast, easy to handle method, but it exhibits relative values of the molar mass distribution and needs calibration. AUC is a rather complicated, time-consuming, and more expensive method, but it can give absolute values of the molar mass distribution because the calibration is inside the method.

Up to now the usual methods for the determination of molar mass distributions (Blair–Williams [1], Fujita [2], and Trautman [3]) take into account a so-called zero time correction which consists of an acceleration effect and a correction for restricted sedimentation near the meniscus. The calculation of the molar mass distribution is usually done by an error-producing nonlinear regression method [1, 2, 4].

The difficulties of this "zero time correction method" can be overcome by the presented "run time integral method," as modern ultracentrifuges directly measure the variable angular velocity ω as a function of time t.

The principles of both methods are given in this paper and the run time integral method is demonstrated with several narrow and broad distributed polymers.

Theory

The zero time correction method

This method is widely used for the determination of molar mass distributions of polymers in solution from sedimentation velocity [1, 2]. It takes into account the effects of concentration C_0, pressure p, diffusion coefficient D and the zero time correction t_0 on the sedimentation coefficient $S = (dR/dt)/(\omega^2 R)$ and the distribution of S, $g(S, p, C_0)$. The equations read:

$$S(p, C_0) = [\ln(R^*/R_{\mathrm{m}})]/[\omega^2(t - t_0)]$$
$$= S(0, C_0)\{1 + B_1[(R^*/R_{\mathrm{m}})^2 - 1]\} \tag{1}$$

$$g(S, p, C_0) = (1/C_0)R^*\omega^2(t - t_0)(R^*/R_m)^2\,(dC/dR)$$
$$\times \{1 - m[(R^*/R_m)^2 - 1]\} \qquad (2)$$

R^* = Distance of boundary of species i from center of rotation

R_m = Radius of the meniscus from center of rotation

$S(0, C_0)$ = Sedimentation coefficient at $p = 0$ bar

$t_0 = t_f + t_r$ = Zero time correction

t_f = Acceleration time, time until the rotor reaches full speed

t_r = Correction for restricted sedimentation near the meniscus

$$B_1 = [k_sC_0 - m(1 + 2k_sC_0)]/[2(1 + k_sC_0)] \qquad (1a)$$

= Parameter for pressure and concentration effect

$$m = \gamma(1/2)\omega^2 R_m^2\,\rho(0) \qquad (2a)$$

γ = Pressure parameter

The determination of the unknown quantities $t_0. = t_f + t_r$, B_1, γ and $S(0, C_0)$ has been done alternatively in the following ways

a) Calculation of t_0 with the equation $t_0 = (2/3)t_f$ [2, 3]

b) Plot of $\ln(R^*/R_m)$ as a function of time t with the intercept $\omega^2\,S(p, C_0)t_0$ and the slope $\omega^2 S(p, C_0)$ [1, 2]

c) Calculation of t_0, B_1 and $S(0, C_0)$ by nonlinear regression methods [1, 2]

The methods a) and b) neglect pressure and concentration effects of the sedimentation. Method c) takes pressure and concentration effects into account, but is error producing and physically uncorrect as at least the factor $\omega^2 t_f$ has to be replaced by the run time integral $\int \omega^2\,dt$ as will be pointed out in the next section.

The run time integral method

The error-producing and physically uncorrect zero time correction procedure, Eqs. (1) and (2), could be overcome as modern ultracentrifuges measure directly the variable angular velocity ω as a function of time. On the other hand, aged analytical ultracentrifuges can easily be equipped with a clock to measure the speed as a function of time up to the acceleration time t_f. Therefore, S and $g(S, p, C_0)$ can be calculated physically exact without defining a zero time correction. Equations (1) and (2) then read

$$S(p, C_0) = [\ln(R^*/R_m)]/\left(\int_0^t \omega^2 dt - A_r\right)$$
$$= S(0, C_0)\{1 + B_1[(R^*/R_m)^2 - 1]\} \qquad (3)$$

$$g(S, p, C_0) = (1/C_0)R^*\left(\int_0^t \omega^2 dt - A_r\right)(R^*/R_m)^2\,(dC/dR)$$
$$\times \{1 - m[(R^*/R_m)^2 - 1]\} \qquad (4)$$

A_r is a parameter for restricted sedimentation near the meniscus and can easily be determined from experimental values as Eq. (3) reduces in the vicinity of the meniscus ($R^* \approx R_m$)

$$\ln(R^*/R_m) = S(0, C_0)\left[\int_0^t \omega^2 dt - A_r\right]. \qquad (5)$$

The procedure for determining the molar mass distribution via the run time integral method is as follows:

1) Measurement of the speed of the AUC as a function of time yields the run time integral $\int \omega^2\,dt$ with $\omega = 2\pi N$ (Fig. 1).

2) Measurement of dn/dR (Schlieren-optics) or concentration C (interference optics, adsorption optics) as a function of time and concentration (Fig. 2).

3) Determination of the parameter A_r. Plot of $\ln(R_{max}/R_m) = f(\int \omega^2 dt)$ yield $\lim_{R_{max} \approx R_m} S(0, C_0)\ A_r$ as intercept and $\lim_{R_{max} \approx R_m} S(0, C_0)$ as slope (Eq. (5), Fig. 3).

4) Determination of the parameter B_1. Plot of $S_{max}(p, C_0) = [\ln(R^*/R_m)]/(\int \omega^2 dt - A_r)$ as a function of $(R_{max}/R_m)^2 - 1$ yields B_1 (Eq. (3), Fig. 4).

5) Determination of $S(0, C_0)$ and $g(S, C_0, t)$ with the help of Eqs. (3) and (4). Calculation of dC/dR from dn/dR with the relation $dC/dR = [(dn/dR) \cdot tg\theta/(l\,dn/dC)] \times K_{cal}\,m_x/E$. θ = Philpotangle, l = cell length, dn/dC = refractive index increment, K_{cal} = calibration constant, m_x = magnification of the Schlieren optics in x-direction, E = magnification of the comparator (Schlieren-optics). The parameter for the pressure effect γ may be either taken

Fig. 1 Rotor speed as a function of run time of the rotor. Acceleration time $t_f = 259$ s

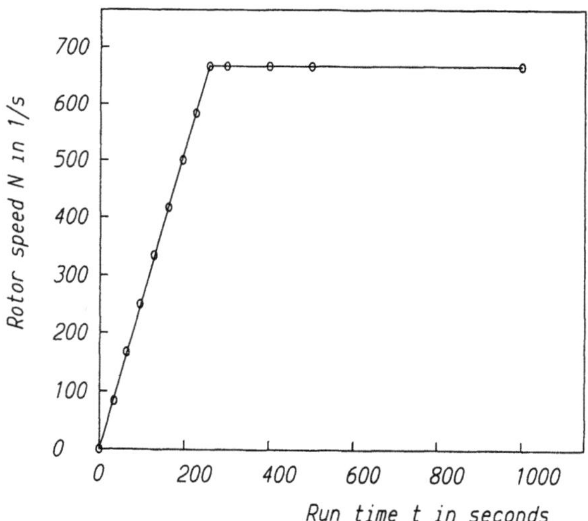

M.D. Lechner and W. Mächtle
Molar mass distribution of polymers

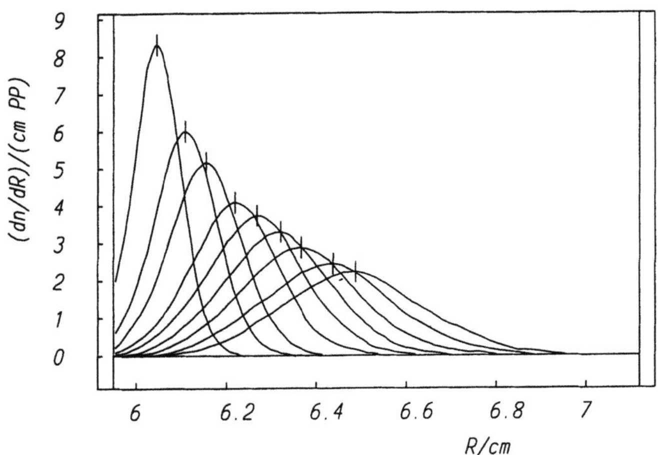

Fig. 2 Dextran T 70 in water. Schlieren optics. $t = 50, 75, 93, 120,$ 138, 158, 178, 205, 226 min (left upper curve to right lower curve). $C_0 = 3.0$ g/l, $T = 25\,°C$, $N = 40\,000$ min^{-1}

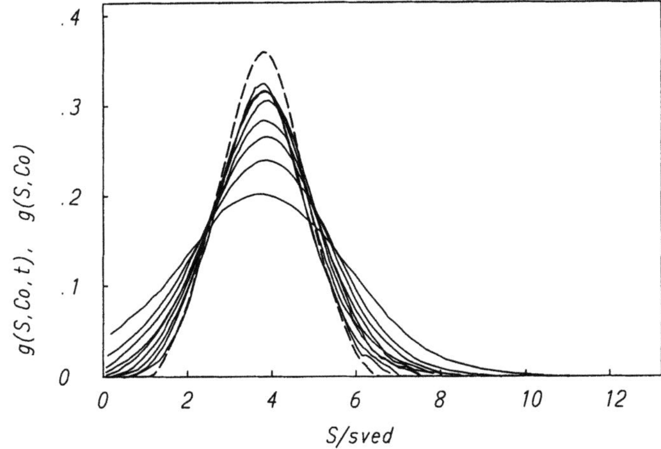

Fig. 5 Dextran T 70 in water. S-Distribution. $t = 50, 75, 93, 120, 138,$ 158, 178, 205, 226 min (lower solid curve to upper solid curve) and $1/t = 0$ min^{-1} (dashed curve). $C_0 = 3.0$ g/l, $T = 25\,°C$, $N = 40\,000$ min^{-1}

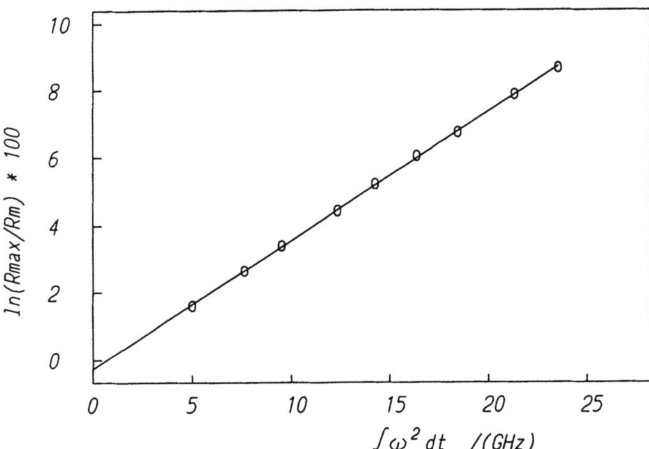

Fig. 3 Dextran T 70 in water. $\ln(R_{max}/R_m)$ as a function of the run time integral. $C_0 = 3.0$ g/l, $N = 40\,000$ min^{-1}

Fig. 4 Dextran T 70 in water. $S_{max}(p, C_0)$ as a function of $(R_{max}/R_m)^2 - 1$. $C_0 = 3.0$ g/l, $N = 40\,000$ min^{-1}

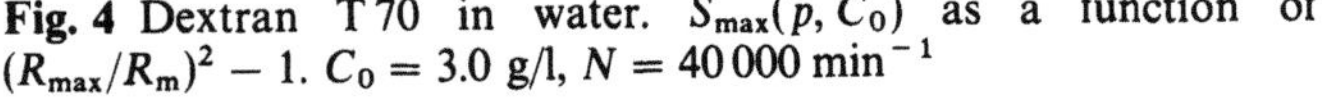

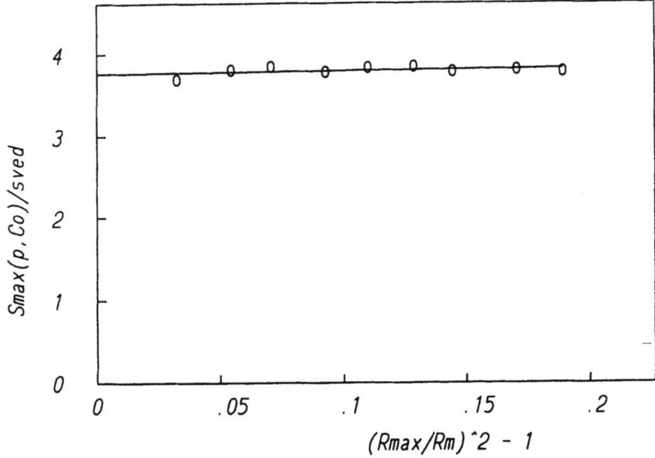

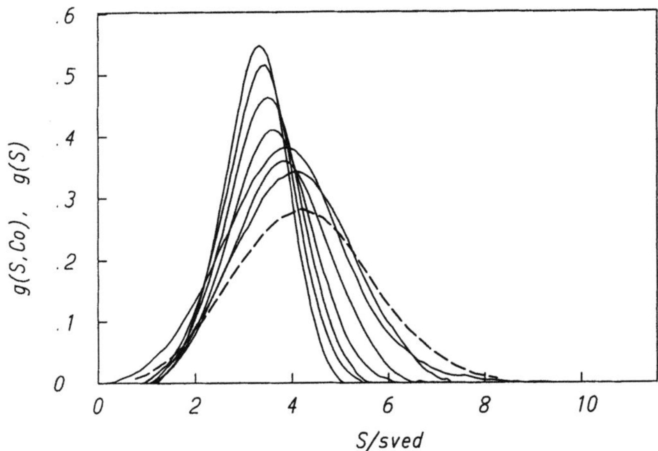

Fig. 6 Dextran T 70 in water. S-Distribution at 7 concentrations (solid curves) and extrapolation to $C_0 = 0$ (dashed curve). $C_0 = 1.01,$ 2.0, 3.0, 4.0, 5.0, 6.0 and 7.0 g/l (lower solid curve to upper solid curve)

from literature ($\gamma = 1,6 \cdot 10^{-9}$ cm g^{-1} s^{-1}) [2] or calculated with the help of Eq. (3) and the definitions of B_1 and m (Fig. 5).

6) Calculation of $g(S, C_0)$ by extrapolating fixed values of $g(S, C_0, t)$ at constant S to $1/t = 0$ (Fig. 5) [2].

7) Calculation of $g(S)$ by extrapolating fixed values of $1/S$ at constant $g(S, C_0)$ to $C_0 = 0$ (Fig. 6) [5].

8) Calculation of $w(M)$ via the relationships $w(M)$ $dM = g(S) \, dS$ and $S = K \, M^a$ (Fig. 7) [2].

Experimental

Materials

Dextran T 70 (Pharmacia Chemicals, Uppsala, Sweden); $M_w = 70\,000$ g/mol; $M_w/M_n = 1.9$.

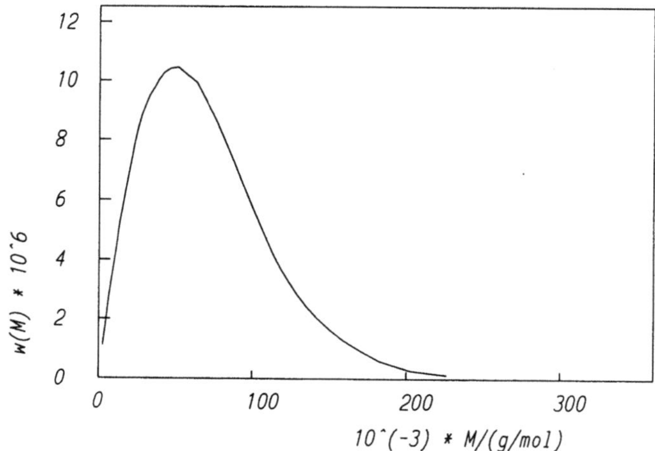

Fig. 7 Dextran T 70 in water. Molar mass distribution. $M_n = 42\,000\,\text{g/mol}$, $M_w = 69\,000\,\text{g/mol}$, $M_z = 89\,000\,\text{g/mol}$, $M_w/M_n = 1.64$

Polystyrene NBS 706 (National Bureau of Standards, Washington DC, USA); $M_w = 258\,000 - 288\,000\,\text{g/mol}$; $M_w/M_n = 1.9 - 2.1$.

Polystyrene PS 90 000 (Polymer Standard Service, Mainz, Germany); $M = 90\,000\,\text{g/mol}$; $M_w/M_n = 1.06$.

Solvents and constants

The solvents used were methylethylketone (MEK, p.a.), toluene (p.a) and water (bidistilled). As values for dn/dc we used $0.216\,\text{cm}^3/\text{g}$ (PS/MEK), $0.110\,\text{cm}^3/\text{g}$ (PS/toluene) and $0.146\,\text{cm}^3/\text{g}$ (Dextran/water).

Methods

Beckman analytical ultracentrifuge, model E, eight-hole rotor, Schlieren optics, multiplexer, 30 mm double-sector cells. The evaluation of the schlieren photos was done with a comparator (enlargement 10-fold), a digitizer and a calculator PC 486.

Results and discussion

The calculation of the S-distribution $g(S, p, C_0)$ – Eq. (4) – needs the pressure parameter γ – Eq. (2a) – which was taken from literature [2]. On the other hand, γ may be calculated by experimental determination of B_1 at different concentrations with the help of Eq. (1a).

Figures 1 to 7 demonstrate in detail the run time integral method with the system Dextran T 70 in water. In Fig. 1 the rotor speed N is plotted against the run time of the rotor. Normally the model E is operated in such a way

that the speed is a linear function of the time t up to the desired speed N. The experimental values allow the calculation of the run time integral for each desired time t.

Figure 2 demonstrates the Schlieren optics curves at nine different times at one concentration for Dextran T 70 in water. The maximum values of the Schlieren peaks are now determined and plotted according to Eq. (5) for the determination of the parameter for restricted sedimentation near the meniscus A_r. The plot is seen in Fig. 3. A_r is slightly different for every concentration and must be determined separately.

The parameter for pressure and concentration effect B_1 is then determined for each concentration according to Eq. (3) and Fig. 4. The figure demonstrates that B_1 is nearly zero for this system.

It is now possible to calculate the sedimentation coefficient S and the S-distribution for each point of the Schlieren curves according to Eqs. (3) and (5). The result is given for one concentration and different times by the solid lines in Fig. 5. These curves are influenced by diffusion of the dissolved macromolecules and must be corrected. The correction is done by plotting fixed values of $g(S, C_0, t)$ at constant S as a function of $1/t$ and extrapolating the $g(S, C_0, t)$-values to $1/t = 0$ [2]. The plot $g(S, C_0, t) = f(1/t)$ is nearly linear if the Schlieren peaks are not in the direct vicinity of the meniscus and the bottom. The result of the extrapolation is shown in Fig. 5 as a dashed line.

Figure 6 summarizes the measurements for seven concentrations as solid lines. The extrapolation to zero concentration is done by plotting the values of $1/S$ at constant $g(S, C_0)$-values as a function of concentration and linear extrapolation to $C_0 = 0$. The result of the extrapolation is given as dashed line in Fig. 6. This line is the true S-distribution, without concentration effect, diffusion effect and Johnson–Ogston effect.

Calculation of the molar mass distribution $w(M)$ is now possible via the relationships $w(M)dM = g(S)dS$ and $S_0 = K_0 \cdot M^{a_0}$. Combination of the two equations give $w(M) = g(S) K_0^{1/a_0} a_0 S^{1 - 1/a_0}$. The S-M relationship for dextran/water is [6]

$$S_0/\text{sved} = 0.012\,M^{0.53}.$$

This relationship allows the calculation of the molar distribution of dextran T 70 which is given in Fig. 7. The result is

$$M_n = 42\,000\,\text{g/mol}, \quad M_w = 69\,000\,\text{g/mol},$$

$$M_z = 89\,000\,\text{g/mol}, \quad M_w/M_n = 1.64$$

in good agreement with values given in the literature [7].

The broad distributed polystyrene NBS 706 in toluene and the narrow distributed polystyrene PS 90 000 in methylethylketone (MEK) has been handled in the same

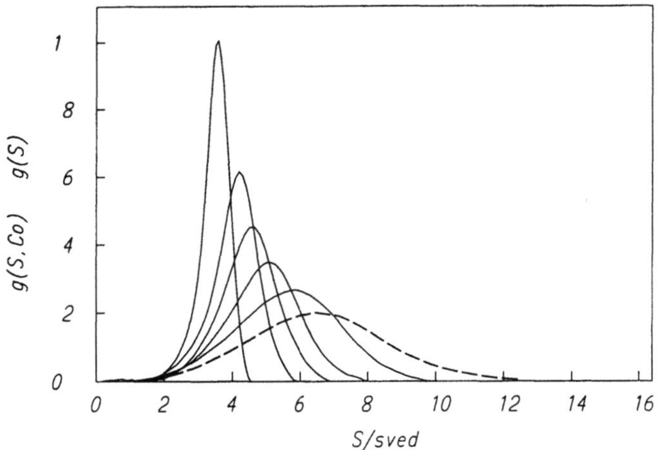

Fig. 8 Polystyrene NBS 706 in toluene. S-Distribution at five concentrations and extrapolation to $C_0 = 0$ (dashed curve). $C_0 = 1.06$, 2.0, 2.98, 4.18 and 6.14 g/l (lower solid curve to upper solid curve)

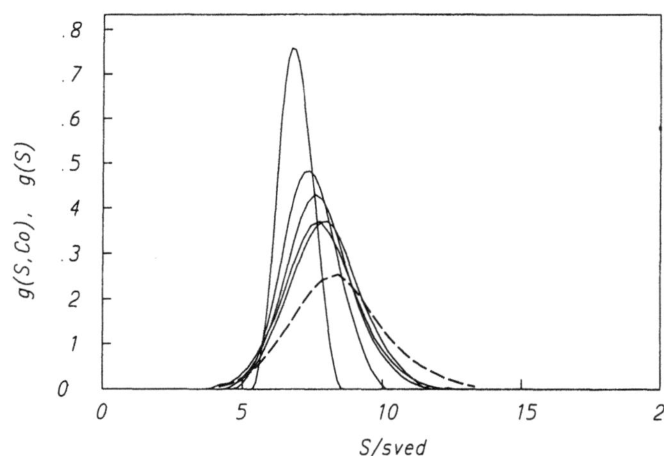

Fig. 9 Polystyrene PS 90 000 in MEK. S-Distribution at five concentrations and extrapolation to $C_0 = 0$ (dashed curve). $C_0 = 1.02$, 1.96, 2.96, 4.0 and 5.96 g/l (lower solid curve to upper solid curve)

way as dextran T 70 except that we have measured only at five concentrations. The results of the system polystyrene NBS 706 in toluene are given in Fig. 8. The dashed line demonstrates the true S-distribution. Concerning the molar mass distribution the problem is that the literature offers a wide variety of S-M relationships [8].

$$K_0 = 0.028 - 0.034; \qquad a_0 = 0.41 - 0.47 .$$

Nonlinear regression calculation with our own experimental values exhibits the following S-M relationship:

$$S_0/\text{sved} = 0.041\, M^{0.41} .$$

Figure 9 summarizes the measurements on polystyrene PS 90 000 in methylethylketone at five concentrations. The S-M relationship for this system, calculated from measure-

ments of six narrow distributed polystyrenes is [9]

$$S_0/\text{sved} = 0.040\, M^{0.472} .$$

With this relationship the molar mass distribution of PS 90 000 is

$$M_n = 73\,000 \text{ g/mol}, \quad M_w = 88\,000 \text{ g/mol},$$

$$M_z = 102\,000 \text{ g/mol}, \quad M_w/M_n = 1.19 .$$

It has been demonstrated that analytical ultracentrifugation is a precise method for the determination of molar mass distribution of polymers. Nevertheless, the method is time consuming and complicated to handle.

Acknowledgements The authors are indebted to the Fonds der Chemischen Industrie and to the BASF AG for financial support. The precision measurements with the AUC were performed by H. Kaiser and H. Roth, Kunststofflaboratorium, BASF AG, Ludwigshafen.

References

1. Blair JE, Williams JW (1964) J Phys Chem 68:161
2. Fujita H (1975) Foundations of Ultracentrifugal Analysis. Wiley, New York
3. Trautman R (1956) J Phys Chem 60:1211
4. Ortlepp B, Panke D (1992) Makromol Chem, Macromol Symp 61:176
5. Baldwin RL, Gosting LJ, Williams, JW, Alberty RA (1955) Discuss Farad Soc 20:13
6. Lechner MD, Mächtle W, unpublished measurements
7. Wan PJ, Adams ET (1974) Am Chem Soc, Pol Prepr 15:509
8. Meyerhoff G (1955) Z phys Chem (Frankf.) 4:335; Appelt B, Meyerhoff G (1980) Macromolecules 13:657; Petrus V, Borsch B, Nystroem B, Sundeloef LO (1982) Macromol Chem Phys 183:1279
9. Lechner MD, Mächtle W, this volume

Progr Colloid Polym Sci (1995) 99:125–131
© Steinkopff Verlag 1995

M.D. Lechner
W. Mächtle

Molar mass distribution from sedimentation fractiometry (SF)-competition with size exclusion chromatography (SEC, GPC)

Received: 6 April 1995
Accepted: 23 May 1995

Prof. Dr. M.D. Lechner (✉)
Physikalische Chemie
Universität Osnabrück
Barbarastraße 7
49069 Osnabrück

W. Mächtle
Kunststofflaboratorium
BASF AG
67056 Ludwigshafen

Abstract The resolutions of size exclusion chromatography (SEC) and sedimentation fractiometry (SF) are compared with mixtures of nearly monodisperse polystyrenes. It has been pointed out that the resolution, i.e., the precision of the molar mass distribution of the SF is as good as for the SEC. A new easy to handle procedure is suggested for the determination of molar mass distribution by SF and demonstrated with narrow and broad distributed polymers.

Keywords Molar mass distribution – sedimentation fractiometry – size exclusion chromatography

Introduction

A precise determination of the molar mass distribution of polymers is only possible by fractionation of the material. This fractionation may be done by classical fractionation, size exclusion chromatography, fractionation by sedimentation or mass spectrometry. Advantages and disadvantages of the four methods are given in Table 1. If one compares the methods for determining the molar mass distribution of polymers by fractionation it seems that fractionation by sedimentation with an analytical ultracentrifuge has some advantages compared with other methods like classical fractionation, SEC, SEC-light scattering and SEC-viscosity coupling techniques. Sedimentation fractiometry does not need a special detector for molar mass distribution. Analytical ultracentrifugation allows fractionation and molar mass determination in one step. What are now the reasons that most workers do not dare to determine the molar mass distribution by ultracentrifugation. The reasons may be that the analytical ultracentrifuge is complicated to handle and the theory of sedimentation is sophisticated and requires more mathematical developments than do classical fractionation and SEC.

Up to now the most common and the most correct method for the determination of molar mass distributions is the zero time correction method of Blair–Williams, Fujita and Trautman [1–3]. The method requires sedimentation measurements at several times and several concentrations. The sedimentation coefficient S and the S-distribution is then calculated according to papers given by Stafford [4] and by Lechner and Mächtle [5]. The authors demonstrate that the following corrections are made to S and $g(S)$: 1) zero time correction t_0 which consists of the acceleration time t_f and a correction for restricted sedimentation near the meniscus t_r; 2) correction for diffusion; 3) correction for pressure; 4) correction for concentration. One can imagine that this is a time-consuming and complicated procedure. It has been pointed out that the zero time correction method may be improved by the so-called run time integral method [4, 5]. Finally, one obtains the molar mass distribution with a $S_0 - M$ relationship $S_0 = K_0 M^{a_0}$ and $w(M)dM = g(S)dS$. In this form the AUC hardly competes with SEC.

Size exclusion chromatography (SEC)

Although a general SEC theory does not yet exist, the SEC handling is very simple. One needs only one concentration of the polymer and then the SEC-people sell us after

Table 1 Determination of molar mass distribution

Classical Fractionation
Time Consuming
Suitable for all Polymers

Size exclusion chromatography (SEC)
Gel permeation chromatography (GPC)
Difficulties with polyelectrolytes
Easy to handle

SEC (GPC) + CLS (classical light scattering)
Difficulties with polyelectrolytes
Favours large molar masses
Easy to handle

Sedimentation fractiometry (SF)
Suitable for all polymers including polyelectrolytes
Complicated theory

Mass Spectrometry (MS)
Suitable for all polymers including polyelectrolytes
Restricted molar mass range

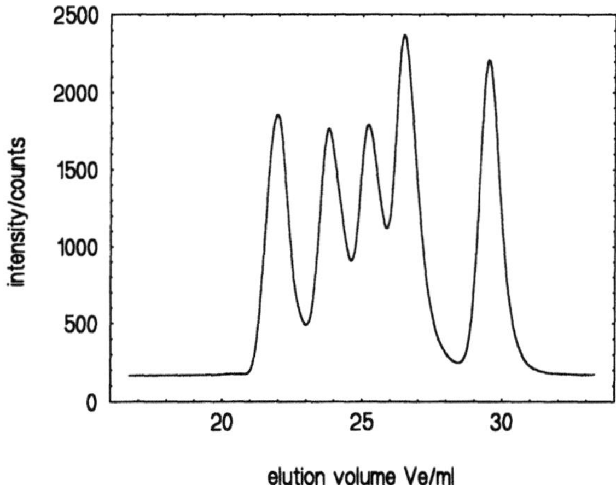

Fig. 1 SEC diagram of five narrow distributed polystyrenes in tetrahydrofurane (see text)

10 min the only true and reliable molar mass distribution and nobody is able to control the SEC-people. What are the principles of SEC and can we learn from them.

The first stage of SEC measurement is the calibration of the SEC configuration with narrow distributed polymers according to the empirical SEC equation

$$\log M = A - BV_e \tag{1}$$

where the coefficients A and B are constants if all other parameters are kept constant (M = molar mass, V_e = elution volume). Figures 1 and 2 give an example of the SEC-calibration with the following configuration:

polystyrene/tetrahydrofurane; concentration: 6 g/l; flow rate: 1.0 cm^3/min; column configuration: 3 columns 10^3 Å, 10^4 Å, 10^5 Å, styrene/divinylbenzene, column length: 300 mm, internal column diameter: 8 mm. calibration standards: narrow distributed polystyrenes; $M = 512\,000$ g/mol, $M_w/M_n = 1.03$; $M_w = 208\,000$ g/mol, $M_w/M_n = 1.04$, $M_w = 92\,300$ g/mol, $M_w/M_n = 1.05$; $M_w = 46\,000$ g/mol, $M_w/M_n = 1.03$; $M_w = 9770$ g/mol; $M_w/M_n = 1.05$.

A plot of $\log M$ as a function of the elution Volume V_e, Fig. 2, results in the SEC-equation

$$\log M = 10.75 - 0.229\,V_e; \quad C_0 = 6 \text{ g/l} . \tag{2}$$

The following equation

$$(M_w/M_n)_{exp} = M_w/M_n + (M_w/M_n)_{dis} , \tag{3}$$

with $(M_w/M_n)_{exp}$ = experimental available polydispersity and $(M_w/M_n)_{dis}$ = polydispersity from axial dispersion takes the fact into account that a fraction of molecules comes too early out of the column, another fraction too

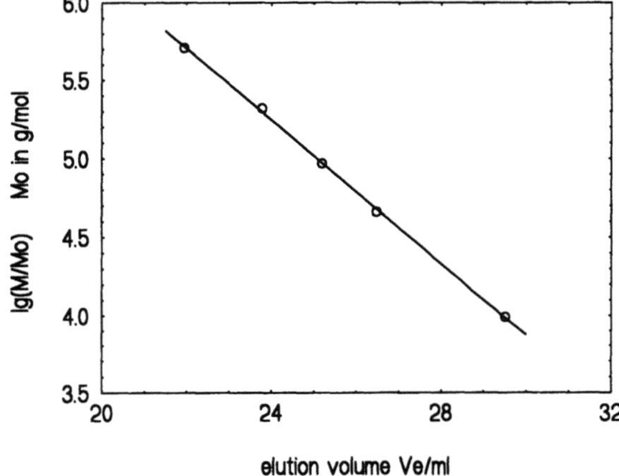

Fig. 2 SEC calibration with five narrow distributed polystyrenes in tetrahydrofurane (see text)

late, and some fractions do not appear at all, rather they disappear in the SEC columns. The SEC equation (2) is only valid for a special concentration, a special flow rate and a special column configuration. If one changes the concentration, the flow rate or the column configuration one has to calibrate the SEC once more.

Sedimentation fractiometry (SF)

Figures 3 and 4 demonstrate that the separation efficiency of an ultracentrifuge is at least as large as that of the SEC. We have fully separated up to four fractions in one solution with the AUC. But in contrast to the SEC, things are more complicated as we have with the SEC only a two-dimensional problem with the variables intensity and elution-volume, and with the AUC we have a three dimensional problem with the variables concentration

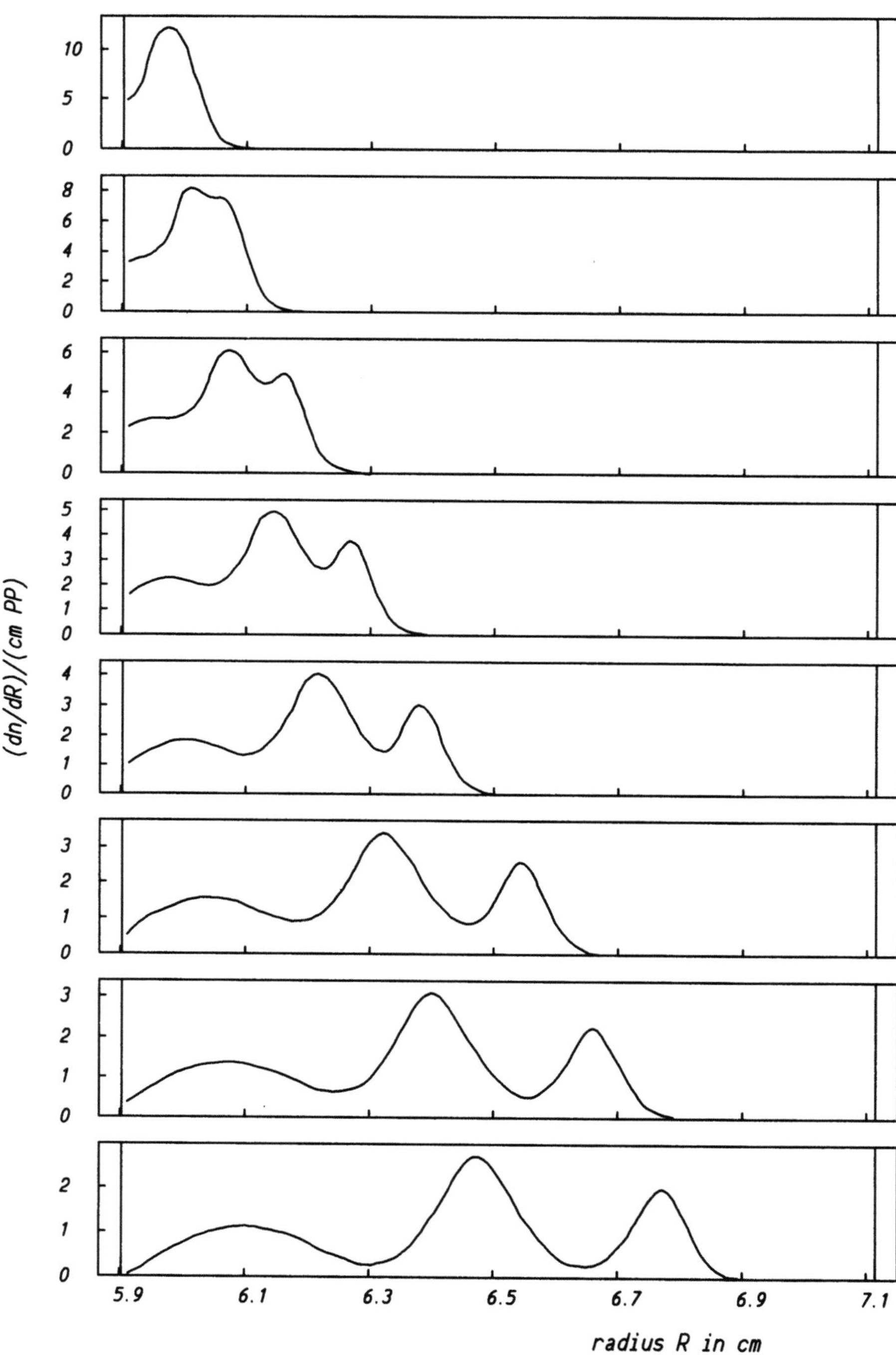

Fig. 3 Schlieren diagram of a mixture of three narrow distributed polystyrenes (M = 20 000; 200 000; 498 000 g/mol) in methylethylketone (C_0 = 3.0 g/l, each component C_0 = 1.0 g/l). t = 17, 23, 32, 42, 52, 67, 78, 88 min

or concentration gradient, distance from center of rotation and time.

To reduce the number of variables and to make measurements with the AUC approximately as easy as with the SEC, we suggest the following procedure for the determination of the molar mass distribution with the analytical ultracentrifuge. The method is somewhat similar to the SEC-procedure and given by the following equations [1–5]

$$S_C = K_C M^{a_C} \tag{4a}$$

$$\log S_C = \log K_C + a_C \log M \tag{4b}$$

$$S_C = [\ln (R^*/R_m)] \bigg/ \left(\int_0^{t'} \omega^2 \, dt - A_r \right) \tag{5}$$

$$g(S_C) = \lim_{1/t \to 0} (1/C_0) R^* \left(\int_0^{t'} \omega^2 \, dt - A_r \right) (R^*/R_m)^2 \, (dC/dR) \tag{6}$$

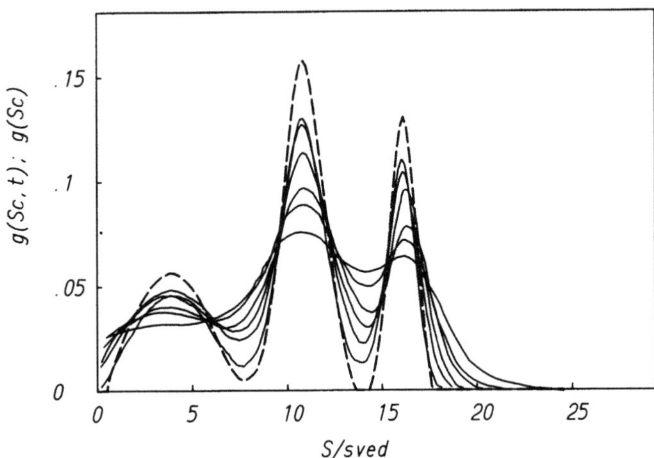

Fig. 4 S-distribution of a mixture of three narrow distributed polystyrenes at different times (solid lines) and extrapolation to $1/t = 0$ min^{-1} (dashed line). See Fig. 3

$$w(M)\,dM = g(S)\,dS \tag{7a}$$

Equations (4a) and (7a) yield

$$w(M) = g(S)\,K_C^{1/a_C}\,a_C\,S_C^{1-1/a_C}\,, \tag{7b}$$

with S_C = sedimentation coefficient at constant concentration; the diffusion effect is extrapolated; the Johnston–Ogston effect and the pressure dependence is included in S_C, M = molar mass, K_C, a_C = constants at constant concentration, R^*, R_m = distance of sedimentation boundary and meniscus from center of rotation, $\int \omega^2\,dt$ = run time integral, A_r = parameter for restricted sedimentation near the meniscus caused by the presence of the air liquid meniscus; surface tension forces at the beginning of the sedimentation, C_0 = initial concentration.

We call this method sedimentation fractiometry (SF). The method is based on the fact that a relationship between sedimentation coefficient S and molar mass M exists not only for infinite dilution but also at distinct concentrations, Eqs. (4a) and (4b). The constants K_c and a_c vary with concentration but are constant at constant concentration. In an arbitrary molar mass distribution it is assumed that Eq. (4a) is valid as long as the sum of the different concentrations of the polymer components $\Sigma C_{0,i}$ remains constant. Equations (5) and (6) give the relationships for calculation of the sedimentation coefficient S and its distribution $g(S)$ from experimental available values. The equations are given in the well known form with the run time integral $\int \omega^2\,dt$ and the parameter of restricted sedimentation near the meniscus A_r. For further explanations see [4, 5].

The diffusion effect has to be taken into account. Among several procedures the extrapolation of $g(S_c)$ to $1/t \to 0$ seems the most reliable one [2]. Another possibil-

ity is to considerate the sedimentation profile as a sum of independent sedimentation and diffusion fluxes and to subtract the diffusion flux from the sedimentation profile according to Lavrenko et al. [6].

The procedure is then as follows:

1) SF calibration with several narrow or one broad distributed polymers of known molar masses and molar mass distributions. The calibration is carried out with only one standard concentration, for example $C_0 = 1$ g/l or $C_0 = 2$ g/l. The S_C-M relationship, Eqs. (4a, b) is then determined in the case of narrow distributed polymers with calibration standards by a plot of $\log S_{\max} = f(M)$ or in the case of one broad distributed polymer by nonlinear regression Eq. (7b).

1.1   Calibration with several narrow distributed polymers

1.1.1 Determination of S_C and A_r from plot of $\ln(R_{\max}/R_m) = f(\int \omega^2\,dt)$ at one concentration according to Eq. (5) with $S_{\max} = S_C$ and $R_{\max} = R^*$; slope $= S_{\max}$; intercept $= S_{\max}A_r$.

1.1.2 Calculation of the constants K_C and a_C from plot of $\log S_{\max} = f(\log M)$, Eqs. (4a, 4b)

1.2   Calibration with one broad distributed polymer.

1.2.1 Determination of A_r from plot of $\ln(R_{\max}/R_m) = f(\int \omega^2\,dt)$ at one concentration according to Eq. (5) with $S_{\max} = S_C$ and $R_{\max} = R^*$; slope $= S_{\max}$; intercept $= S_{\max}A_r$.

1.2.2 Determination of $S_C(t)$ and $g(S_C, t)$ according to Eqs. (5) and (6) from experimental values of R^*, R_m, ω, t, C_0 and dC/dR.

1.2.3 Extrapolation of $g(S_C, t)$ at constant S-values to $1/t = 0$.

1.2.4 Calculation of the constants K_C and a_C according to Eq. (7b) by nonlinear regression.

2) Determination of molar mass distribution

2.1.1 Repetition of step 1.2.1 from calibration with the unknown polymer

2.1.2 Repetition of step 1.2.2 from calibration with the unknown polymer

2.1.3 Repetition of step 1.2.3 from calibration with the unknown polymer

2.1.4 Calculation of $w(M)$ according to Eqs. (7a, 7b) from experimentally determined values of S_C and $g(S_C)$.

Experimental

Materials

Dextran T 70 (Pharmacia Chemicals, Uppsala, Sweden); $M_w = 70\,000$ g/mol; $M_w/M_n = 1.9$

Dextran T 40 (Pharmacia Chemicals, Uppsala, Sweden);
$M_w = 40\,000$ g/mol
Polystyrene NBS 706 (National Bureau of Standards, Washington DC, USA); $M_w = 258\,000{-}288\,000$ g/mol; $M_w/M_n = 1.9{-}2.1$

6 narrow distributed polystyrenes (Pressure Chemical Company, PCC; Pittsburgh, USA):

PS20000	$M = 20\,000$ g/mol	$M_w/M_n < 1.06$
PS90000	$M = 90\,000$ g/mol	$M_w/M_n < 1.04$
PS200000	$M = 200\,000$ g/mol	$M_w/M_n < 1.06$
PS300000	$M = 300\,000$ g/mol	$M_w/M_n < 1.06$
PS500000	$M = 500\,000$ g/mol	$M_w/M_n < 1.2$
PS1800000	$M = 1\,800\,000$ g/mol	$M_w/M_n < 1.3$

The given molar masses M refer to the fact that for narrow distributed polymers $M \approx M_n \approx M_w \approx M_z$. M and M_w/M_n are given by PCC and controlled by us with light scattering measurements, osmotic pressure measurements, sedimentation velocity and sedimentation equilibrium (unpublished results).

Solvents

The solvents used were methylethylketone (MEK, p.a.) and water (bidistilled).

Methods

Beckman analytical ultracentrifuge, model E, 8 hole rotor, Schlieren optics, multiplexer, 30 mm single sector cells. The evaluation of the schlieren photos was done with a comparator (enlargement 10 fold), a digitizer and a calculator PC 486.

Results and discussion

As examples for the demonstration of the SF we choose polystyrene in MEK and dextran in water. In the case of polystyrene/MEK 6 narrow distributed polystyrenes (given in the experimental section) at a concentration of 2 g/l were used for calibration. The constants and operational conditions for this system are: density solvent $\rho_1 = 0.7994$ g/cm^3; partial specific volume $v_2 = 0.910$ cm^3/g; refractive index increment $dn/dC = 0.216$ cm^3/g; speed $= 40\,000$ min^{-1}. Figure 5 demonstrates the calibration curve. Additionally the calibration curves for concentrations of $C_0 = 1.0$ g/l and $C_0 = 0.0$ g/l are given. Evaluation of the calibration curve yields the calibration equation

$$S_C = 0.12 M^{0.37}; \qquad C_0 = 2.0 \text{ g/l}.$$

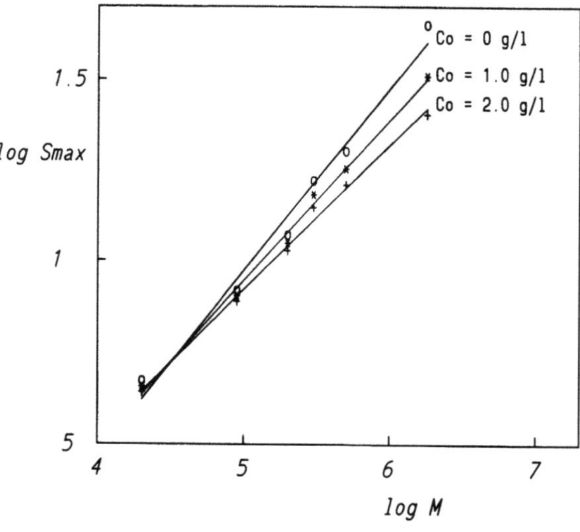

Fig. 5 SF calibration with six narrow distributed polystyrenes in methylethylketone at concentrations $C_0 = 0$ g/l, $C_0 = 1.0$ g/l and $C_0 = 2.0$ g/l. S in sved; M in g/mol (see text)

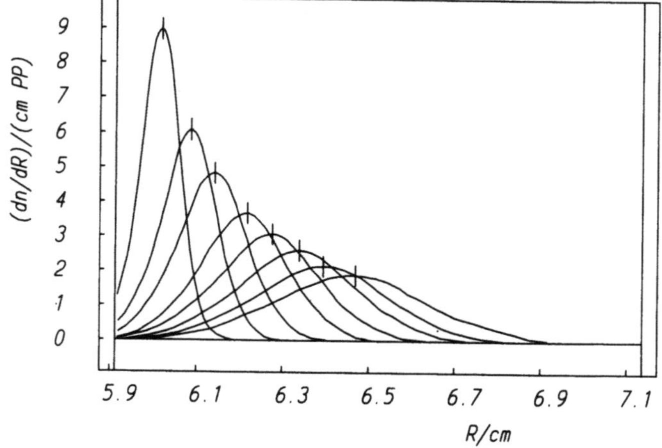

Fig. 6 Schlieren diagram of polystyrene NBS706 in methylethylketone. $t = 19, 28, 35, 45, 53, 62, 70, 79$ min; $C_0 = 2.06$ g/l; $T = 25\,°$C; $N = 40\,000$ min^{-1}

Figures 6 and 9 demonstrate the evaluation procedure of polystyrene NBS 706 in MEK according to steps 2.1.1 to 2.1.4 of section III. The results of the determination of the molar mass distribution by SF is $M_n = 135\,000$ g/mol; $M_w = 264\,000$ g/mol; $M_z = 441\,000$ g/mol; $M_w/M_n = 1.96$. A comparison of these values with those from literature is easy because polystyrene NBS 706 is a well characterized polymer [7a, 7b]. The results of the molar mass distribution of this polymer are in good agreement with values from literature [7a, 7b]. The values given by Budd [7a] are: $M_n = 133\,000$ g/mol; $M_w = 261\,000$ g/mol; $M_z = 512\,000$ g/mol. The values given by Kehrhahn et al. [7b]

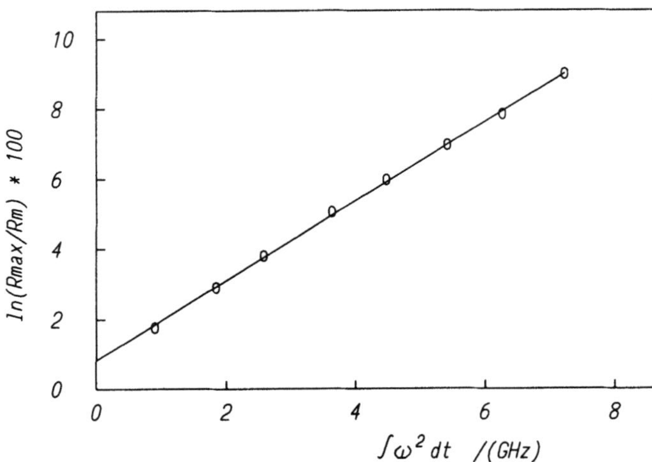

Fig. 7 $\ln(R_{max}/R_m)$ as a function of the run time integral. Polystyrene NBS706 in methylethylketone; $C_0 = 2.06$ g/l; $T = 25\,°C$; $N = 40\,000$ min^{-1}; $A_r = 1,1$ GHz

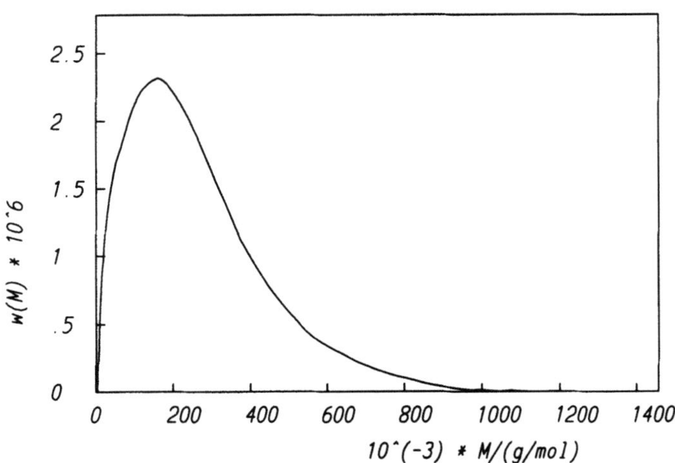

Fig. 9 Molar mass distribution of polystyrene NBS706. $M_n = 135\,000$ g/mol; $M_w = 264\,000$ g/mol; $M_z = 441\,000$ g/mol; $M_w/M_n = 1.96$

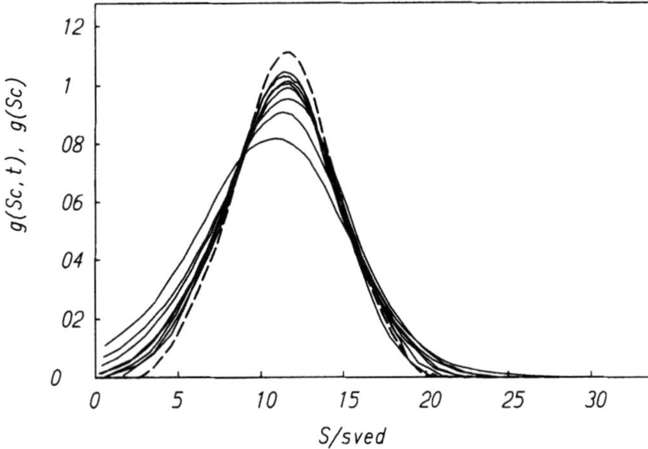

Fig. 8 S-distribution of polystyrene NBS706 in methylethylketone. $t = 19, 28, 35, 45, 53, 62, 70, 79$ min (solid lines) and extrapolation to $1/t = 0$ min^{-1} (dashed line). $C_0 = 2.06$ g/l; $T = 25\,°C$; $N = 40\,000$ min^{-1}

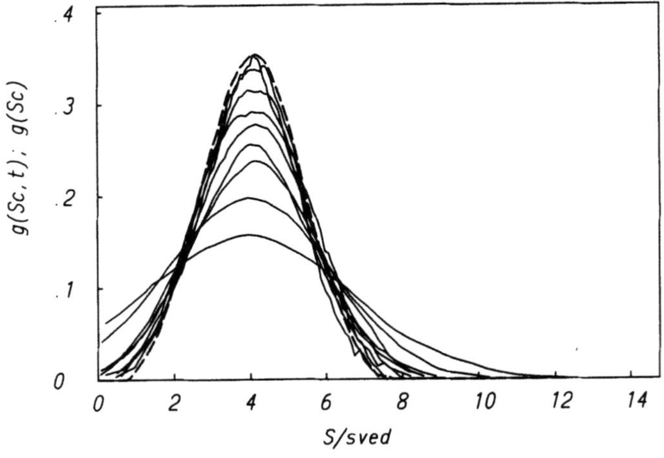

Fig. 10 S-distribution of dextran T 70 in water. $t = 32, 48, 74, 91, 114, 136, 156, 176, 200$ min (solid lines) and extrapolation to $1/t = 0$ min^{-1} (dashed line). $C_0 = 1.01$ g/l; $T = 25\,°C$; $N = 40\,000$ min^{-1}

are: $M_n = 124\,000$ g/mol; $M_w = 265\,000$ g/mol; $M_z = 445\,000$ g/mol.

For the system dextran water, we choose the well known and characterized dextran T 70 at a concentration of 1.0 g/l for calibration [8]. The molar masses and the other constants of the system are given in the literature [8] and controlled by us (unpublished measurements): $M_n = 42\,000$ g/mol; $M_w = 68\,000$ g/mol; $M_w/M_n = 1.62$; density solvent $\rho_1 = 0.997$ g/cm^3; partial specific volume $v_2 = 0.607$ cm^3/g; refractive index increment $dn/dC = 0.146$ cm^3/g. The operational conditions were: speed $N = 40\,000$ min^{-1}; temperature $T = 25\,°C$; wavelength $\lambda = 546$ nm. Figure 10 demonstrates the S-distribution of

dextran T 70 in water. Evaluation of this S-distribution according to step 1.2.1 to 1.2.4, section III yields the calibration equation

$$S_C = 0.0204\,M^{0.48}; \qquad C_0 = 1.0 \text{ g/l} .$$

Figure 11 shows the S-distribution of dextran T 40 in water. The evaluation procedure according to step 2.1.1 to 2.1.4 of section III yields the molar mass distribution of dextran T 40:

$M_n = 18\,000$ g/mol; $\qquad M_w = 39\,000$ g/mol;

$M_z = 60\,000$ g/mol; $\qquad M_w/M_n = 2.16$

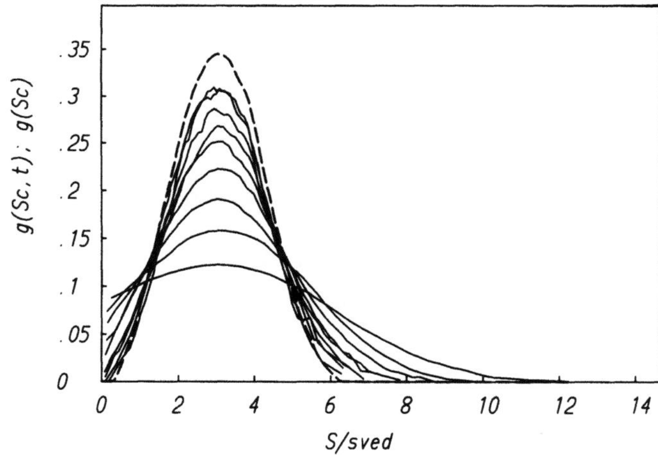

Fig. 11 S-distribution of dextran T 40 in water. $t = 43, 60, 78, 100, 120, 140, 160, 180, 200$ min (solid lines) and extrapolation to $1/t = 0$ min^{-1} (dashed line). $C_0 = 1.04$ g/l; $T = 25\,°C$; $N = 40\,000$ min^{-1}

Regarding the value of $M_w = 40\,000$ g/mol given by Pharmacia and similar reaction conditions concerning the production of dextran T 70 and dextran T 40 this is a reasonable result.

We hope that it is demonstrated by this procedure and the two examples that SF is a competition to SEC concerning the determination of molar mass distribution of polymers.

Acknowledgements The authors are indebted to the Fonds der Chemischen Industrie and to the BASF AG for financial support. The measurements with the AUC were performed by M. Kaiser and H. Roth, Kunststofflaboratorium, BASF AG, Ludwigshafen.

References

1. Blair JE, Williams JW (1964) J Phys Chem 68:161
2. Fujita H (1975) Foundations of Ultracentrifugal Analysis, Wiley, New York
3. Trautman R (1956) J Phys Chem 60:1211
4. a) Stafford WF in Harding SE, Rowe AJ, Horton JC (Eds.), "Analytical Ultracentrifugation in Biochemistry and Polymer Science", p. 359 ff, The Royal Society, Cambridge 1992. b) Stafford WF in Schuster TM, Laue ThM (Eds.), "Modern Analytical Ultracentrifugation", p. 119 ff, Birkhäuser, Boston 1994
5. a) Lechner MD, Mächtle W, this volume b) Mächtle W (1984) Makromol Chem 185:1025 c) Mächtle W (1988) Angew Makromol Chem 162:35
6. Lavrenko PN (1994) Polymer 35:2133
7. a) Budd PM (1988) J Polym Sci B26:1143 b) Kehrhahn J-H, Lechner MD, Mächtle W (1993) Polymer 34:2447
8. Wan PJ, Adams ET (1974) Am Chem Soc Pol Prepr 15:509

Progr Colloid Polym Sci (1995) 99:132–135
© Steinkopff Verlag 1995

P. Beyer
M.D. Lechner

Reaction kinetic and molar mass distribution of the polycation poly[(dimethyleneimino)ethylene-(dimethyleneimino)-methylene-1,4-phenylenemethylenedichloride]

Received: 27 March 1995
Accepted: 6 June 1995

P. Beyer (✉) · Prof. Dr. M.D. Lechner
Physikalische Chemie
Universität Osnabrück
Barbarastraße 7
49069 Osnabrück, Germany

Abstract The polycation Poly[(dimethyleneimino)ethylene-(dimethyleneimino)-methylene-1,4-phenylene-methylenedichloride], Poly(XDC/TED), was synthesized of the monomers Dichloro-p-Xylylene (XDC) and Tetramethyl-ethylenediamine (TED). The molar mass averages M_n, M_w and M_z, and the molar mass distribution (MMD), $w(M)$, were determined by equilibrium-ultra-centrifugation. The resulting MMD was compared with the theoretical expected distribution. By considering the reaction kinetics we found a distribution which predicts the MMD of any Poly(XDC/TED) synthesis.

Key words Polycation – reaction kinetic – equilibrium ultracentrifugation – molar mass distribution

Introduction

Poly(XDC/TED) is a polycation of the ionene-type. Ionene polymers are polyammonium salts with two positive nitrogens per repeating unit in the backbone. They are unbranched and of relatively low molar masses. The degree of polymerization is controllable by well-defined reaction conditions.

Ionenes are industrially used as flocculation and electrostatic agent. Medical studies investigate the use of ionenes for their neurobiological effect, as an anthiheparin agent and in the treatment of cancer.

We synthesized Poly(XDC/TED) to investigate polyelectrolyte complexes between synthetic polyanions and polycations. The formation of water-soluble interpolymer complexes was the aim of our work. The characterization of the polycation, as presented here, is a necessary step for the description of the polyelectrolyte complexes.

Kinetics

The polymerization of Poly(XDC/TED) proceeds by nucleophilic bimolecular substitution (S_N2)[1]. Tsuchida [2] describes the exact mechanism as:

Equimolar amounts of XDC and TED react to the dimer (XDC/TED). The transition state is stabilizised by the solvent Dimethylformamide (DMF) or Dimethyl Sulfoxide (DMSO) and the aromatic ring. The fast start reaction consumes nearly all monomers so that only dimer-dimer- and oligomer-dimer reactions are responsible for a much slower chain growth. The reaction rate constant of the dimer formation, k_{start}, is larger than the combined rate constants, k_{growth}, of the chain growth (Fig. 1).

In this case, we cannot expect the classical Schulz–Flory–MMD [3] for polycondensation-products. The kinetic model of Kumar et al. [4] takes into account the different rate-constants of monomers and higher homologues. They found that the polydispersity deviates from the most probable value of 2; it is less than 2 when the reactivity increases with chain length and larger than 2 when it decreases.

Analytical ultracentrifugation

The ultracentrifuge Spinco Model E has been used for the sedimentation equilibrium measurements. It is equipped

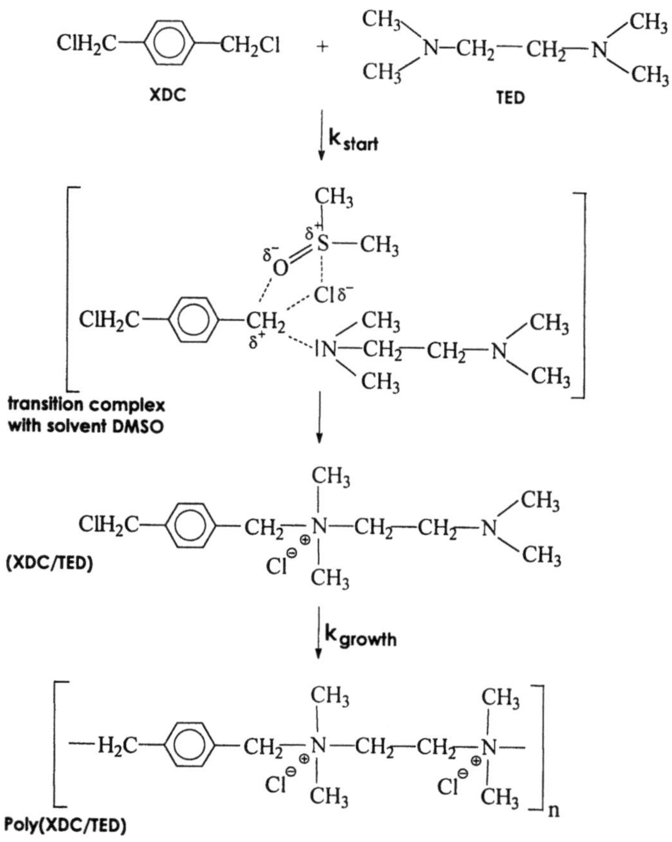

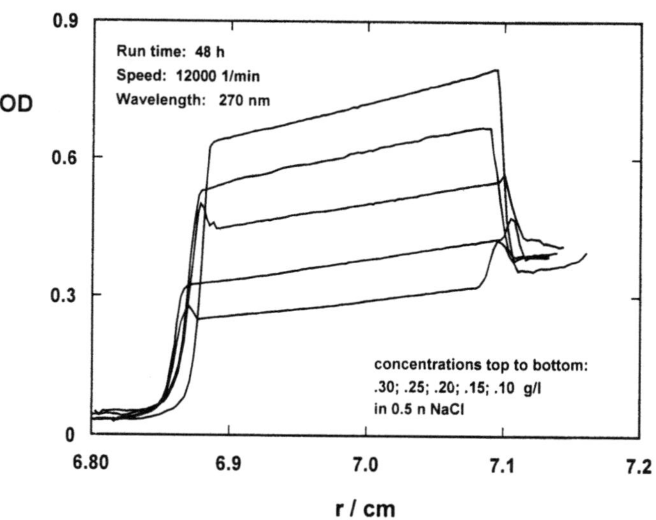

Fig. 2 Equilibrium measurement of synthesis 3

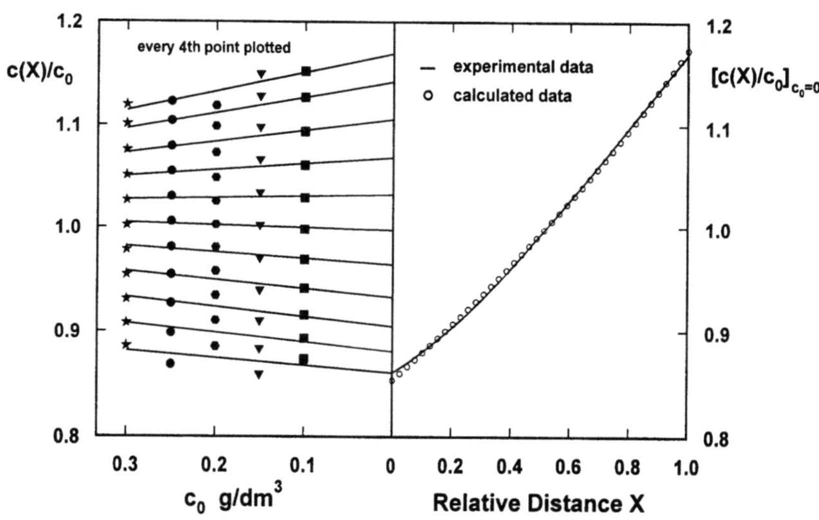

Fig. 1 Mechanism of Poly(XDC/TED) synthesis. The reaction rate constant k_{start} is larger than k_{growth}

with a Flossdorf optic [5], UV/VIS detector, and a multiplexer.

We measured different polymer concentrations (below 0.5 g/dm³) in 0.5 molar NaCl solutions at 25 °C in double-sector cells (12 mm length) with velocities between 12 000 and 20 000 rpm.

The data-aquisition is done by linking the Beckman Dynograph with a PC and digitalization of the plotter dc-output. The evaluation of the x, y-ASCII-files is performed by rearanging the data to the XL-A-file format and calculating the results with programs which were developed for the new analytical ultracentrifuge Optima XL-A (Beckman, Palo Alto, California, USA).

The signal-to-noise ratio is very high, due to the beam focusing with a Flossdorf optic and, if necessary, further increased by averaging of some repeated cell-scans. This method was not applied to the present investigations due to the almost noiseless measurements (Fig. 2).

The procedure for the determination of the number-, weight- and z-average molar masses M_n, M_w, M_z, and the molar mass distribution $w(M)$ is based on the determination of the experimentally measured reduced concentra-

Fig. 3 Left: Extrapolation of $c(X)/c_0$ data from equilibrium run (synthesis 3) to $c_0 = 0$. Right: The extrapolated data $[c(X)/c_0]_{c_0=0}$ plotted versus relative distance X. This concentration profile (—) is compared with the calculated data (○)

tion profile $c(X)/c_0$, extrapolation to zero concentration $c_0 = 0$, and direct calculation of the molar mass distribution by nonlinear regression of the integral equation according to the Simplex procedure [6, 7] (Fig. 3).

The three-parameter function of Hosemann and Schramek [8] is used as MMD.

$$w(M) = a d^{(k+1)/a} \Gamma^{-1}((k+1)/a) M^k e^{(-dM^a)} \tag{1}$$

The parameters d and k are calculated and the parameter a is used to approximate different distribution models (see Table 1).

The P_N-value is used to calculate the density function $H(P)$ of the degree of polymerization P. It holds for the Schulz–Flory–MMD:

$$H(P) = \left(1 - \left(1 - \frac{1}{P_N}\right)\right)^2 P \left(1 - \frac{1}{P_N}\right)^P . \tag{3}$$

The Wesslau-distribution [10] is obtained with the following equation:

$$H(P) = \frac{1}{\sigma(2\pi)^{0.5}} P^{-1} \exp\left(-\frac{(\ln P - \ln P_N)^2}{2\sigma^2}\right) . \tag{4}$$

Calculating the MMD with kinetic data

There are no specific kinetic data available for the start and chain growth reaction of the Poly(XDC/TED) synthesis. Therefore, we used the classical model of polycondensation to estimate the expected molar mass. For condensation reactions which consume the catalyst (XDC), the number average of the degree of polymerization, P_N, can be obtained from the reaction rate constant, k, the start concentration, c_0, and the reaction time t by the following equation:

$$P_N^2 = 1 + 2k c_0^2 t \qquad P_w/P_n = 1.5 . \tag{2}$$

We assume that c_0 is the dimer concentration (XDC/TED). Hence, the rate constant from Tsuchida [9] belongs to the chain growth reaction.

Table 1 Parameter a of Eq. (1) with the corresponding molar mass distributions.

Parameter a	Distribution function
0.1–0.5	Wesslau- and square-root-MMD
1	Schulz–Flory–MMD
2	Poisson–MMD

Results

The P_N-values (Eq. (2)) calculated with kinetic data and the calculated M_w-data have been checked with static light scattering (SLS) and AUC (Table 2). The evaluation of sedimentation equilibrium measurements leads to the

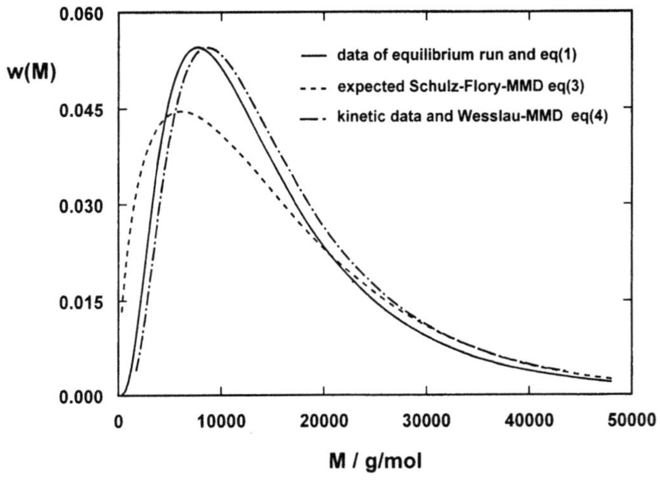

Fig. 4 Molar mass distribution of Poly(XDC/TED) synthesis 3

Table 2 Kinetic data and molar mass of different Poly(XDC/TED) synthesis

Poly(XDC/TED)				Kinetic					SLS[1]	AUC	
	$k/(\frac{mol}{l\,min})$	$c/(mol/l)$	t/h	$\vartheta/°C$	Solvent	P_N	$M_w/(\frac{g}{mol})^2$	$M_w/(\frac{g}{mol})$	$M_w/(\frac{g}{mol})$	$U = M_w/M_n$
Synthesis 1	0.10	0.30	24	40	DMF[3]	5	2200	2600	–	–
Synthesis 2	0.42	0.75	24	40	DMSO[4]	26	11300	12000	12300	3
Synthesis 3	0.19	0.75	96	20	DMSO	34	15200	16000	15000	2

[1] Static light scattering
[2] calculated from P_N
[3] Dimethylformamide
[4] Dimethylsulfoxide

Wesslau–MMD. It is more polydisperse then the expected Schulz–Flory distribution (Eq. (3)).

This corresponds with the described reaction mechanism and the kinetic model of Kumar. Furthermore, we used the logarithmic distribution function (Eq. (4)) together with kinetic data to calculate the theoretical MMD of any Poly(XDC/TED) synthesis (Fig. 4).

Acknowledgement Support of this work by *Deutsche Forschungsgemeinschaft, Fonds der Chemischen Industrie* and Dr. W. Mächtle, BASF AG, is gratefully acknowledged.

References

1. Wang J, Meyer WH, Wegner G (1994) Macromol Chem Phys 195:1777
2. Tsuchida E, Sanada K, Moribe K (1972) Makromol Chem 151:207
3. Schülz GV (1939) Z Phys Chem B43:25
4. Gupta SK, Kumar A, Bhargava A (1979) Polymer 20:305
5. Floßdorf J, Schillig H (1979) Feinwerktechnik & Messtechnik 87:93
6. Lechner MD, Mächtle W (1992) Makromol Chem Rapid Commun 13:555
7. Lechner MD, Mächtle W (1991) Progr Colloid Polym Sci 88:62
8. Hosemann R, Schramek W (1962) J Polym Sci 59:20
9. Tsuchida E, Kaneko M, Nishide H (1972) Makromol Chem 151:221
10. Wesslau H (1956) Makromol Chem 20:129

Progr Colloid Polym Sci (1995) 99:136–143
© Steinkopff Verlag 1995

U. Sedlack
M.D. Lechner

Direct digital data capture for sedimentation velocity experiments using UV/VIS optics

Received: 30 March 1995
Accepted: 5 June 1995

U. Sedlack (✉) · M.D. Lechner
Physikalische Chemie
Universität Osnabrück
Barbarastraße 7
49069 Osnabrück, Germany

Abstract We developed a substitute for the dynograph/multiplexer unit of the Beckman Model E Analytical Ultracentrifuge (AUC) consisting of a high-speed analog-to-digital converter and a personal computer. With this combination, we determine directly the intensity of light pulses measured at the photomultiplier. The computer program is written in the C + + programming language under DOS. It measures peak heights, relates them to the cell numbers they belong to, calculates optical densities from sample and reference sector, and displays the optical densities as a function of radius on the computer screen. Thus, it is possible to measure simultaneously all five concentrations that can be carried by the six-hole rotor of the Beckman Model E AUC during one scan. A multiplexer or flash lamp is no longer necessary. After a short description of the method, some general comments on requirements of sedimentation velocity experiments and on noise problems are made. The hardware and software for direct digitizing is presented and a further enhancement is proposed using a high-speed, two-channel recording system and one additional photomultiplier recording the primary intensity.

Key words Analytical ultracentrifuge – absorption optics – sedimentation velocity experiments – noise problems – direct digital data capture

Introduction

The absorption optical system for the analytical ultracentrifuge (AUC) was originally developed by Svedberg and Pedersen [1]. The design of photoelectric scanners at the Berkeley University of California [2] made this method quite popular in the 1970s and 1980s. Compared with Schlieren and Rayleigh Interference systems which incorporated a time-consuming photographic method of evaluation, the absorption optical method with photoelectric scanners enabled experimenters to directly view the sedimentation process. The principle of the first photoelectric scanners was improved stepwise, yielding enhanced linearity and accuracy of the measured curves that show optical density as a function of radius. The most important of these was the introduction of the "pseudo" double-beam method [3]. It simulated double-beam operation by employing the rotor as a light chopper, causing first one compartment of the cell containing the solvent to appear in the light path and then the other, which was filled with solution. By this means imperfections of the optical system could be recognized as well as fluctuations of the lamp intensity.

A further development concerned the use of rotors that could carry more than one sample. Schachman [2] reported a modification of his scanning unit that was able to distinguish between two different samples. This was accomplished by a special electronic switching circuitry which was triggered by an external source attached to the

Progr Colloid Polym Sci (1995) 99:136–143
© Steinkopff Verlag 1995

rotor, selecting one of the two cells for examination. Cheng and Littlepage [4] applied and modified the principle of this scanning unit for the construction of a photoelectric scanner capable of analyzing six samples in a single run. This scanner can be seen as a prototype for the dynograph/multiplexer unit of the Beckman Spinco Model E which has been the state of the art for more than 20 years.

Recent developments at Beckman Instruments Inc. lead to the Optima XL-A AUC which presents an enhanced design of the UV/VIS optics and a fully computer-controlled operation including the ability to scan the wavelength of light. A four-hole rotor operation is made possible by the use of a flash lamp that illuminates only one cell during a scan. This substitutes the old switching circuitry of the Spinco Model E which has proved to be the source of trouble in many cases (photopickup defects).

Today, the use of pulsed lasers [5], CCD cameras, and digital image processing equipment [6] made Schlieren and interference optical measurements, now applied as well to multi-hole rotors, as fast and reliable as those carried out with UV/VIS optics, without the need of a UV/VIS absorbing sample. But one should always keep in mind the technical effort necessary therefore, and moreover, absorbance optics can achieve the greater sensitivity by principle, yielding in lower sample concentrations necessary for examination.

So, it is still reasonable to improve the technical development of the UV/VIS optical AUC, especially to enhance the signal-to-noise ratio of scans and to speed up multihole rotor operation. One attempt to do so, aiming at the special needs of molar mass distribution (MMD) determination in sedimentation velocity experiments, is the scope of this article. For our work, which is not yet finished, the Spinco Model E with six-hole rotor served as a test base.

General description

Let us consider an UV/VIS optical AUC with a six-hole rotor carrying double-sector cells. Connecting an oscilloscope to the output of the photodetector of the AUC, one will see a train of double peaks (Fig. 1) at the screen of the scope. Zooming into one of those double peaks, one finds the intensity transmitted through the reference sector of the corresponding cell represented in the amplitude of the first peak, and the intensity transmitted through the sample sector as the second. This configuration is according to the split beam spectrophotometer principle that was mentioned earlier. Any electronic equipment that is to be used for recording of optical densities (OD) has to deter-

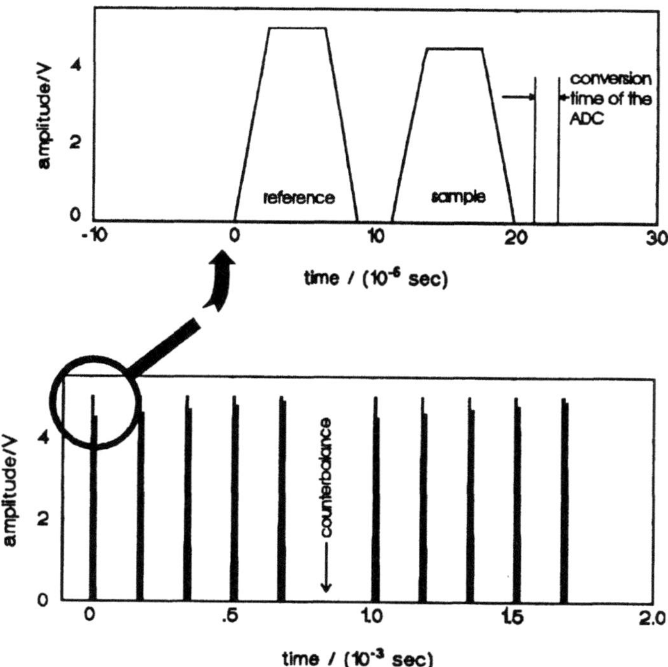

Fig. 1 Typical signal timing that could be measured at the photodetector of an analytical ultracentrifuge with UV/VIS optics under "no-noise conditions" (hypothetical) A rotational speed of 60 000 rpm is assumed. In the zoomed diagram the conversion time of the analog to digital converter ADS-7819, which was used for our experiments, is shown additionally

mine the heights of those peaks, and to relate them to the sectors of the AUC rotor that are responsible for them.

To emphasize the difference between direct and indirect digital data capture, we should first describe equipment that performs indirect digitization: In this case an analog-to-digital converter (ADC) unit is just a substitute for the pen writer which was originally used to record UV/VIS scans, as known from the Beckman Model E AUC. The rest of the original electronic equipment is unchanged. As there are computer programs to evaluate AUC measurements available in almost any laboratory, this avoids the time-consuming and inaccurate translation from paper scan to digital data. We call this "indirect digitization" since the light pulses measured at the photomultiplier are processed in an analog manner, passing analog electronic gates, logarithmic amplifiers and an analog differential amplifier which gives an analog $OD(r)$ curve. The task of the digitizing program is to pick up these preprocessed optical densities. For that purpose, an ADC capable of 10 conversions per second is sufficient.

In contrast to this, direct digitizing connects directly to the photomultiplier output. So, suitable programs do have to handle pulses. Therefore, the corresponding hardware must consist of a transient recorder with a high speed

ADC, typically capable for 10^6 conversions per second (1 MHz sampling rate) at 12-to 14-bit resolution, giving a division of the interesting range of light intensity into 4096 to 16384 quantization steps. Briefly, a data capture program for direct digitizing has to perform three major tasks:

1) Communication with the ADC, i.e., starting of data conversion, reading data values, etc.

2) Sieve peaks out of the raw data. Peak height and peak distance are to be determined, the first for determination of transmitted intensities and the latter for revolution counting.

3) Online charting and storage of data.

Additionally the program has to care for the radial position of the detector. Either it has to control a vertical translation stage, or it must calculate the actual radius by a radius time relation if a non-computer-controlled scanning unit moves the detector with constant velocity. In the latter case, top and bottom positions have to be determined from the observation of vanishing and appearing sector peaks. The advantages of direct digitizing are:

- All sectors that are carried with the AUC rotor can be examined simultaneously. Therefore, no multiplexer or flash lamp is necessary.
- The total scanning time is reduced, i.e., five concentrations can be examined in the same period of time that is needed for one concentration using a multiplexer.
- No restrictions apply to the maximum number of cells in a rotor. If available, rotors with eight and more holes can be handled.
- Analog electronic wiring can be reduced and thereby the noise due to electromagnetic induction.
- Only industry standard or consumer electronics is involved. This results in easier servicing and cheaper equipment.
- The procedure is open for further enhancement: If a faster or more accurate ADC is available, it can substitute the old one with only minor changes in hardware and software.

Direct digitizing for sedimentation velocity experiments

Evaluating sedimentation velocity data, one has to carry out two extrapolations, the time extrapolation $(1/t \to 0)$ and the concentration extrapolation $(C \to 0)$ (see [7] in this volume). The more concentrations measured and the shorter the time between two measurements of one concentration, the more accurate the results will be:

- More data points diminish the effect of noisy data.
- The method of extrapolation, linear or nonlinear, can be

chosen more reliably if there are more data points. An estimation about the type of interaction that causes concentration dependence could be possible, too.

Using conventional recording equipment to distinguish between several concentrations that are carried in a multi-hole rotor, one is confronted with a dilemma: Available methods (multiplexer or flash lamp) can record only one concentration at one scan. So, one has to decide whether a large number of scans for each concentration is carried out for only a few concentrations, or whether a large number of concentrations is used with less data points on the time axis.

Maybe this dilemma does not exist for users of 4-hole rotors since they can examine only four/three concentrations during one experiment, but using an eight or more hole rotor the problem becomes serious. Direct digitizing avoids the problem by recording all concentrations during one scan.

The extension of the digitizing equipment using a second photomulitplier for tracing primary intensity, which was proposed in the preceding section, enhances the sample capacity by decreasing the number of reference sectors to one. Hence, the sectors that are no longer used as a reference can be used for additional concentrations: An eight-hole rotor would be capable of 15 concentrations if there is no counterbalance necessary.

Some considerations on noise

Any experimenter using the AUC wishes to obtain $OD(r)$ curves that are noise free. As experience tells, this is not possible. But, nevertheless, one should control the sources of noise in order to avoid those which are avoidable and thereby improve the signal-to-noise ratio (SNR).

1) Brownian motion of particles that absorb the light gives a contribution to the noise that is found. As this is of random noise type, it can be diminished by overlaying two or more scanned curves when an equilibrium run is carried out. In velocity runs it is possible to interpolate two curves that were taken within a short period of time.

2) If the light source is an arc lamp, xenon or mercury, one has to heed the fluctuation of light intensity. This starts with the selection of a certain brand of a lamp. Recently, some manufacturers began to offer high stability type arc lamps. These lamps are preferable to conventional ones.

Also, it has to be avoided that the following optics focus too narrowly on the light arc. This would cause strong intensity fluctuations if the arc is moving.

Another contribution to noise is the intensity fluctuation inside the light arc. There were several attempts made to incorporate a regulation mechanism into the power supply of the lamp, but all those which are known to the authors do not work properly. Thus, the remaining fluctuations have to be recognized in some way: The first possibility to do this was already mentioned with the split beam method. By this method all contributions to intensity fluctuations are recognized that take longer than the time Δt between the appearance of reference and sample sector. The fluctuations during Δt cannot be taken into account. If this is the desire, one has to monitor the primary intensity with a second photomultiplier.

3) The components of the optical pathway, mirrors and lenses, are subjected to oil and dust which accumulate on their surface during prolonged operation. These deposits cause additional absorbtion or scattering of light leading to fluctuations of intensity if the radius of observation changes. Spatial fluctuations of intensity are considered by the reference sector when the split beam method is applied.

4) A sort of spatial fluctuation that cannot be considered by the split beam method is that due to scratches on cell windows. These scratches may specifically appear at the reference sector or at the sample sector of a cell. The perturbation of the light beam by scratches can be seen as a contribution to the baseline of the resulting $OD(r)$ curve. Thus, they can be corrected by a baseline correction method as it was proposed with the time derivative method [8].

5) Photomultiplier noise contributes to intensity fluctuations as well. External magnetic fields cause photoelectrons and secondary electrons to deviate from their normal trajectories inside the photomultiplier giving a non-constant gain. Thus, the photomultiplier should be mounted inside a special housing that incorporates a shielding layer; most effective is a mu metal layer. Such housings are commercially available. Advanced models of such housings even allow a temperature control which gives the additional advantage of reduction of thermally based noise.

A factor that contributes to photomultiplier noise is the photomultiplier dynode voltage. If a certain limit of dynode voltage is exceeded, the photomultiplier noise increases drastically. This presents a special problem if a regulation mechanism is attached to the photomultiplier power supply that keeps the output voltage for the reference sector constant, for example, at 5 V as with the Model E: If the primary intensity is low due to poor focusing of the optics or due to an old lamp being used, the dynode voltage will be increased to the limit mentioned and thereby a new source for noise arises.

6) Electromagnetic fields do not only disturb the photodetector operation, but also cause induction in the following analog electronic equipment. The recipe against induction is, like in the case of the photomultiplier, an extensive shielding of all analog parts. Additionally the analog electronic pathways should be kept short. Where this is not possible the impedance of this path should be kept low.

7) If a digitizer is used, attention has to be paid to the design of this unit: The analog and the digital side of this unit should be separated from each other as well as possible. A separate analog and digital power supply and ground is mandatory. If this is not taken into account, an interference between the analog and digital parts will be the source for additional noise.

Hardware of the digitizer

A hardware block diagram for a direct digitizing unit is presented in Fig. (2). The photomultiplier produces a voltage over the resistor R. We use the 9257 QB photomultiplier by Thorn-EMI. The voltage over R is amplified by an operational amplifier of type OPA671 by Burr-Brown. In the next step, the signal is filtered by a 50 Hz notch filter. During a first test of the digitizer without the notch filter it could be seen that 50 Hz gives the main contribution to baseline noise. The filter is of a hybrid type, manufactured on customer specifications by PTEK, Mainz.

Behind the notch filter the signal is shifted by an active subtracting unit in order to match the input range of ± 2.5 V of the used ADC. The active part of this offset adjusting circuitry is a TLE2037 operational amplifier from Texas Instruments. The ADC is an ADS7819 by Burr Brown, which is capable of a 800 kHz sample rate at 12-bit resolution. In our case, it is operated at 750 kHz. The ADS7819 encourages a proper design by offering separate analog and digital grounds. Furthermore, it provides a sample and hold circuit on the analog input side and a three-state latch register at the digital output side integrated on the chip, thus minimizing the amount of electronic circuitry needed.

Data conversion is controlled by a timer/buffer board consisting of 32 K-word storage and a timer/counter that is responsible for creating the conversion clock and for successive addressing of the target address inside the storage buffer. A new converted data value is transferred from the ADC's latch register to the actual address of the buffer, thereupon the address is incremented and a new conversion is started. For setting the timer/buffer board status, it has a status register which can hold, among others, information, about whether data shall be written to the buffer from the ADC side or shall be read from the computer side. The status register can be set from the personal computer side via a data port, and thus, by software. Data transfer to the computer is carried out via two other data ports, one for the high and one for the low byte of the 12

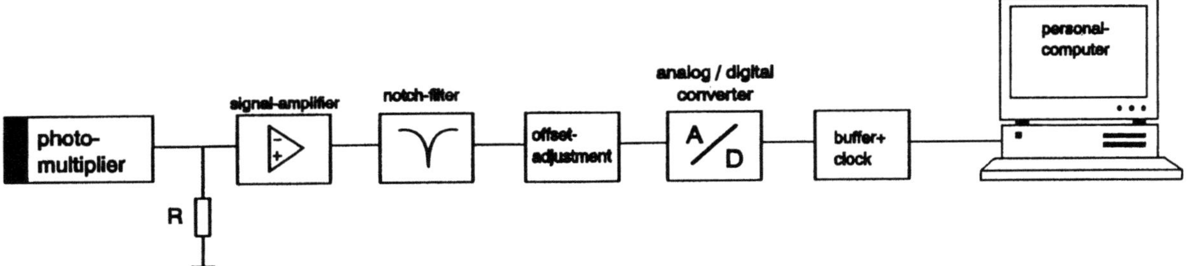

Fig. 2 Block diagram of the hardware for direct digitizing

bits of one data word. When reading data, the data address counter is incremented after addressing the low byte port.

Software

All the software was developed in the C + + programming language under DOS. C + + is a widely spread computer language that is favourable for larger software projects, mainly where modular structured programs are wanted. The operating system DOS is suitable for the project since it is a single tasking system (the digitizing program does not have to share resources with other tasks), 16-bit data are sufficient in our case (ADC data are 12-bits wide), and the amount of data that has to be stored in RAM during one scan is only a few Ks. Moreover, it is quite easy to handle self-made hardware under DOS.

As stated in the introduction, a program for direct digitization has to perform three tasks. The first two of them, communication with the ADC and sieving for pulses, have been implemented in one class of the C + + programming language which we called UZDDD. The corresponding functions (or methods) are named *startconversion*, *startreading*, and *readsingle* for communication, and *sieve* for sieving.

Since the *sieve* function is the heart of the software, it is appropriate to give a sketch of its interior: The function works through a filled digitizer buffer, 32 000 data words that have to be condensed to one peak height for each sector, and this at least 150 to 300 times during one radial scan. Thereby it has to detect the peaks, store their heights and the distance between them. For peak detection two different strategies can be applied: The first one is the treshold method, where a peak is detected by the exceeding of a certain treshold that has been defined by the user. The treshold value has to be selected above the maximum baseline noise level. The second is the gradient method, where a peak is detected by the average slope of three or four data points. Since the gradient method appeared to be too time consuming, we chose the first one.

During the first tests there occurred problems with signal disturbances called spikes, i.e., short-needle-like pulses that fooled the sieve function. So, as a further criterion the width of the peak has to be considered. If the peak is shorter than a certain expected time, it is considered as a spike.

The height of any peak that is detected is added to a store which is referenced by the peak number. Also the distance to the last appearance of this sector and the gap between the actual sector number and the preceding one is stored. By the gap a relation from the apparent sector numbers to the real ones can be made if no triggering source is used, like it is in our case. Then, a counterbalance cell can be used to identify sector number zero. The gap preceding sector zero is twice as wide as the others.

After working through the content of the digitizer buffer, the stores for peak heights, distances and gaps are divided by the number of appearances of the related sector. A condensed structogram of the *sieve* function is presented in Fig. (3).

After sieving the obtained raw data, the main program can access the resulting peak heights by the *data* function, which needs as a parameter the index of the sector for which the transmitted intensity is desired. Also, the actual speed of rotation can be accessed with the *rpm* function or via the *frequency* function.

Online charting and storage of data is managed by a C + + class (module) called *diagram* which was designed in order to minimize the time needed for graphical output during a scan.

The screen image is built up before scanning; during the scan only a few operations are necessary to insert new data points into the diagram windows. *Diagram* is also provided with a *store* method that stores all data points that have been plotted in a diagram. The storing routine was overloaded for our special purpose in order to produce files that are in conformance with the XL-A output format.

Therefore, the main program has to handle the graphics and digitizing modules. In the beginning it reads

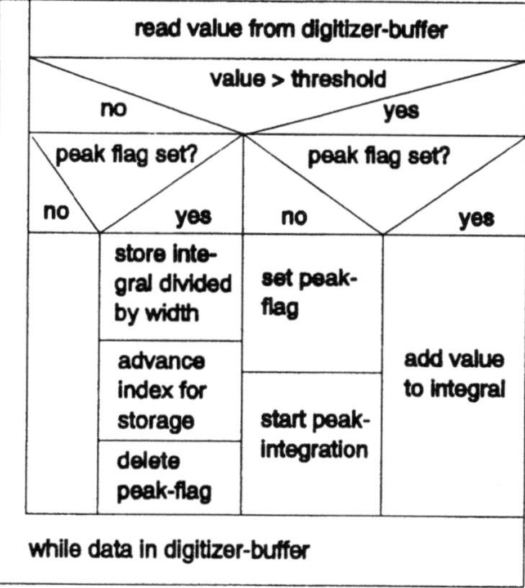

Fig. 3 Nassi–Shneiderman structogram of the key algorithm for peak detection and height determination

general information from a parameter file, such as type of rotor and sample names. Then it initializes the graphics and digitizer hardware. The following measurement loop is repeated until the user presses the ESC-key at the end of a scan. From the beginning of the program to its end the integral $\int \omega^2 dt$ is recorded and displayed on the screen.

First tests and conclusions

The first tests were carried out with a digitizer version that is connected to the sampler input testpoint located at the front of the dynograph of the Model E. So, we use the

preamplifier stage of the dynograph which is followed by the OPA671 inside the new digitizer. This gives the advantage of a still unmodified Model E AUC as a base for testing algorithms and ADC components, while the scanning system and HV power supply is still operated by the dynograph. The disadvantage is that we have the same sources for signal fluctuation as with the original dynograph: The main source of these is the long coaxial cable leading the weak photomultiplier signals to the preamplifier. It functions as well as an antenna for electromagnetic fields that are produced by the electric contactors, relais, and heavy electric motors in its vicinity. Due to the antenna function of this cable, spikes occur in the measured raw data which mix up the *sieve* routine. This can be seen as dropouts of OD values during a scan. The old dynograph of the Model E seems to be quite insensitive to these disturbances.

In the next version of the software, spikes will be recognized by an "expected width of real peaks" criterion. Since this will slow down the operation of the program, the further aim will have to be the avoiding of the spike sources in a planned version of a scanner that is especially designed for the needs of direct digitizing.

In a word, at the present state the quality of the scans is not as good as it is needed for MMD determination. An example of those scans is presented as a screen dump of the program in Fig. (4). It can be seen that the digitizing program is able to distinguish between the different sectors, recording correct scans.

A test of the digitizer with a better shielding of cable and using the new version of software could not be carried out yet, but will be made soon. We hope to be able to present complete sedimentation velocity runs in a subsequent paper.

Fig. 4 Screen dump of the running program. The picture was made during one of the first tests. Due to the limitations of our present test equipment the scans are still very noisy

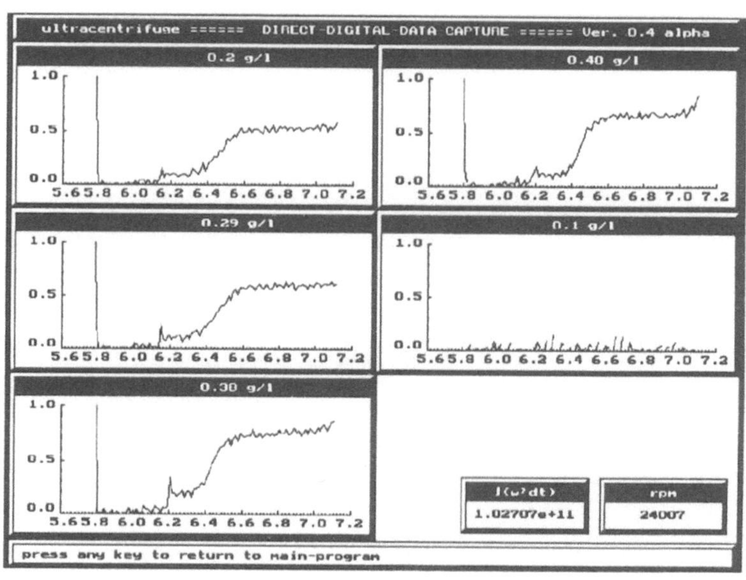

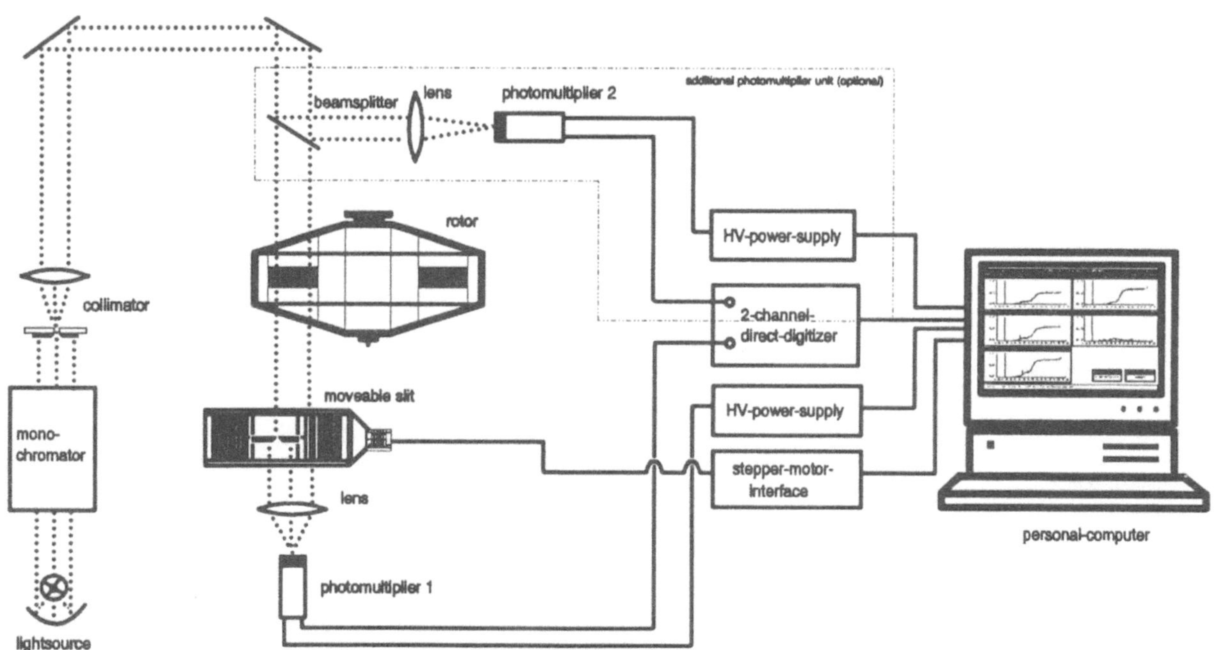

Fig. 5 Block diagram of a digital AUC with UV/VIS optics, including an additional photomultiplier unit for tracing primary intensity

Next steps of development

At the actual state of development, our digitizer is not able to operate without the dynograph unit of the Model E AUC: The radial scanning system and the high voltage (HV) power for the photomultiplier is still supplied by the dynograph. So, the next step has to be the installation of a computer-controlled scanning system and a computer-driven HV supply.

The scanning system will consist of a vertical translation stage which is driven by a five-phase stepper motor. There are units of this type commercially available that can resolve less than 1 μm of positioning accuracy. On this stage a slit will be mounted, the detector being fixed in save distance to the stepper motor. This arrangement is necessary since the stepper motor is operated with rectangular pulses of high current, yielding high electromagnetic induction in its vicinity. On the other hand, mounting the photomultiplier on the translation stage results in noise due to dynode jittering. Behind the translation stage a large lens is positioned that focuses the light that was transmitted through the slit onto the photomultiplier input window. The whole unit, translation stage, lens, and detector, will be mounted on an optical bench.

The HV power supply will be a programmable power module from Thorn-EMI that enables us to control the photomultiplier voltage by the computer.

In the next step we plan to use a two-channel recording system and one additional photomultiplier recording primary intensity in the same space of time as the original multiplier. This will result in greater sample capacity and in lower noise:

- If all samples are dissolved in the same solvent it will be sufficient to fill only one sector with this solvent. The other sectors can be used for additional concentrations of the sample.
- The reference intensity I_0, which is by definition that part of the primary intensity I_0^* of the light source that was transmitted through the pure solvent and the optical components, can be corrected for fluctuation of I_0^* over time. Thus, I_0 at the time where a sample sector is measured can be determined from the intensity measured for the reference cell, I_0', multiplied with the fraction I_{mon}/I_{mon}', where I_{mon}' is the intensity monitored by the second photomultiplier at the time when the reference sector passed the light beam, and I_{mon} is the monitored intensity when the sample is measured.

Progr Colloid Polym Sci (1995) 99:136–143
© Steinkopff Verlag 1995

References

1. Svedberg T, Pedersen KO (1940) The Ultracentrifuge, New York, Oxford University Press
2. Schachman HK, Edelstein SJ (1966) Biochemistry 5:2681–2705
3. Lamers K, Putney F, Steinberg IZ, Schachman HK (1963) Arch Biochem Biophys 103:379ff
4. Cheng PY, Littlepage JL (1966) Anal Biochem 15:211–231
5. Mächtle W, Klodwig U (1979) Makromol Chem 180:2507–2511
6. Laue TM, Anderson AL, Demaine PD (1994) Progr Colloid Polym Sci 94:74–81
7. Lechner MD, Mächtle W, Steinmeier D (1995) this volume
8. Stafford WF (1992) Anal Biochem 203:295–301

Progr Colloid Polym Sci (1995) 99:144–153
© Steinkopff Verlag 1995

W. Mächtle
G. Ley
J. Streib

Studies of microgel formation in aqueous and organic solvents by light scattering and analytical ultracentrifugation

Received: 23 May 1995
Accepted: 16 June 1995

Dr. W. Mächtle (✉) · G. Ley · J. Streib
Kunststofflaboratorium
BASF Aktiengesellschaft
67056 Ludwigshafen, Germany

Abstract Fourteen nearly monodisperse aqueous poly-n-butylmethacrylate (PBMA) dispersions with diameters of about 60 nm were prepared by emulsion polymerization. These PBMA-dispersions contain different amounts of the crosslinker methallylmethacrylate (MAMA), $p_{MAMA} = 0/0, 02/0, 04/0, 06/0, 08/0, 1/0, 2/.../10$ wt.%.

First, these aqueous dispersions were characterized by light scattering (LS) and analytical ultracentrifugation (AUC). Particle size distributions, diffusion coefficients, sedimentation coefficients and particle densities were determined.

Then these PBMA/MAMA-particles were transferred directly into the organic PBMA-solvent tetrahydrofuran (THF). Particles with small amounts of MAMA are completely dissolved, whereas particles with high amounts of MAMA are completely crosslinked and exist as swollen microgel particles with varying volume swelling ratios, q. Particles with medium amounts of MAMA consist of both dissolved macromolecules and crosslinked, extremely swollen microgels. Also by LS and AUC, the molecular masses, diffusion coefficients, sedimentation coefficients and densities of the dissolved macromolecules and the microgel particles were measured. Further, both the wt.%-ratio of dissolved macromolecules/microgel particles and the swelling ratio q of the microgel particles were determined.

The measured quantities are discussed in relation to existing crosslinking theories. The hydrodynamic behavior of microgel particles in the interesting transition region from dissolved macromolecules to crosslinked particles is also discussed.

Key words Microgel – analytical ultracentrifugation – light scattering – particle size distribution – density gradient – swelling degree – crosslinking theory – hydrodynamic behavior – polymer dispersion

Introduction

Microgel dispersions consist of dispersed particles in the diameter range $20 < D < 2000$ nm, in which all linear primary macromolecules within one single particle are cross-linked. That means that every particle is one single giant globular macromolecule (see scheme in Fig. 1). If the dispersion medium is a good solvent for the polymer it will swell.

In recent years there has been an increasing interest in such microgel particles [1]. First, there is a scientific

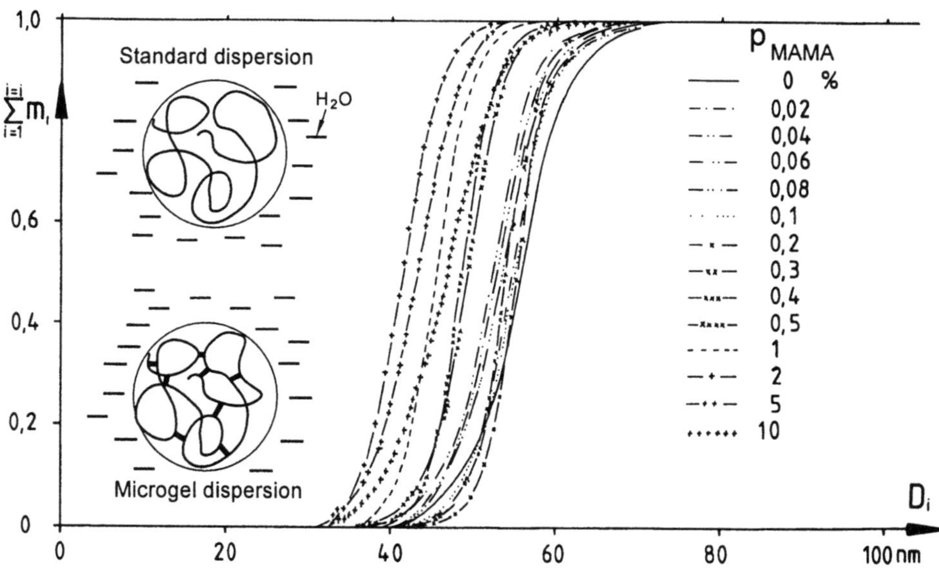

Fig. 1 AUC-particle size distributions of 14 aqueous PBMA-dispersions with different concentrations of the crosslinker MAMA

interest, as microgel dispersions are suitable models for studies of

a) film-forming processes from polymer dispersions [2, 3],

b) the glass transition process in simple, quasiatomic liquids [4], and

c) the ion exchange in polyelectrolytes [5].

Second, there is also an economic interest in microgels, because they are used, for example, as adhesives, lacquers, ion exchangers, drug reduction agents, super absorber and filling material of chromatographic columns.

In this paper we present a systematic study of the formation process of microgel particles in dependence of the crosslinker content, with an emphasis on the first crosslinking steps, in contrast to two earlier AUC-papers [6, 7] where mainly completely crosslinked microgels are analyzed. The first crosslinking steps, that means the continuous transition from uncrosslinked particles with only linear primary macromolecules inside to completely intraparticular crosslinked particles. We also study the volume swelling ratio q of these microgel particles as a function of the crosslinker percentage and their hydrodynamic behavior in this transition region. Another intention of our paper is to demonstrate that the density gradient technique in an AUC, first used by [8], is an excellent tool to study microgels.

Experimental

Materials

With a standard emulsion polymerization procedure, given in ref. [2], we prepared 14 aqueous poly-*n*-butyl-

methacrylate dispersions (abbreviated PBMA), which approximately have the same unimodal, narrow particle size distribution (PSD), presented in Fig. 1. The average diameter D of all 14 samples is nearly the same, about 60 nm, which means that if these particles are completely crosslinked, the corresponding microgels will have a molecular mass of about $M = 70 \cdot 10^6$ g/mol. The solid content of all aqueous dispersions was 30 wt.%.

The difference between these 14 samples is the different amounts of crosslinker added to the BMA during the polymerization. We use methallylmethacrylate (MAMA) as the crosslinker. The 14 different percentages p of MAMA, relative to BMA, varied from 0 up to 10 wt.%. These p_{MAMA}-values are given in Fig. 1. Within this column the microgel character of these particles increases with MAMA content.

Instrumentation

Most of our measurements to characterize the microgel dispersions are done with an analytical ultracentrifuge (AUC), Model E from Beckman, Palo Alto, California, USA, with Schlieren-optics. We used an eight-hole rotor, single-sector cells (thickness 12 or 30 mm) and concentrations in the range of $c = 1 - 8$ g/l. Different AUC-techniques are employed (detailed descriptions can be found in the cited references): PSD-measurements [9], sedimentation runs [10] to fractionate according to size (i.e., to separate dissolved macromolecules from microgels) and density gradient runs [11] to fractionate according to macromolecule or particle density ρ.

Static (SLS) and quasielastic light scattering (QELS) are performed with an ALV-Goniometer SP 86 from

Peters, Langen, FRG, using an argon laser (514 nm) and an ALV-Correlator 3000. These yield (via Zimm-Plots and Stokes–Einstein-relation) the weight average molecular mass M_w, the radius of gyration R_g, the diffusion coefficient $\tilde{D}$ and the hydrodynamic radius R_h.

The auxiliary parameters specific refractive index increment, dn/dc and partial specific volume, $(\bar{v}) = 1/\rho_{PM}$, needed to interpret LS- and AUC-measurements, were measured with a Shimadzu differential refractometer and the density balance Paar DMA 60 + DMA 602 from Anton Paar K.G., Graz, Austria. For all 14 samples we assume the same values, measured on the $p_{MAMA} = 0\%$-sample, $dn/dc = 0.140 \ cm^3/g$ $(\lambda = 546 \ nm)$ and $(\bar{v}) = 0.940 \ cm^3/g$ in H_2O and $dn/dc = 0.077 \ cm^3/g$ $(\lambda = 546 \ nm)$ and $(\bar{v}) = 0.933 \ cm^3/g$ in the organic solvent tetrahydrofuran (THF). This asumption is well fullfilled for the first 11 samples up to 1% MAMA. For the last three samples, 2–10% MAMA, there is a small deviation (see, for example, the density gradients in Figs. 2 and 3), which we neglect.

Also, some membrane osmometer (MO) measurements were performed to estimate the number average molecular mass M_n, using an Acetate 10 000-membrane and a digital-MO from Knauer, Berlin, FRG. Solvent was toluene.

All measurements are done at a temperature of 25 °C.

Results

Figure 2 shows AUC-Schlieren photos of aqueous H_2O/metrizamide density gradients (DG). For the sake of clarity, we show only five of the total 14 gradients. We selected the samples with 0/0, 1/0, 2/5 and 10% MAMA.

Under the lowest Schlieren photo the radial density distribution inside the AUC-cell is indicated as ρ-axis. As expected, all particles in Fig. 2 show a very narrow turbidity band with uniform particle densities of about $\rho = 1.05$–$1.06 \ g/cm^3$. We see a small particle density increase with increasing MAMA-content, because MAMA has a higher density than BMA.

Now, we transfer these dispersed particles from the *aqueous* medium (in Fig. 2) to an *organic* medium. We do that (see Fig. 4) by diluting the original concentrated (30 wt.%) aqueous dispersions by about 1 to 100, with the organic medium THF which is a good solvent for PBMA-molecules and completely miscible with water. This means that all non-crosslinked particles will be dissolved completely in single primary PBMA-macromolecules and all completely crosslinked particles will only swell (the swelling degree q will depend on the degree of crosslinking). In

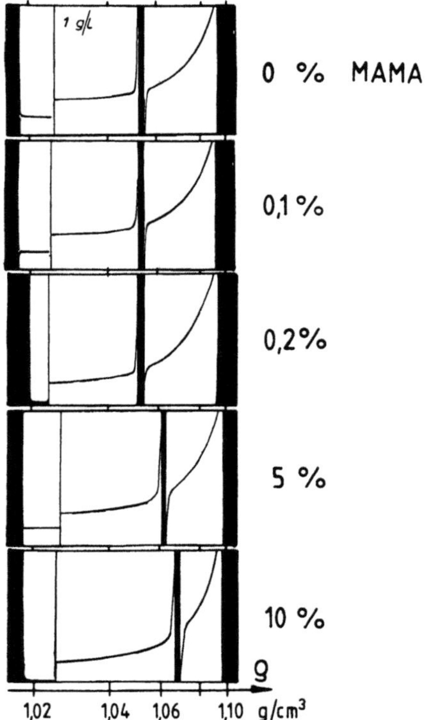

Fig. 2 Schlieren-photos of (aqueous) 90 H_2O/10 metrizamide density gradients in an AUC of five PBMA-microgel dispersions with different concentrations of the crosslinker MAMA

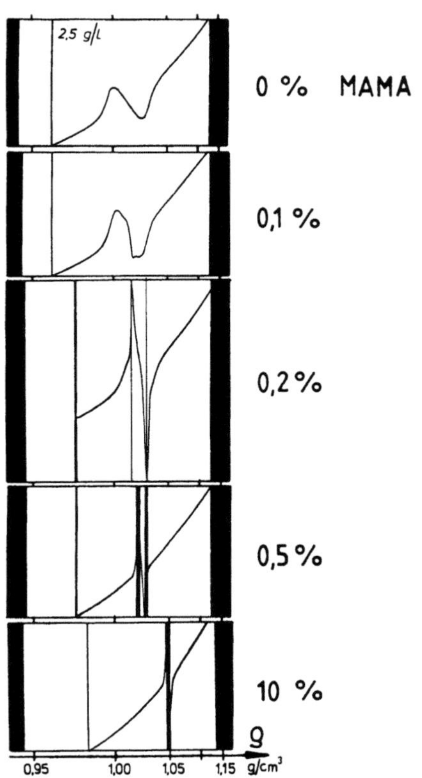

Fig. 3 Schlieren-photos of (organic) 80 tetrahydrofuran/20 diiodomethane density gradients in an AUC of five PBMA-microgel dispersions with different concentrations of the crosslinker MAMA

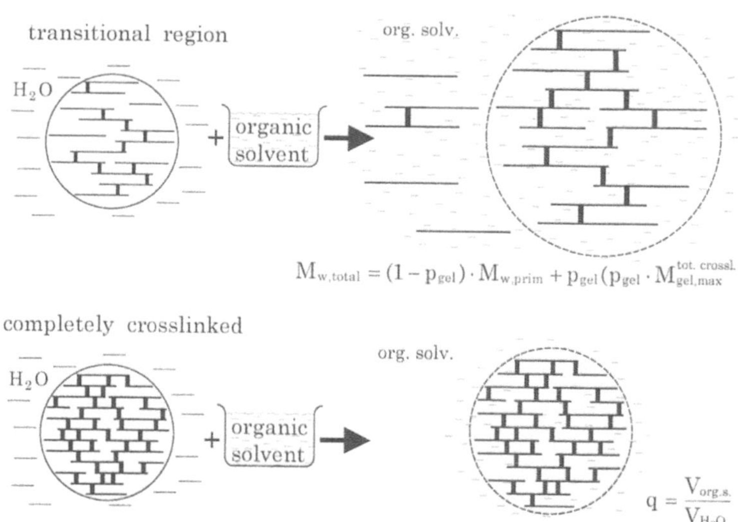

Fig. 4 Scheme of the transfer of microgel-dispersions from the aqueous in an organic solvent medium. Upper part: very weakly crosslinked microgels. Lower part: completely crosslinked microgels

a transitional region (with small amounts of the cross-linker MAMA, upper part of Fig. 4) we will see both highly swollen microgel *particles* (microgel percentage in wt.% is p_{gel}) and dissolved *macromolecules* (macromolecule percentage is $p_{Molec} = 1 - p_{gel}$). Possible macromolecules are monomer, dimer, trimer, a.s.o. of the primary linear PBMA-macromolecules and other branched molecules.

Figure 3 shows the result of this transfer experiment, the Schlieren-photos (again only five of the total 14) of *organic* THF/diiodomethane (DJM) density gradients.

The uppermost Schlieren-photo of the sample 0% MAMA shows only a double Schlieren peak of dissolved primary macro*molecules*, with a molecular mass of about one million (from the broadness of this peak). In contrast to this, sample 10% MAMA (the lowest Schlieren-photo) shows only highly crosslinked microgel particles in the form of a narrow turbidity band. Or, in other words, we see giant globular macromolecules with a molecular mass of only about $70 \cdot 10^6$.

Between these two extremes there is an interesting transitional region. In sample 0.1% MAMA dissolved (linear?) macromolecules dominate. However, in the middle of the double Schlieren peak we see a small deviation which represents the first appearance of about 5% microgel particles. In sample 0.2% MAMA the ratio is already 50:50. First, we see about 50% of highly swollen microgel particles, shown as a transparent gel-band, and second, we see about 50% of dissolved macromolecules shown as a superimposed double Schlieren peak. In sample 0.5% MAMA only medium swollen microgel particles are observed.

Let us now quantify these first qualitative results by means of sedimentation runs of THF-solutions. Figure 5 shows only one *S*-run-Schlieren-photo at a concentra-

tion of $c = 6$ g/l from every sample. In reality, we took about 40 photos of each of the 14 samples at four to six different concentrations (see Fig. 6) and at eight different sedimentation times.

These photos yield two values for each Schlieren peak, first the sedimentation velocity (or coefficient) s, a measure of the molecular mass (only if the shape of all particles is constant what is fulfilled here), and second from the peak area the mass percentage p of the related macromolecules and of the associated microgel particles respectively.

For the samples 0% and 0.1% MAMA in Fig. 5 we see only the slow Schlieren peak of dissolved macromolecules with a sedimentation velocity $s = 4.3$ and 4.8 Svedberg-units (Sved). In sample 0.2% MAMA we see *two* peaks: first a slow peak of dissolved macromolecules with $s = 5.0$ Sved and a mass percentage of $p_{Molec} = 55\%$, and second we see a fast peak of highly swollen microgel particles with $s = 26$ Sved and $p_{gel} = 45\%$. In sample 0.5% MAMA we see a trace of about 5% macromolecules on the left-hand side at the meniscus and in the middle the dominating 95 wt%-peak of swollen microgel particles with $s = 116$ Sved. In sample 10% MAMA we see only the 100%-peak of very slightly swollen, fast microgel particles with $s = 520$ Sved. This increase from 116 to 520 Sved is the result of a drastic decrease in the swelling ratio q of the microgel with increasing MAMA-content only. It is interesting to note that all visible Schlieren peaks of dissolved macromolecules are uniform, small in shape and have nearly the same s-values (s_c (6 g/l) = 4.3–5.0 Sved, see Fig. 5 and $s_0(c \to 0) = 7.7$–9.3 Sved, see Figs. 6 and 7b). That means we see mainly primary linear PBMA-macro-molecules. Dimers, trimers etc. must be rare, if present.

Figure 6 shows the concentration dependence of the s-values for the microgel Schlieren peaks. This plot, $1/s$ as

W. Mächtle et al.
Microgel formation studies

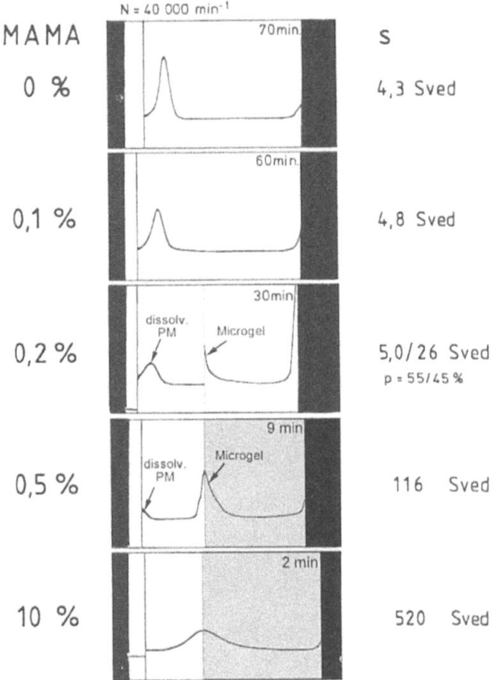

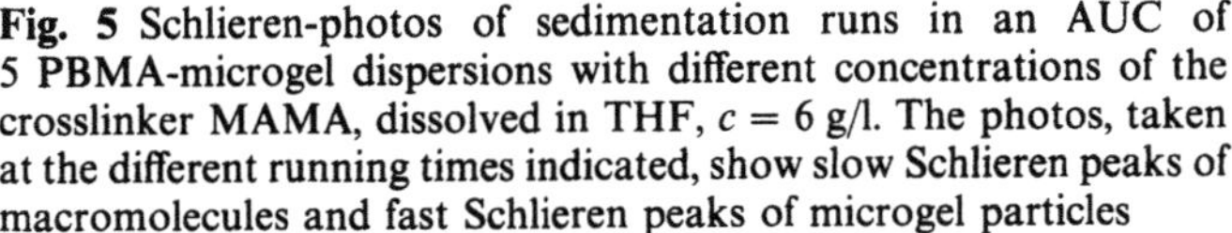

Fig. 5 Schlieren-photos of sedimentation runs in an AUC of 5 PBMA-microgel dispersions with different concentrations of the crosslinker MAMA, dissolved in THF, $c = 6$ g/l. The photos, taken at the different running times indicated, show slow Schlieren peaks of macromolecules and fast Schlieren peaks of microgel particles

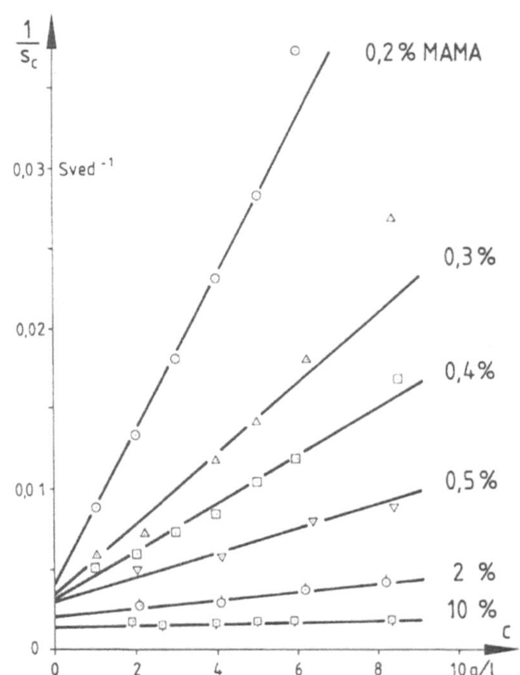

Fig. 6 Plot of the reciprocal sedimentation coefficient s as function of concentration c of the microgel Schlieren peaks of 6 PBMA-microgel dispersions prepared with different concentrations of the crosslinker MAMA, dissolved in THF

a function of c, is used to extrapolate to zero concentration, $c \to 0$, which yields the s_0-values, needed for theoretical calculations as shown in Figs. 7 and 8.

Figure 7 shows a summary of all our single measurements in THF. In Fig. 7a we show the mass percentage of microgel p_{gel}, calculated from the Schlieren peak areas, plotted as function of the crosslinker fraction p_{MAMA}. In Fig. 7b is shown s_0 of the microgel particles as well as of the dissolved macromolecules, plotted also as function of p_{MAMA}. This diagram shows clearly that the crucial transitional region from pure linear macromolecules ($M \approx 10^6$) to pure completely crosslinked microgel particles ($M \approx 70 \cdot 10^6$) occurs within a crosslinker range of $0.1 < p_{MAMA} < 0.5\%$ MAMA. Within this transitional region we find simultaneously *both*, dissolved macromolecules (mainly primary linear PBMA-macromolecules) and dispersed/dissolved microgel particles.

In addition to the two upper AUC-diagrams Fig. 7 shows two LS-diagrams. The first one (Fig. 7c) shows the diffusion coefficient $\tilde{D}_0$ of the total sample (macromolecules plus microgel particles) measured by QELS in THF as a function of p_{MAMA} (for $p_{MAMA} \geq 0.2\%$ MAMA this $\tilde{D}_0$-value is nearly identical with the value of the pure

microgel particles because of their high LS-power). The second LS-diagram (Fig. 7d) shows the weight average molar mass $M_{w,total}$ of the total sample measured via SLS in THF (via Zimm-plots, Fig. 9) as function of p_{MAMA}. This last diagram also indicates a transitional region within $0.1 < p_{MAMA} < 0.6\%$ MAMA by a continuous increase within this range of $M_{w,total}$ from 700 000 up to $\approx 60 \cdot 10^6$ g/mol with increasing p_{MAMA}.

Using the s_0-values of the dissolved (linear) *macromolecules* in Fig. 7b and a scaling relation

$$s_0 = K \cdot M^a = 0.12 \,\text{Sved} \cdot M_{s_0}^{0.31} \tag{1}$$

established in our laboratory for linear n-PBMA-molecules in THF at 25 °C, we calculated their M_{s_0}-values indicated in Fig. 7d. The numerical values, $800\,000 < M_{s_0} < 1.2 \cdot 10^6$ g/mol, are in good agreement with the SLS-value M_w (0% MAMA) = 700 000 g/mol. This again indicates that we see in all samples mainly primary linear PBMA-macromolecules.

Using the s_0-values and the $\tilde{D}_0$-values (for $p_{MAMA} \geq 0.2\%$ MAMA) of the dispersed/dissolved microgel particles in Figs. 7b and 7c and the known Svedberg-

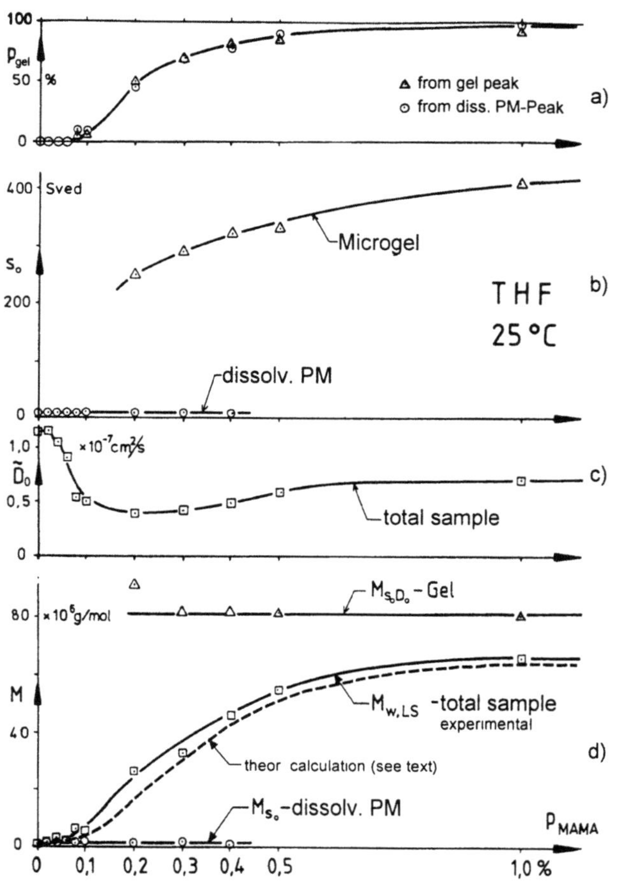

Fig. 7 Plots of different measuring parameters of PBMA-microgel dispersions, dissolved in THF, always as function of the crosslinker concentration p_{MAMA}. The parameters are, 7a: gel percentage inside the original particles p_{gel}, 7b: sedimentation coefficient s_0 of microgels and (primary) macromolecules, 7c: QELS-diffusion coefficient $\tilde{D}_0$ of the total samples (microgel + dissolved macromolecules), 7d: molecular mass $M_{s_0 D_0}$ of the microgels according to the Svedberg-equation, SLS-molecular mass $M_{w,total}$ of the total sample and molecular mass M_{s_0} of the dissolved macromolecules (broken line theoretical calculation, see text).

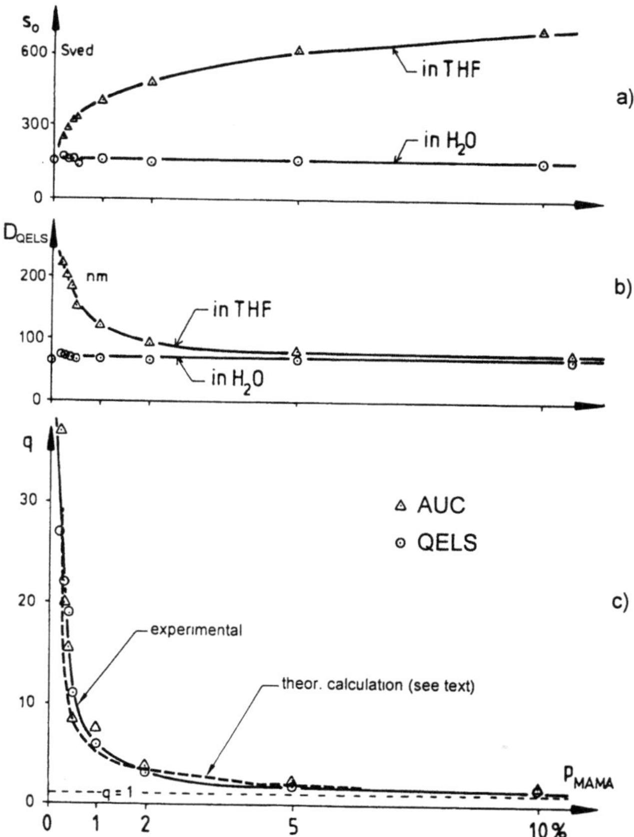

Fig. 8 Plots of different measuring parameters of the microgel parts of PBMA-microgel dispersions, in H_2O and dissolved in THF, as function of the crosslinker concentration p_{MAMA}. The parameters are, 8a: sedimentation coefficient s_0 in THF and H_2O, 8b: hydrodynamic QELS-diameters $D_h = 2 \cdot R_h$ in THF and H_2O, 8c: volume swelling ratios q calculated from the above AUC and QELS data (broken line theoretical calculation, see text).

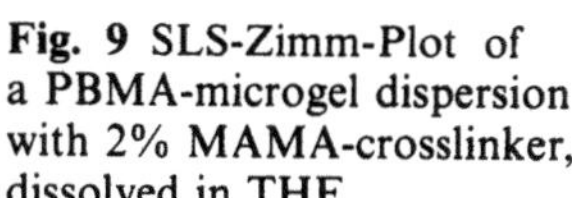

Fig. 9 SLS-Zimm-Plot of a PBMA-microgel dispersion with 2% MAMA-crosslinker, dissolved in THF

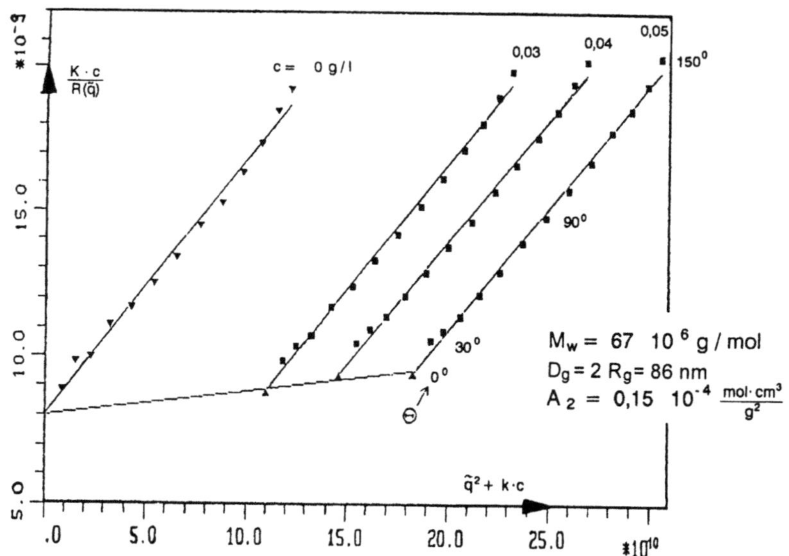

relation

$$M_{s_0 \tilde{D}_0} = \frac{R \cdot T \cdot s_0}{\tilde{D}_0 (1 - (\bar{v}) \cdot \rho_s)}, \tag{2}$$

which is valid for macromolecules/particles of every form (coil, rod, sphere, etc.), we calculated their $M_{s_0 \tilde{D}_0}$-values given in Fig. 7d. All numerical values are about $M_{s_0 \tilde{D}_0} \approx 80 \cdot 10^6$ g/mol, which agree roughly with the theoretically expected value $70 \cdot 10^6$.

Let us now look at the volume swelling ratio q of the microgel particles in THF in Fig. 8, which is analogous to Fig. 7, but the p_{MAMA}-range is increased from 0–1% to 0–10% MAMA.

In Fig. 8a you see not only the s_0-values of the swollen microgel particles in THF. You also see their s_0-values, non-swollen, measured in H_2O. From the ratio $s_{0,THF}/s_{0,H_2O}$ we calculated the volume swelling ratio

$$q = \frac{V(\text{swollen})}{V(\text{non-swollen})} = \frac{V(\text{THF})}{V(H_2O)} = \left(\frac{D(\text{THF})}{D(H_2O)}\right)^3 \tag{3}$$

of the different microgels, indicated in Fig. 8c, in the following manner. We first use Stokes'-law for spherical particles (fulfilled for the investigated microgels)

$$D = \sqrt{\frac{18 \cdot \eta_s \cdot s_0}{\rho_p - \rho_s}}, \tag{4}$$

valid in THF and H_2O, which relates particle diameter D with s_0, viscosity η_s and density ρ_s of the solvent and particle density ρ_p.

Then we postulate volume additivity of polymer and solvent inside a swollen particle, which results in the equation

$$\rho_p(q) = \left[1 - \frac{1}{q}\left(1 - \frac{\rho_p(q=0)}{\rho_s}\right)\right]\rho_s. \tag{5}$$

Introduction of (5) and (4) into the q-definition, Eq. (3), yields

$$q = \left(\frac{\rho_p(q=0) - \rho_{H_2O}}{\rho_p(q=0) - \rho_{THF}} \cdot \frac{\eta_{THF}}{\eta_{H_2O}} \cdot \frac{s_{0,THF}}{s_{0,H_2O}}\right)^{-3}. \tag{6}$$

The q-ratios calculated in this manner are indicated as triangular points in Fig. 8c.

Additionally, q was measured independently with the QELS-method. This was performed, as indicated in Fig. 8b, by measuring the hydrodynamic diameters ($D_h = 2 \cdot R_h$) of the microgel particles separately, swollen in THF and non-swollen in H_2O. The resulting QELS-swelling ratios $q = (D_{h,THF}/D_{h,H_2O})^3$ are indicated in Fig. 8c as circular points.

The agreement between both methods, AUC and QELS, is good. The swelling ratio q varies greatly between 35 for the most weakly crosslinked microgel particles with 0.2% MAMA and 1.5 for the strongest crosslinked particles with 10% MAMA. That means even the 10% MAMA-particles are weakly swollen: THF is a good solvent for PBMA!

Discussion

First crosslinking steps

Figure 7a shows that the critical crosslinker concentration, that is, the minimum concentration for 50% gel is experimentally p_{MAMA} ($p_{gel} = 50\%$) $\approx 0.2\%$ MAMA. A theory of Flory [12, page 384] for *macroscopic* gels roughly allows to estimate this value. According to Flory for this critical crosslinking density of beginning crosslinking of all (linear) primary macromolecules the following relation is valid,

$$p_{\text{crossl.critic}} = \frac{N_{\text{crossl.}}}{N_{\text{mon.}}} = \frac{1}{P-1} \approx \frac{1}{P_w} \approx \frac{M_{\text{mon}}}{M_{w,\text{prim}}}$$

$$= 0.021\% \text{ MAMA}, \tag{7}$$

where N_{crossl} and N_{mon} are the numbers of crosslinker (MAMA) and monomer (BMA) molecules. $P = M/M_{\text{mon}}$ is the polymerization degree of the primary macromolecules. The calculated numerical value 0.021% MAMA follows from our special case, $M_{w,\text{prim}} = M_{w,\text{PBMA}} = 700\,000$ g/mol and $M_{\text{mon}} = M_{\text{BMA}} = 142$ g/mol. This calculated low value seems not unrealistic because of the lower reactivity of the methallylic group in copolymerization and side reactions like degradative transfer and intramolecular loop formation.

Using the simple model of step-by-step crosslinking of linear primary macromolecules, illustrated in the upper half of Fig. 4 it is possible to theoretically calculate the molecular mass $M_{w,\text{total}}$ of our different samples as a function of p_{MAMA}. This can then be correlated to the corresponding experimental curve in Fig. 7d. From Fig. 4 we find both a mass percentage $p_{primM} = 1 - p_{gel}$ of primary macromolecules with $M_{w,\text{prim}} \approx 700\,000$ g/mol (dimer, trimer a.s.o, as already stated, are seldom so that they can be neglected) and a mass percentage p_{gel} of microgel particles with a molecular mass of $M_{w,\text{gel}} = p_{gel} \cdot M_{w,\text{gel,max}}^{\text{tot.crossl.}} \approx p_{gel} \cdot 70 \cdot 10^6$ g/mol. The weight average molecular mass of both is

$$M_{w,\text{total}} = f(p_{gel}) = (1 - p_{gel}) \cdot M_{w,\text{prim}} + p_{gel}(p_{gel} \cdot M_{w,\text{gel}}^{\text{tot.crossl.}})$$

$$\approx M_{w,\text{LS,total}}$$

$$= (1 - p_{gel}) \cdot 700\,000 \text{ g/mol} + p_{gel}^2 \cdot 70 \cdot 10^6 \text{ g/mol}. \tag{8}$$

Using the experimental results of Fig. 7a, the relation $M_{w,\text{total}}$ as a function of p_{MAMA} can be calculated. The broken line in Fig. 7d is the result of this calculation. There is a good agreement between $M_{w,\text{total}}$-values (continuous line) from light scattering and our calculation (broken line). This means that the simple model works surprisingly well.

Swelling ratio q of completely crosslinked microgel particles

According to Flory–Huggins' crosslinking theory of *macroscopic* gels [12, page 576], the following relation between q and the number average molecular mass M_c of the molecular chain between two crosslinking points is valid:

$$q^{5/3} - \frac{q}{2} = \frac{0.5 - \chi}{\rho_p V_s \left(\dfrac{1}{M_c} - \dfrac{2}{M_{n,\text{prim}}} \right)}, \tag{9}$$

where χ is the Flory–Huggins interaction parameter between polymer chain and solvent within the swollen gel, ρ_p the density of the non-swollen polymer ($= 1.064$ g/cm³ in our case for PBMA), V_s the molar volume of the solvent ($= 81.7$ cm³/mol for THF) and $M_{n,\text{prim}}$ the number average molar mass of the primary macromolecules before crosslinking ($= 125\,000$ g/mol, measured via membrane osmometry for sample 0% MAMA). Also according to [12]

$$M_c = \frac{N_{\text{total}} \cdot \bar{M}_{\text{mon}}}{2 N_{\text{crossl.}}} = \frac{\bar{M}_{\text{mon}}}{2 x_{\text{crossl.}}} \approx \frac{M_{\text{BMA}}}{2 p_{\text{MAMA}}} \tag{10}$$

is valid, where $\bar{M}_{\text{mon}}$ is the average molar mass of the two monomers (BMA and MAMA), N_{total} and $N_{\text{crossl.}}$ are the numbers of all monomers and of the crosslinker and $x_{\text{crossl.}}$ is the molar fraction of crosslinker. Because $M_{\text{BMA}} \approx M_{\text{MAMA}}$ is valid, $\bar{M}_{\text{mon}} \approx M_{\text{BMA}} = 142$ g/mol and $x_{\text{crossl.}} \approx p_{\text{crossl.}} = p_{\text{MAMA}}$ (mass fraction) is valid too. Introduction of this M_c-relation (10) into Eq. (9) yields

$$q^{5/3} - \frac{q}{2} = \frac{(0.5 - \chi) M_{\text{BMA}}}{2 \rho_p V_s \left(p_{\text{MAMA}} - \dfrac{M_{\text{BMA}}}{M_{n,\text{prim}}} \right)}. \tag{11}$$

All numerical values in this Eq. (11) are known, with exception of χ for our special BMA/MAMA-microgel system. The best fit between our measuring points and the theoretical line, given by Eq. (11), yields a value of $\chi = 0.44$. This value is physically very reasonable and in accord with [13] and especially with [7], where the swelling ratios q of different microgel systems were determined via measurements of the intrinsic viscosity $[\eta]$. The final q-p_{MAMA}-relationship, calculated with this set of numerical parameters, is shown as a broken line in Fig. 8c in comparison to the experimental AUC and QELS results (con-

tinuous line). The agreement is surprisingly good. In this case it would appear that the Flory–Huggins' crosslinking theory for macroscopic gel is also valid for microgels down to 60 nm. The same result was found in [7] for different microgel systems.

Hydrodynamic behavior, transition from linear coil to "hairy ball" to hard sphere

The 14 samples of Fig. 1, dissolved in the organic solvent THF, are an interesting series for which, with increasing crosslinker content p_{MAMA}, we find a continuous transition from linear macromolecules with only weakly crosslinked microgel to densely crosslinked spherical microgel particles or, in other words, from linear random coils via "hairy balls" to hard spheres, as illustrated in Fig. 10.

It is therefore interesting to ask what measurements are able to distinguish between coils, hairy balls and hard spheres and which provide the possibility to quantify the degree of hairy ball character? We think all parameters are useful which are sensitive to hydrodynamic behavior, i.e., to movements of the dissolved/dispersed macromolecules/microparticles relative to the surrounding sol-

Fig. 10 Scheme of the transition from linear coils (macromolecules) over hairy balls to hard spheres of polymer dispersions with increasing concentration of crosslinker in an organic solvent

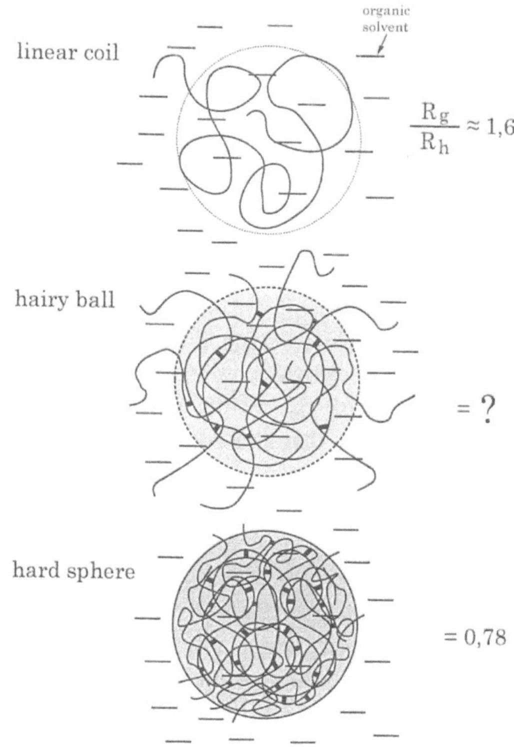

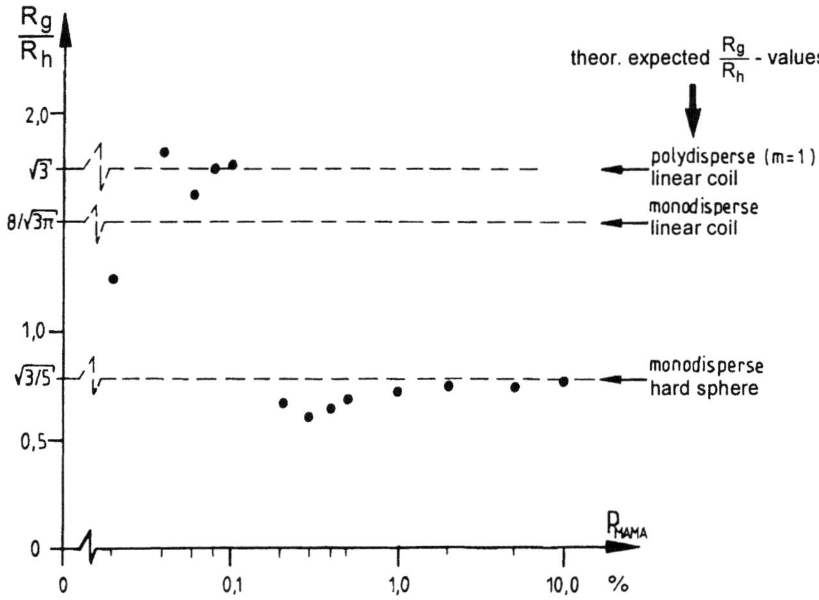

Fig. 11 Plot of the molecular form factor R_g/R_h of different PBMA-dispersions, dissolved in THF, as function of the crosslinker concentration p_{MAMA}

vent. These are diffusion and sedimentation movements. We will discuss both using the AUC results of Fig. 6 and the QELS-measurements of Fig. 11.

QELS and SLS (Fig. 9) also yield the hydrodynamic radius R_h and the radius of gyration R_g of the total sample in THF. The ratio of both, called the molecular form factor $F = R_g/R_h$, is plotted in Fig. 11 as a function of p_{MAMA}. The plot shows a transition from coils to hard spheres with increasing p_{MAMA}.

According to theory [14] and as illustrated in Fig. 10 the numerical values of these ratios are about 1.6 (1.7 for polydisperse and 1.5 for monodisperse coils) for the limiting case of a linear random coil and 0.78 for the limiting hard sphere case. The values for the intermediate "hairy ball" structure, should be between them, depending on the degree of hairy ball character. Figure 11 shows our experimental results, R_g/R_h as function of p_{MAMA}. The horizontal broken lines represent the theoretically expected numerical limiting case values. With increasing crosslinker content there is a clear transition from linear random coils to hard spheres near a MAMA-content around 0.2% MAMA, in agreement with the results in Fig. 7.

Figure 6 demonstrates that the concentration dependence parameter, k_s, of the sedimentation coefficient s, defined by

$$\frac{1}{s} = \frac{1}{s_0} + k_s \cdot c, \tag{12}$$

is another useful parameter, which allows differentiation between linear coils, hairy balls and hard spheres. Only the most crosslinked microgel with 10% MAMA in Fig. 6 shows a hydrodynamic behavior which we expect for an ideal hard sphere: That is, the sedimentation coefficient s is independent on c. All other microgels in Fig. 6, which are not so densely crosslinked, show a concentration dependence of s. This is very systematic: The weaker the crosslinking, the stronger the dependence; in other words, the less crosslinked, the more pronounced is the hairy ball character. The weakest crosslinked microgel with 0.2% MAMA, 35-fold swollen in THF, shows the largest concentration dependence of s, which is nearly the same as that of a linear random coil.

The transition from linear coil to hard sphere is softer, according to the AUC-measurements in Fig. 6, than the light-scattering measurements in Fig. 11. We think for our BMA/MAMA-system the AUC is closer to the reality and the concentration dependence parameter k_s of the sedimentation coefficient is a better measure for the degree of the hairy ball character than the form factor R_g/R_h. The reason for this is that the errors in R_g and R_h are very high in the region $0.1 < p_{MAMA} < 0.5\%$, where both, macromolecules and microgel particles are present.

Conclusions

Our studies demonstrate that light-scattering techniques (SLS and QELS) and especially the different AUC-methods, which are *fractionating* methods, yield a good characterization and a deep insight into the molecular structure of colloidal microgel systems. With increasing crosslinker content the first steps of crosslinking inside the latex particles during emulsion polymerization are well described by a simple model of step-by-step crosslinking of primary linear macromolecules. Flory–Huggins' crosslinking theory for macroscopic gels, describing the swelling ratio q as a function of crosslinker content, seems to be valid for microgels down to 60 nm diameter. The concentration dependence parameter k_s of the sedimentation coefficient s is a good measure to quantify the transition from linear random coil via hairy ball to hard sphere. In order to interpret correctly all kinds of scattering measurements (LS, SANS, SAXS) of weakly crosslinked microgel particles in organic solvents it is important to consider, that most of such systems simultaneously contain both dissolved (linear) macromolecules and highly swollen microgel particles. For improving crosslinking theories of very weakly crosslinked gels investigations on microgels should be very helpful, as they are experimentally easier to handle than macroscopic gels.

References

1. Antonietti A, Bremser W, Schmidt M (1990) Macromol 23:3796–3805
2. Hahn K, Ley G, Schuller H, Oberthür R (1986) Colloid Polym Sci 264:1092–1096 ibid. (1988) 266:631–639
3. Zosel A, Ley G (1993) Macromol 26:2222–2227
4. Bartsch E, Frenz V, Möller S, Sillescu H (1993) Physica A 201:363–371
5. Mächtle W, Ley G, Rieger J (1995) Colloid Polym Sci 273:708–716
6. Shashoua VE, Van Holde KE (1958) J Polym Sci 28:395–411
7. Shashoua VE, Beaman RG (1958) J Polym Sci 33:101–117
8. Buchdahl R, Ende HA, Peebles LH (1963) J Polym Sci, Part C, 1:143–152
9. Mächtle W (1992) Ang Makromol Chem 162:35–52
10. Mächtle W (1992) in Harding SE (Ed), Analytical Ultracentrifugation in Biochemistry and Polymer Science, Royal Society of Chemistry, Cambridge, England, Chap. 10
11. Mächtle W (1984) Colloid Polym Sci 262:270–282
12. Flory PJ (1953) Principles of Polymer Chemistry, Cornell Univ Press, Ithaka, New York, pp 384, 576
13. Hoffman M (1974) Makromol Chem 175:613–639
14. Burchard W (1983) Advances in Polymer Science 48, Light scattering from Polymers, Springer, Berlin

Progr Colloid Polym Sci (1995) 99:154–161
© Steinkopff Verlag 1995

Formation of reversible concentration gradients during the centrifugation of gels

H. Hinsken
E. Selic
W. Borchard

Received: 27 March 1995
Accepted: 3 July 1995

Dedicated to Prof. Dr. Joachim Klein on the occasion of his 60th birthday

H. Hinsken · E. Selic
Prof. Dr. W. Borchard (✉)
Angewandte Physikalische Chemie
Gerhard-Mercator-Universität-
Gesamthochschule Duisburg
47048 Duisburg, Germany

Abstract A new method for the investigation of gels in the Analytical Ultracentrifuge is proposed and used. During the measurements only small changes in the concentration profiles of the gel phases occur. Therefore, a differential relation can be used for the evaluation of the Schlieren patterns. Real equilibria could be shown by the path independence of the obtained gradients.

Key words Gels – Schlieren patterns – continuous equilibria – gradient method

Introduction

The application of gels is gaining increasing importance in industry as well as in private households. There is a special interest in the investigation of water-swellable polymers because their application is widespread.

The elasticity of a gel, which can be considered as an elastic mixture, is achieved by a crosslinking of the polymer component. For our purpose, it is important that the crosslinks are stable during the measurements.

There exist different thermodynamic methods for the characterisation of gels, some of them are mentioned here, where equilibria of swelling [1–7], swelling pressure [8–15], deswelling of a maximum swollen gel [16–18], vapor pressure [19] and of sedimentation-diffusion [20–22] or elastic moduli [23–26] have been determined. By investigating gels with the AUC a combination of these methods has been used [22, 27–32].

Under isothermal conditions a gel in the centrifugal field of the AUC is a continuous system which is determined by the gradient of the potential of this field. The action of the external field leads to a radial concentration gradient inside the gel phase by sedimentation, deformation and diffusion. Near the meniscus gel/vapor the con-centration of the crosslinked polymer will be lowered compared with the initial concentration, whereas the polymer concentration near the cell bottom must increase due to the mass balance.

In general, there are two cases to be distinguished [27]. In the first case the effective centrifugal field is kept low by applying low rotational speeds. The polymer concentration ρ_2 at the meniscus gel/vapor is lower than the initial value $\rho_{2,t=0}$ at the time $t = 0$, but still higher than the concentration in the maximum swollen gel $\rho_{2,s}$. No movement of the gel meniscus can be observed. In case of incompressible gels its movement is not expected if there is no volume change on mixing in the system. This is valid as long as the partial specific volumes of the components are constants in the considered concentration range.

In the second case a movement of the gel meniscus is forced by the application of relatively high centrifugal fields. The polymer concentration at the phase boundary gel/vapor cannot become lower than $\rho_{2,s}$. As soon as this concentration is reached, solvent will be excluded from the gel and the appearing meniscus gel/solvent begins to move in the direction of the cell bottom due to the condition of the mass conservation of the crosslinked polymer in the gel phase.

Whereas the second case was investigated in our group during recent years, assuming the polymer density in the gel phase to be a linear function of the radial distance r, the first case described above was only predicted theoretically. It has been shown for the first time that it is possible to obtain a radial concentration gradient inside the gel phase without a perceptible movement of the meniscus gel/vapor by means of an optical detection system. For this case a more appropriate type of evaluation, which is a differential method, is given. With the knowledge of the refractive index increment $dn/d\rho_2$ it is possible to determine the polymer concentration dependent on the distance to the axis of rotation as in the case of a polymer solution by integration of the measurable Schlieren curve. Therefore, it has to be required that the concentration gradient can be determined along the gel phase completely. In earlier investigations this was not always possible because of the turbidity of the higher concentrated gels. But meanwhile more powerful components are used in the optical system of the AUC so that these problems do not occur anymore [30, 33].

Theory

The deformation–diffusion equilibrium

McBain and Stuewer were the first to observe, in 1935, that the behavior of a gel in the gravitational field was quite different from that of a solution [34]. Svedberg investigated gels in the ultracentrifuge in 1940 [35]. The sedimentation of a gel was defined as a movement of the meniscus gel/solvent with the respect to the laboratory system. This definition was also used by other groups later [36]. But it has to be stressed that a concentration gradient may be formed inside the gel although a movement of the meniscus of the gel does not yet take place.

A schematic representation of a gel in a sector-shaped cell of the ultracentrifuge is given in Fig. 1 [27]. In the upper part a) the case is demonstrated where a gel is run in the cell at lower rotational speed then a critical one, ω_{crit}, which is necessary for the sedimentation of the meniscus gel/vapor. In the lower part b) the gel meniscus $r_m^{g/v}$ has shifted to higher r-values, where the meniscus solvent/vapor $r_m^{s/v}$ appears.

It was predicted a few years ago that it is possible to obtain changes in the concentration profile of the gels either with an exclusion of solvent at the meniscus of the gel but as well without the exclusion of solvent. This can be explained by the schematic representation of concentration profiles within the gel phase at different values of the angular velocity ω_i in Fig. 2.

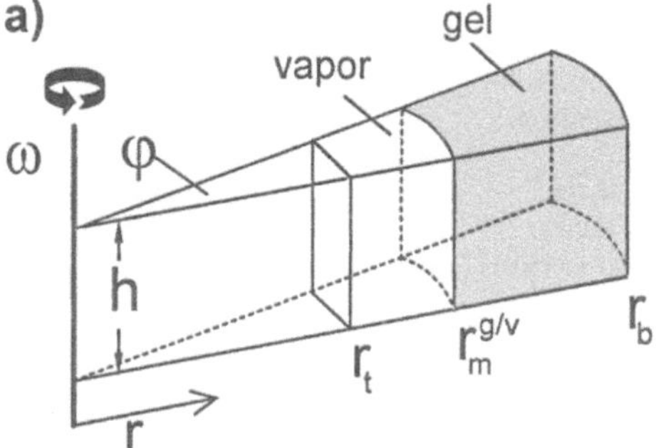

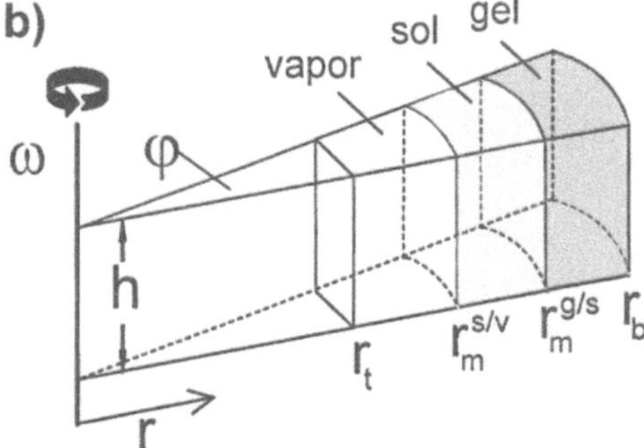

Fig. 1 Schematic representation of a gel in a sector-shaped cell of an Analytical Ultracentrifuge: a) $\omega < \omega_{\mathrm{crit}}$, b) $\omega > \omega_{\mathrm{crit}}$ [27]. r = radial distance from the axis of rotation, ω = angular velocity, φ = sector angle of the centerpiece, h = height of the centerpiece, indices t = top of the cell, m = meniscus, b = bottom of the cell. Phase boundaries: g/v = gel/vapor, s/v = solution/vapor, g/s = gel/solution; see text

At the beginning the concentration of the polymer is constant throughout the gel between the bottom of the cell and the meniscus gel/vapor.

Thermoreversible gels which are formed by physical crosslinking reaction of the polymer, like a coil-helix transition with subsequent aggregation of the helices, may correspond to a state of swelling which is not identical to the maximum swollen state. This means the polymer concentration in the gel ρ_2 is higher than in the maximum swollen gel given by $\rho_{2,\mathrm{s}}$.

In the centrifugal field the osmotically active pressure difference will cause a concentration gradient inside the gel phase, as is demonstrated for ω_1. If ω is increased the equilibrium concentration at the bottom of the cell increases further and that at the boundary gel/vapor decreases.

As the concentration at the phase boundary cannot drop below $\rho_{2,\mathrm{s}}$, the meniscus has to move towards the bottom for rotational speeds $\omega > \omega_2$. ω_2 is the limiting case with the concentration of the maximum swollen gel at the meniscus. For higher speeds the gel boundary has to move from the original position $r_m^{g/v}$ to $r_m^{g/s}$, which means that the solvent or sometimes a solution is excluded.

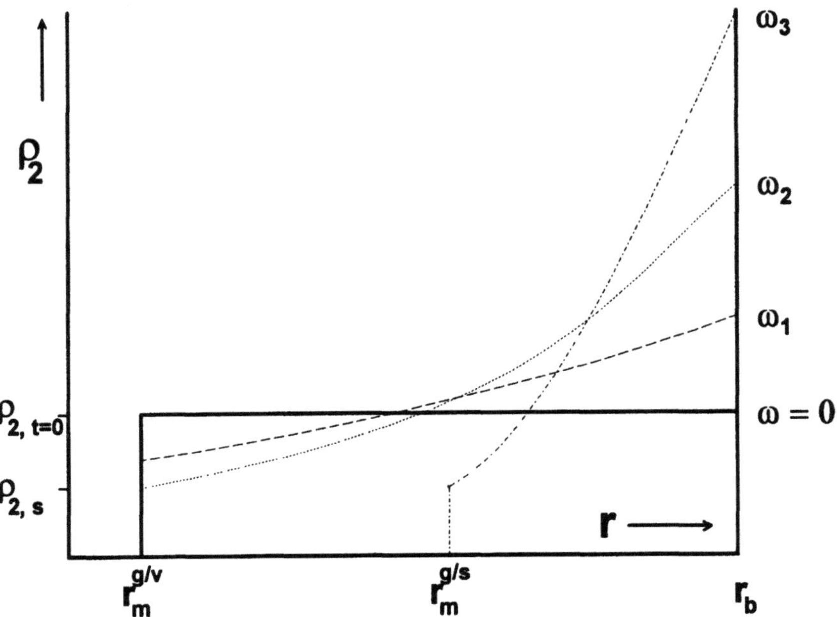

Fig. 2 Schematic representation of concentration profiles inside the gel phase for different rotational speeds ω_i; $(\omega_1 < \omega_2 < \omega_3)$ [27]. ρ_2 = partial density of the polymer, indices: t = time, s = saturation value at swelling equilibrium

In our earlier investigations, forcing a movement of the gel surface by the application of high centrifugal fields, the concentration of the network at every point of the gel was calculated by an approximation because it was sometimes difficult to determine the gradients in turbid systems at higher polymer concentrations. During the deformation of an isotropic gel under isothermal conditions a continuous thermodynamic equilibrium was presupposed and proved [22, 27–32]. Based on the cell geometry and the mass balance, a linear concentration gradient of the crosslinked polymer in the gel phase was assumed with the concentration of the maximum swollen gel at the meniscus gel solvent (corresponding the curve for ω_3 in Fig. 2) [28, 29].

The generalized Svedberg–Pedersen equation for the calculation of the swelling pressure π_s was derived and used for the evaluation of the experimental data

$$\pi_s = \omega^2 \int_{r=r_m^{g/s}}^{r} \frac{\rho_2}{\rho_1 \tilde{V}_1} (1 - \tilde{V}_2 \rho) r \, dr \; , \tag{1}$$

where ρ is the density of the gel and ρ_1 the partial density of the solvent inside the gel both depending on r. This equation is called generalized because it is valid generally under the circumstances mentioned, not only, for example, for highly swollen gels [35].

Whereas this case was investigated in our group during recent years and described in detail in several papers [27–32], we now want to take into consideration the first case, namely the appearance of a concentration profile inside the gel phase without exclusion of the solvent corresponding the curves for ω_1 and ω_2 in Fig. 2.

The most important advantage is that the deformations which occur in the gel phase are very small ($\lambda \approx 1$),

whereas the deformations in our former measurements were much higher ($\lambda \ll 1$).

Without a movement of the gel surface, it is possible that we do not yet have the maximum swollen gel at the meniscus gel/vapor. But this is not a necessary assumption anymore because we want to calculate the concentration – as in case of polymer solutions – with the knowledge of the refractive index increment $dn/d\rho_2$ by integration of the Schlieren curve without additional assumptions concerning the concentration gradient.

For a given gel we only have small changes around the initial concentration during the measurement. The concentration near the cell bottom increases, whereas the concentration near the meniscus decreases slightly, but is still higher than in the maximum swollen gel, see ω_1 in Fig. 2. According to the mass conservation of the crosslinked polymer, the total concentration inside the gel cannot change if volume changes on mixing are disregarded.

Considering these small changes in the concentration range, we now use a differential method for the construction of the swelling pressure-concentration curve.

The differential of the swelling pressure π_s to the distance of the axis of rotation r can be calculated by means of the generalized Svedberg–Pedersen equation, Eq. (2), in the differential form

$$\frac{d\pi_s}{dr} = \omega^2 \frac{\rho_2}{\rho_1 \tilde{V}_1} (1 - \tilde{V}_2 \rho) r \; . \tag{2}$$

This relation can be used for the calculation of the differential coefficient of the chemical potential of the solvent with respect to r, namely $d\tilde{\mu}_1/dr$. The swelling pressure can be considered as an osmotically active pressure where the gel is acting as a semipermeable membrane. Therefore the following relation is valid

$$\frac{d\tilde{\mu}_1}{dr} = -\frac{d(\pi_s \tilde{V}_1)}{dr} . \tag{3a}$$

In many systems the partial specific volume of the solvent can approximately be considered as being independent of the distance to the axis of rotation so that we have for an incompressible gel

$$\frac{d\tilde{\mu}_1}{dr} \approx -\tilde{V}_1 \left(\frac{d\pi_s}{dr}\right) . \tag{3b}$$

The differential change of the chemical potential of the solvent with the polymer concentration $d\tilde{\mu}_1/d\rho_2$ is expressed by the following relationship

$$\frac{d\tilde{\mu}_1}{d\rho_2} = \frac{d\tilde{\mu}_1}{dr} \bigg/ \frac{d\rho_2}{dr} . \tag{4}$$

We know $d\tilde{\mu}_1/dr$ from the measurements combining Eqs. (2) and (3). From the Schlieren pattern the gradient of the refractive index dn/dr is deduced. With the refractive index increment for a given system $dn/d\rho_2$ the differential coefficient of the partial density of the polymer with respect to the distance to the axis of rotation $d\rho_2/dr$ is obtained from the following relationship

$$\frac{d\rho_2}{dr} = \frac{dn}{dr} \bigg/ \frac{dn}{d\rho_2} . \tag{5}$$

Instead of the differential form of the generalized Svedberg–Pedersen relation, the integral form reads

$$-\Delta\tilde{\mu}_1 = \omega^2 \int_{r_1}^{r_2} \frac{\rho_2}{\rho_1} (1 - \tilde{V}_2 \rho) r \, dr , \tag{6}$$

which is essentially Eq. (1) but with different integration limits. Equations (2) and (6) are advantageous in describing $\Delta\tilde{\mu}_1$ at a certain concentration or a very narrow concentration range. Therefore the lowest limit of $\Delta\tilde{\mu}_1 = 0$ does not have to be known. This corresponds to the swelling equilibrium. The advantage for such a small range of concentration is that the partial specific volumes are independent of the density, $\tilde{V}_i \neq f(\rho_2)$. This means furthermore that the assumption that no anisotropy of the gels occurs is fulfilled here. It is therefore not necessary to presuppose it.

In Fig. 3a a schematic representation of a swelling pressure-concentration curve is given. w_2^0 is the initial polymer concentration at which the thermoreversible gel has been set. The different parts (A, B, C) of the swelling pressure-concentration curve are obtained by drying this gel to different polymer concentrations after gelation. The whole swelling pressure-concentration curve can be con-

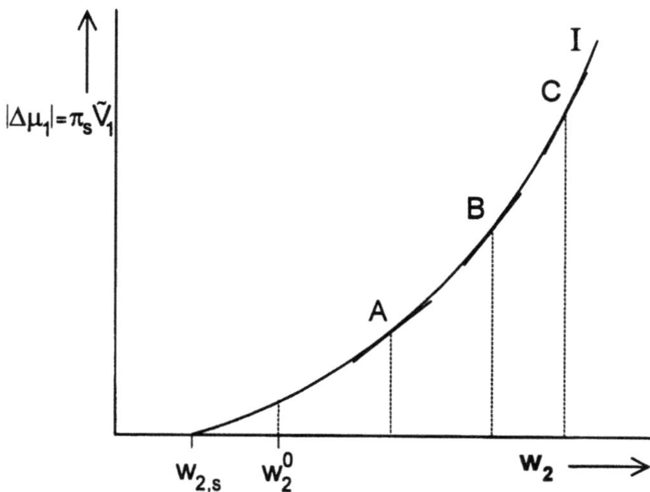

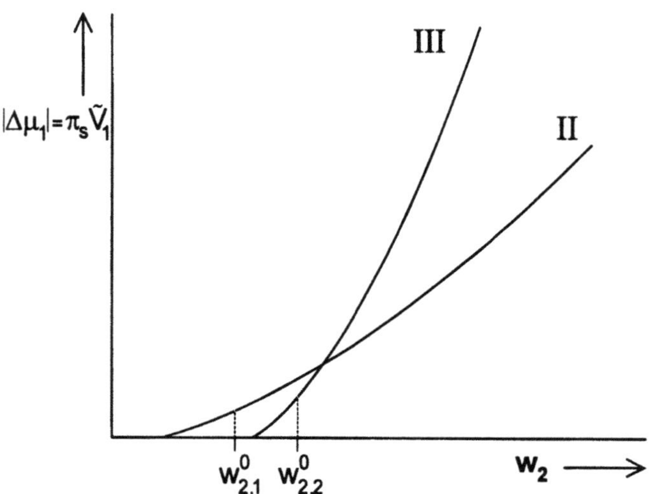

Fig. 3A, B Schematic representation of swelling pressure-concentration curves (I, II and III). $w_{2,i}^0 \equiv$ initial concentration of the cross-linked polymer, $w_{2,s} \equiv$ concentration of saturation, $w_2 \equiv$ polymer concentration; see text

structed from the slopes at different values of r in a single run. Therefore it is called a differential method. We only have small changes in the polymer concentration within the range of one measurement. If during the drying no change in the network structure occurs, a single swelling pressure-concentration curve is expected, as has been shown in Fig. 3a. It must be mentioned that the gel slices have to be brought into the shape of the corresponding ultracentrifugal cell.

The difference $\Delta\tilde{\mu}_1$ can be obtained by the values for r_1 and r_2, which are two positions on the curve within the small area of one single measurement

$$\Delta\tilde{\mu}_1 = \tilde{\mu}_1(r_2) - \tilde{\mu}_1(r_1) . \tag{7}$$

Strictly speaking, it is not yet proved that the structure of a physically network remains unchanged if a change of a polymer concentration is caused either by a partial drying of the gel (isotropic deformation) or by an actually anisotropic deformation in the centrifugal field of the AUC. This is explained in detail by means of an additional schematic representation of an $\pi_s(w_2)$-curve in Fig. 3b. The differentiation between the influence of the initial polymer concentration during the network formation w_2^0 and the change of the polymer concentration during the measurements w_2 should be emphasised by this figure. In Fig. 3a we actually have one gel with the polymer concentration of w_2^0 at the moment of network formation and we get the gels with the other concentrations by drying of this one. The single measurements only include small concentration ranges (A, B, C). The whole swelling pressure-concentration curve is built up from these parts. In Fig. 3b we have different initial concentrations $w_{2,i}^0$, this means we have gels with different polymer contents at the point of network formation. This may induce different network structures leading to the intersection of the curves II and III as shown in Fig. 3b. In one measurement we cover a much wider range of concentration because the applied forces are much higher if we use relatively high rotational speeds. This leads to the exclusion of solvent and, correspondingly, to an increase of the polymer concentration at the cell bottom (see Fig. 2). But as soon as a displacement of the gel meniscus takes place there is the danger of implying high non-linear deformations. Therefore, we now prefer the differential method with smaller concentration ranges but very small deformations.

The results of the differential method may be used to test statistical theories which connect the mixing term and the deformation term of the chemical potential of the solvent in the gel with molecular parameters. For example, the so-called Flory–Huggins–Staverman equation [37, 38, 39] has been used in a semiempirical way [27–32].

$$\frac{\Delta\tilde{\mu}_1}{RT} = -\frac{\pi_s V_1}{RT} = \underbrace{\ln w_1 + w_2 + \chi_w w_2^2}_{\text{mixing term}} + \underbrace{C_w w_2^{1/3}}_{\text{network term}} \quad . \tag{8}$$

For different investigated systems the interaction parameter χ_w was found to be dependent on the polymer concentration

$$\chi_w(w_2) = \chi_{w,0} + \chi_{w,1} w_2 \ . \tag{9}$$

The network term C_w is composed of different quantities characteristic for the network

$$C_w = \frac{M_1}{\bar{M}_c}\left(1 - \frac{2}{f}\right)\left[\frac{\overline{r^2}}{\overline{\overline{r_0^2}}}\right] . \tag{10}$$

M_1 is the molar mass of the solvent, $\bar{M}_c$ the number average molar mass of the chains between two junction points of the network, f is the functionality of the junction

points, $\overline{r^2}$ is the mean square end-to-end distance of the network chains in the undeformed state and $\overline{r_0^2}$ the mean square end-to-end distance in the reference state of the dry polymer.

This statistical theory has been worked out for non-electrolyte gels and can be applied to polyelectrolyte gels in a semiempirical way. Because of the lack of information on some systems more refined theories could not be proven. Network defects and other kinds of irregularity are not taken into account. With the differential method, which we want to use now, in principle it should be possible to check the statistical theories and, if necessary, to modify them.

As will be shown later in the discussion, the change of the concentration profile of a crosslinked polymer dependent on time differs from that of an uncrosslinked polymer.

Referring to case a) mentioned above and illustrated in Fig. 2, there is no movement of a peak maximum – as in case of solutions – nor any movements of the meniscus as a criterion for the sedimentation velocity. But as the external field will always lead to continuous changes in the concentration of the polymer, which is a sedimentation of the crosslinked polymer, the movement of the meniscus at the gel surface cannot be used for the description of the sedimentation velocity. A suggestion to define the sedimentation of a gel based on the relative movement of the center of mass of the crosslinked polymer has been made recently [40].

Results and discussion

In Figs. 4a–e some Schlieren patterns are shown which have been measured by investigating gels at not too high fields, this means at rotational speeds between 3000 and 12 000 rpm. These Schlieren patterns are corrected by histogram, brightness and contrast, and are on a larger scale in the direction of the gradient. Therefore, the gradient is a little broad. Its middle section will be used for the later evaluation.

These pictures prove that real equilibria are achieved inside the gel phase. This can be demonstrated by the path independence of the obtained gradients. For one selected rotational speed it makes no difference whether the equilibrium concentration profile is reached coming from higher or lower speeds to the chosen one.

The first picture (Fig. 4a) was taken 6 min after the run was started at 6000 rpm. The example is a κ-carrageenan/water gel with a initial polymer concentration of 3% by weight. Nearly nothing has happened inside the gel so far. The concentration of the polymer is the same at every point in the gel phase.

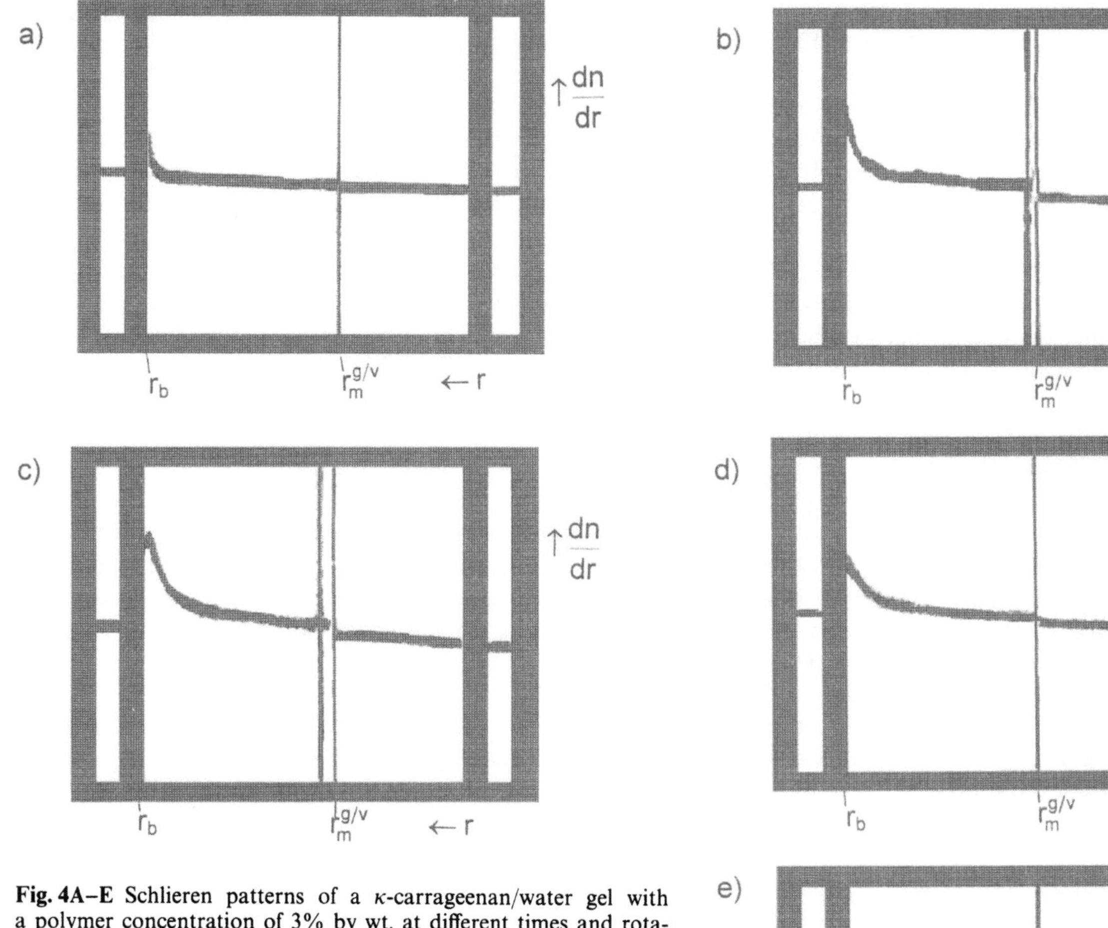

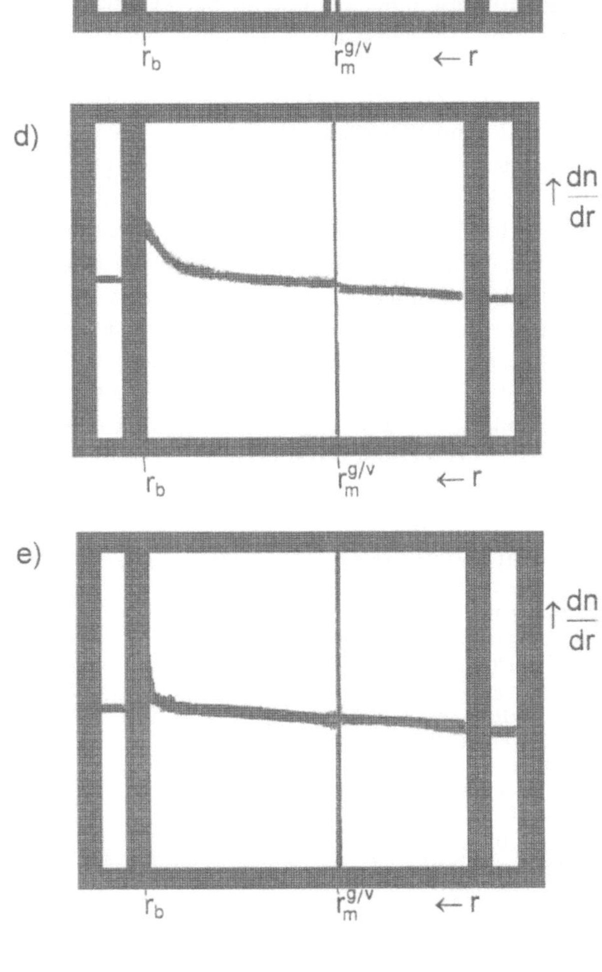

Fig. 4A–E Schlieren patterns of a κ-carrageenan/water gel with a polymer concentration of 3% by wt. at different times and rotational speeds for the proof of real equilibria by means of the path independence of the gradients. a) after 6 min at 6000 rpm, b) after 5 days at 8000 rpm, c) after 1 week at 12 000 rpm, d) after the speed was decreased again to 8000 rpm for 6 days, e) after the speed was decreased again to 3000 rpm for 3 days; see text

The next Schlieren pattern (Fig. 4b) shows the same sample after 5 days at a rotational speed of 8000 rpm. The gradient has become steeper near the cell bottom. The continuous equilibrium corresponding to that speed has been reached here.

In Fig. 4c a Schlieren pattern of the same sample but after 1 week at a rotational speed of 12 000 rpm is shown. It is obvious that the concentration profile has changed further. There is a broad peak near the cell bottom. But in contrast to the Schlieren patterns of polymer solutions, the typical movement of a peak maximum or concentration changes near the meniscus of the solutions, which is the gel surface here, cannot be found. At the beginning the gradient is becoming steep at the cell bottom like in Fig. 4b, and then declines a little. No changes could be observed during the 3 days before this picture was taken.

In Figs. 4b and 4c the second vertical line in the middle section of the Schlieren pattern, appearing on the left of the very sharp meniscus gel/vapor, is not an additional meniscus. This blurred line is perhaps caused by small gel pieces sticking to the cell walls. In Fig. 4c it is becoming obvious that the concentration gradient of the polymer in the gel phase passes right through this line without any bend or kink. This irregularity is reversible whatever its origin; in Fig. 4d it has disappeared again.

During our earlier investigations including movements of the meniscus, we often observed irreversible changes in the network beyond a critical value of angular velocity. So it was of special interest to see what happened to the gel after decreasing the speed again.

In the next Schlieren pattern of Fig. 4d it can easily be seen that the gradient of the polymer concentration is getting flatter again by decreasing the speed. It was first decreased from 12 000 to 8000 rpm where it was kept for 6 days.

Afterwards the speed was decreased to 3000 rpm (Fig. 4e) for 3 days. One can see that the obtained Schlieren pattern greatly resembles the one in Fig. 4a. The polymer concentration is the same along the whole gel again. This shows that there are no irreversible changes in the network during the measurements at those low rotational speeds and that real equilibria were achieved.

A more detailed evaluation of the Schlieren patterns of Figs 4b), 4c), 4d) and a Schlieren pattern taken after 3 days at 6000 rpm has been carried out by means of a calibration cell, where the gradient of the optical pathway is known. First of all, the dependence of the partial density ρ_2 on the distance from the axis of rotation r has been determined from Schlieren curves by use of the law of mass conservation. The polymer concentration is identical to the initial concentration at the value $r_{1/2}$, where the area below the Schlieren curve is 50% of the total area.

In all cases it was found that $\rho_2(r)$ is approximately a linear function of r with a correlation coefficient better than 0.92 and that the slopes of the curves increase with increasing ω. Secondly, $(d\pi_s/dr)$ has been calculated for a κ-carrageenan/water gel with a polymer concentration of $\rho_2 = 3.005 \cdot 10^{-2}$ gml^{-1} starting from Eq. (2) and using the earlier result that the partial specific volume of water $\tilde{V}_1$ is identical to the specific volume of water $\tilde{V}_{01}$, the partial density of water ρ_1 has been taken from the density-concentration relation [31, 41]. The density of the gel is given by $\rho = \rho_1 + \rho_2$. Finally, $(d\pi_s/d\rho_2)$ has been calculated combining Eqs. (3b) and (4):

$$\left(\frac{d\pi_s}{d\rho_2}\right) = \frac{(d\pi_s/dr)}{(d\rho_2/dr)} . \tag{11}$$

For all experiments it is clearly shown that the polymer concentration increases from the meniscus to the bottom and that the average concentration difference with respect to the initial concentration is in a range of 9% to 33.5% corresponding to the different speeds. The values of $(d\pi_s/d\rho_2)$ increase from the meniscus to the bottom.

From Eq. (2) it can be expected that $(d\pi_s/d\rho_2)$ has the same value for all runs at the initial concentration of the gel which is given by ρ_2^0. In Fig. 5 the results are plotted. It can be seen that the $(d\pi_s/d\rho_2)$-values at ρ_2^0 differ in a range

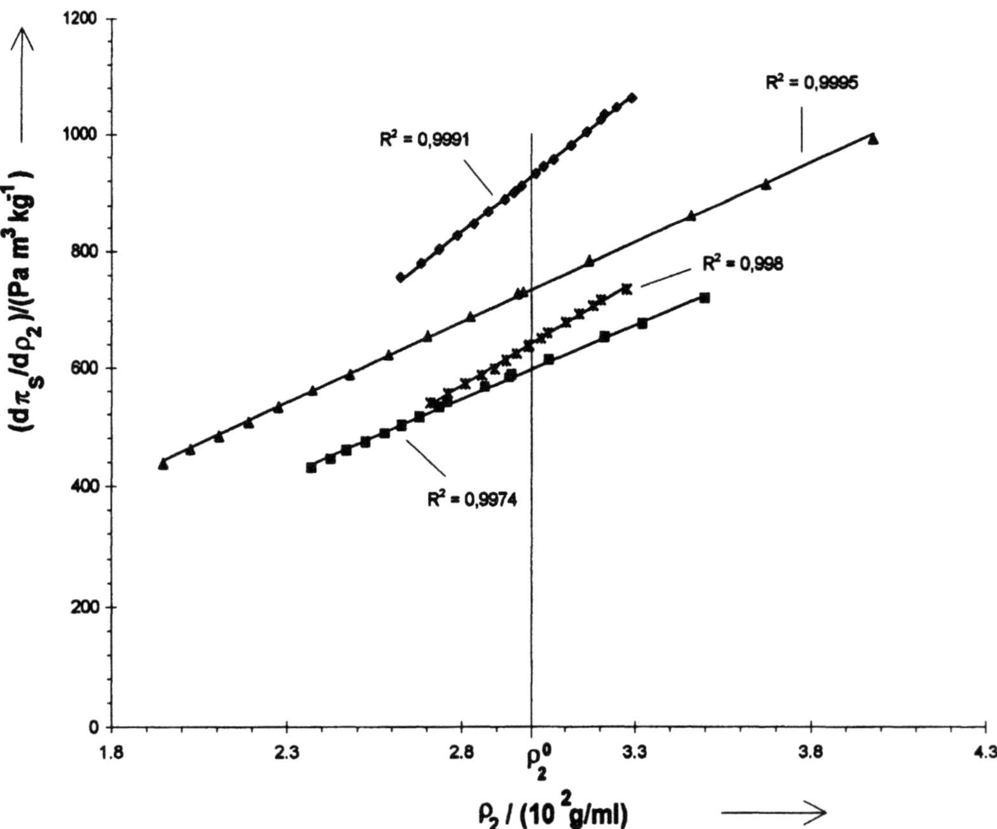

Fig. 5 Quantity $(d\pi_s/d\rho_2)$ vs ρ_2 for four different runs of a κ-carrageenan/water gel corresponding to the conditions of to the Schlieren patterns presented in Figs. 4b) to d) and a Schlieren pattern taken after 3 days at 6000 rpm; 4b) ◆ (calculated for $\omega = 8000$ rpm), 4c) ▲, 4d) ■, 3 days at 6000 rpm ✳; R^2 = correlation coefficients of the lines, ρ_2^0 = initial concentration of the polymer; see text

of 585 to 903 $Pa\,m^3\,kg^{-1}$. Coming from low rotational speeds to higher rotational speeds the denominator of Eq. (11) increases and therefore $(d\pi_s/d\rho_2)$ has to decrease. This is the reason why the $(d\pi_s/d\rho_2)$-values are too high if the continuous equilibrium is not yet reached. From this we expect that with an approach to the equilibrium the curves representing the dependence of $(d\pi_s/d\rho_2)$ on ρ_2 have to shift to lower values. Correspondingly, with an approach to the equilibrium starting from relatively higher rotational speeds the concentration gradients have to decrease. With respect to Eq. (11) the $(d\pi_s/d\rho_2)$-values have to increase, if the system reaches the new equilibrium. Therefore the corresponding curve in Fig. 5 has to shift to higher values.

The equilibrium value of $(d\pi_s/d\rho_2)$ at ρ_2^0 corresponds to the average of two values, which have been obtained via two different equilibrium approaches coming from relatively lower and higher speeds. From these considerations we conclude that the equilibrium value of $(d\pi_s/d\rho_2)$ is given by $590\,Pa\,m^3\,kg^{-1}$. This has been taken from the nearly identical curves of Fig. 5 corresponding to the runs at 6000 from 0 rpm and 8000 from 12 000 rpm. In all other cases the continuous equilibrium has not yet been established.

Additionally, we note that differential evaluation methods always have larger margins of errors because of their sensitivity as compared to a non differential method.

Figure 5 clearly shows that $(d\pi_s/d\rho_2)$ is a linear function of ρ_2 with a correlation coefficient better than 0.99. From the slopes of the curves we obtain the second derivative of the $\pi_s(\rho_2)$-curve which is a proof for the high sensitivity of the gradient method.

Furthermore, it has been shown for the first time that the concentration gradients in gels are approximately constant at not too high external centrifugal fields. This condition was an assumption in all earlier investigations where the movement of the gel meniscus was measured [22, 29–32].

Acknowledgement We thank the Deutsche Forschungsgemeinschaft and the Max-Buchner-Stiftung for financial support of the project. We thank Steinkopff Verlag for permission to use Figs. 1 and 2, and S. Frahn for her assistance.

References

1. Rehage G (1964) Kolloid-Zeitschrift & Zeitschrift für Polymere 194:1
2. Rehage G (1964) Kolloid-Zeitschrift & Zeitschrift für Polymere 196:97
3. Rehage G (1964) Kolloid-Zeitschrift & Zeitschrift für Polymere 199:1
4. Gnanou Y, Hild G, Rempp P (1987) Macromolecules 20:1662
5. Rempp P, Herz JE, Borchard W (1978) Adv Polym Sci 26:105
6. Graham NB, Nwachuku NE, Walsh DJ (1982) Polymer 23:1345
7. Petrovic ZS, MacKnight WJ, Koningsveld R, Dusek K (1987) Macromol 20:1088
8. van de Kraats EJ (1968) Rec Trav Chim, Pays-Bas 87:1137
9. van de Kraats EJ, Winkeler MAM, Potters JM, Prins W (1969) Rec Trav Chim, Pays-Bas 88:449
10. Emberger A (1975) Diplomarbeit, Clausthal
11. Borchard W, Emberger A, Schwarz J (1978) Angew Makromol Chemie, Bd 66, 986:43
12. Borchard W (1993) Colloid Polym Sci 271:843
13. Borchard W (1966) Dissertation, Aachen
14. Posnjak E, Freundlich H (1912) Kolloid-Beih 3:417
15. Prins W, Pennings AJ (1961) J Polym Sci 49:507
16. Steinbrecht U (1991) Dissertation, Duisburg
17. Borchard W, Steinbrecht U (1991) Colloid Polym Sci 269:95
18. Boyer RF, Spencer RS (1948) J Polym Sci 3:97
19. Tomka I (1973) Dissertation, Bern
20. Johnson P, Metcalfe JC (1967) Europ Polym J 3:423
21. Johnson P (1968) Chem Soc Special Publication 23:243, Burlington House, London
22. Holtus G, Cölfen H, Borchard W (1991) Progr Colloid Polym Sci 86:92
23. Oppermann W (1978) Dissertation, Clausthal
24. Oppermann W, Rose S, Rehage G (1985) Br Polymer J 17:175
25. Flory PJ (1953) Principles of Polymer Chemistry, Cornell University Press, London
26. de Gennes PG (1979) Scaling Concepts in Polymer Physics, Cornell University Press, London
27. Borchard W (1991) Progr Colloid Polym Sci 86:84
28. Cölfen H, Borchard W (1991) Progr Colloid Polym Sci 86:102
29. Borchard W, Cölfen H (1992) Makromol Chem, Macromol Symp 61:143
30. Cölfen H (1993) Dissertation, Duisburg
31. Hinsken H (1992) Diplomarbeit, Duisburg
32. Hinsken H, Borchard W to be published in Progr Colloid Polym Sci (1995) "Continuous swelling pressure equilibria of the system κ-carrageenan/water"
33. Cölfen H, Borchard W (1994) Progr Colloid Polym Sci 94:90
34. Mc Bain JW, Stuewer RF (1936) Kolloid-Zeitschrift 74, Heft 1:10
35. Svedberg T, Pedersen KO (1940) Die Ultrazentrifuge, Steinkopff Verlag, Dresden, Leipzig p 26
36. Johnson P (1970) in: Photographic Gelatin I, Academic Press, London, p 13
37. Staverman AJ, van Santen JM (1941) Rec Trav Chim Pay-Bas 60:76; Staverman AJ (1941) Rec Trav Chim Pay-Bas 60:640
38. Huggins ML (1942) J Phys Chem 46:151
39. Flory PJ (1942) J Chem Phys 10:51
40. Hinsken H, Borchard W, presented on the 8th Symposium of Analytical Ultracentrifugation (2.3.–3.3.1995), Berlin, to be published separately
41. Cölfen H, Borchard W (1994) Macromol Chem Phys 195:1165

Progr Colloid Polym Sci (1995) 99:162–166
© Steinkopff Verlag 1995

P. Voelker

Measurement of the extinction coefficient of prostate specific antigen using interference and absorbance optics in the Optima XL-A analytical ultracentrifuge

Received: 27 April 1995
Accepted: 19 May 1995

P. Voelker (✉)
Beckman Instruments Inc.
1050 Page Mill Road
Palo Alto, California 94304, U.S.A.

Abstract Establishment of an accurate extinction coefficient is often of interest when a precise amount of material is required for a biochemical reaction or when converting an association constant from absorbance units into concentration units. Although a gravimetric approach can provide a quick estimate of the concentration, the presence of salts or impurities can adversely affect its accuracy. Analytical ultracentrifugation is often used when more accurate estimates of the extinction coefficient are required. This technique has been applied to prostate specific antigen (PSA), and the value of the extinction coefficient estimated by amino acid analysis is confirmed by sedimentation analysis.

The extinction coefficient of PSA was measured by analytical ultracentrifugation using the method of sedimentation boundary analysis. This method utilizes centrifugal force to form a solvent boundary by layering a reference buffer onto a sample via a thin capillary connecting both sectors of the centerpiece i.e., a synthetic boundary centerpiece. The plateau formed by the boundary is then optically measured as it decreases over time due to diffusional effects. By using a combination of absorbance and interference optics, the OD and concentration (from the fringe displacement) can be sequentially measured and used in Beer's Law to directly solve for the extinction coefficient. A value for the extinction coefficient of PSA at 280 nm of 1.71 L/g-cm was obtained. This is in close agreement with a literature value of 1.84 L/g-cm using a concentration estimated by amino acid analysis.

Key words Extinction coefficient – interference optics – fringe displacement – analytical ultracentrifugation

Introduction

The Optima XL-A analytical ultracentrifuge provides a means of evaluating the size, shape, purity and associative behavior of a wide range of macromolecules that absorb between 190 and 800 nm. Equipping the XL-A with an interferometer allows an even wider range of macromolecules to be studied since following the samples behavior depends solely on a change in its refractive index. Combining both absorbance and interference optics can facilitate existing applications as well as create new ones. In this study a combination of absorbance and interference optics is used to estimate the extinction coefficient of prostate specific antigen (PSA) using the method of sedimentation boundary analysis. A description of the XL-A interferometer is also presented.

Theory

Interference optics

When light passes through a pair of evenly spaced slits, two wave fronts are formed, which result in either constructive or destructive interference. This image appears as a series of parallel, alternating dark and light lines called fringes. If a reference buffer and a sample solution are placed behind the two slits, the wave fronts moving through both samples create a fringe displacement or shift due to the difference in the refractive index between the two solutions (see Fig. 1). The extent of the fringe shift is directly proportional to the concentration of the sample. The basic equation for the fringe displacement (ΔJ) is:

$$\Delta J = \Delta n l/\lambda \, , \tag{1}$$

where λ is the wavelength of the light used (670 nm) and l is the pathlength of the cell (1.2 cm). The change in refractive index Δn, is described by the following equation:

$$\Delta n = c \, (dn/dc) \, , \tag{2}$$

where c is the protein concentration in mg/ml, and dn/dc is the specific refractive index increment in ml/g. The specific refractive index increment is independent of concentration [1] and is found to be relatively constant for most proteins. In this study, a specific refractive index increment of 0.186 ml/g for BSA (at 25 °C and 578 nm) was used [2]. As a general rule, a ΔJ of 1 corresponds to a concentration of about 0.25 mg/ml for most proteins in a 1.2 cm cell. Substituting dn/dc and c from Eq. (2) into Eq. (1) directly relates concentration to fringe displacement.

$$\Delta J/c = (dn/dc) \, l/\lambda \tag{3}$$

After inserting the appropriate constants, Eq. (3) becomes:

$$\Delta J/c = 3.33 \text{ ml/mg} \tag{4}$$

Eq. (4) provides a rapid method of determining the concentration of a protein from its fringe displacement at 25 °C and 670 nm.

XL-A Interferometer

The interferometric accessory of the Optima XL-A was designed to run either independently or in parallel with the standard absorbance optical system. The interferometer contains its own separate light source, optical track and detector (see Fig. 2). The light source consists of a 10-mwatt laser diode operating at 670 nm and is attached directly to the monochromator. The two slits (currently set at a distance of 4 mm and a width of 0.152 mm) are located at the base of the laser. Light passes through the cell containing the reference and the sample and is picked up by the optical track located below the refrigeration can. The entire optical track is kept under vacuum to minimize convective air currents. The optical track includes a condenser and camera lens which operate in concert to image the 2/3 plane of the cell onto the detector. (The 2/3 plane is

Fig. 1 The fringe pattern of a single-component system as part of a velocity run. This scan represents one of five images that collectively spans the radial dimensions of the cell

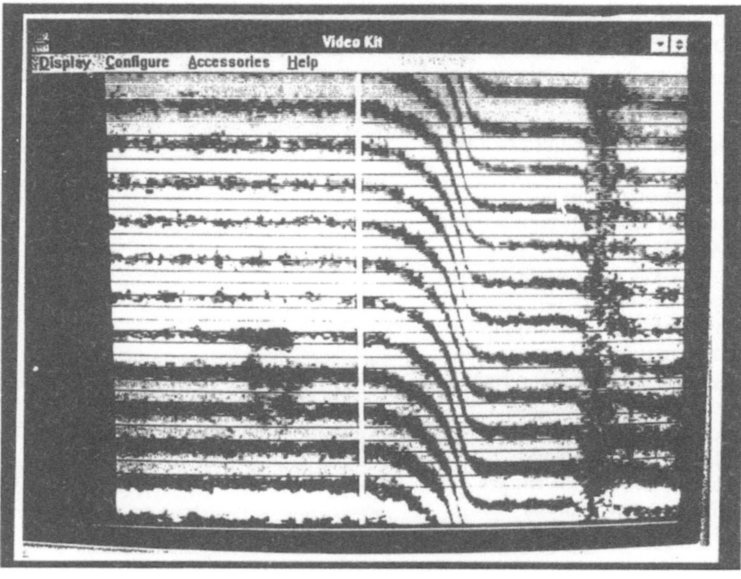

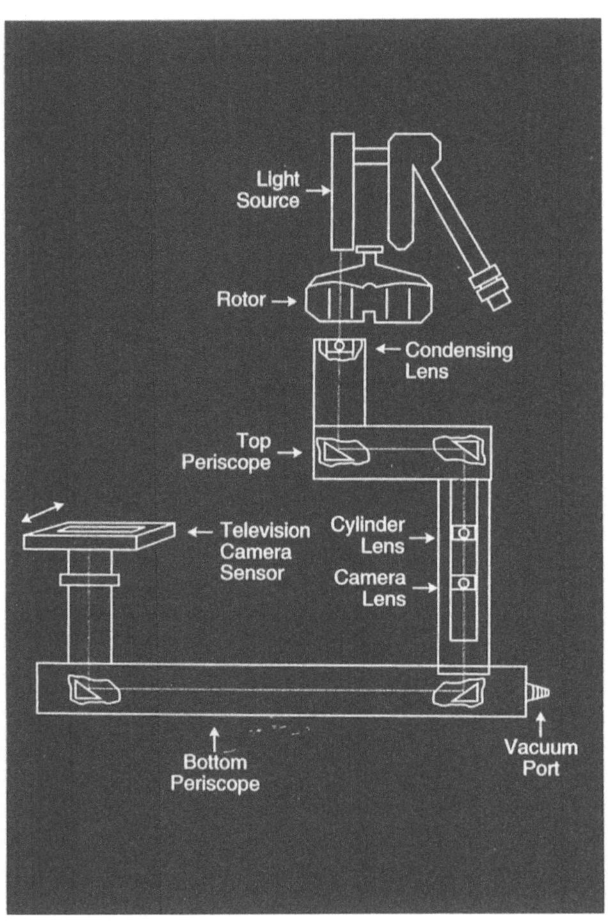

Fig. 2 An overview of the XL-A interference optical system [4]. The laser light source is attached to the monochomator. The two slits are located at the basis of the laser. The optical track is below the refrigeration can. The convoluted optical path is necessary in order to avoid other hardware that is present in this area. The detector is a high resolution CCD camera

a position that is measured from the inner cell radius and is considered to be the object plane [3]). The cylinder lens serves to overlap the image from the reference cell with the corresponding image from the sample cell to create an interference fringe pattern. The fringe pattern is read by a detector, a high resolution CCD camera. The fringe pattern is converted to a relative vertical fringe displacement by means of a Fourier analysis conducted at the spacial frequency of the fringes (see Fig. 3). A scan and a plot of the smoothed data takes about 15 s.

The radial image of a standard two-channel centerpiece (1.4 cm) is magnified to a least 1700 pixels on a CCD array. This translates to a resolution of about 8 μm/pixel. Early camera configurations used a 492 pixels per column $\times$ 512 pixels per row CCD array, which meant that the camera had to move five times in order to capture the entire image [4]. An example of a single image is shown in Fig. 1. The number of fringes that are displayed are about 11 (measured vertically). This translates to a resolution of about 40 pixels/fringe. This value exceeds our earlier expectations of 20–25 pixels/fringe. Although the moving camera worked reasonably well, there were occasional problems in aligning the five images. Alternatively, an instrument is being developed using a CCD pixel array of 96×2048. This allows capturing the entire cell in a single image, which means faster acquisition times and more reliable Fourier transforms of the fringe pattern. The resulting image from this design displays about four to six vertical fringes (not shown).

Experimental

A combination of interference and absorbance optics was used to determine the extinction coefficient of the glycoprotein PSA. Absorbance optics were used to measure the material at 280 nm. Interference optics were used to determine the concentration from the fringe shift using Eq. 4. The absorptivity of PSA was determined by solving Beer's Law using a cell pathlength of 1.2 cm. The extinction coefficient was calculated by multiplying the absorptivity

Fig. 3 A fringe pattern of a two component system (a 1:1 mixture of a 2 mg/ml cyclohexane solution of 106 and 300 KDa monodisperse polystyrene macromolecules) as a part of velocity run. The image on the right displays the fringe pattern. (Note: the inner radius is shown on the right hand side of the image.) The plot on the left displays the relative vertical fringe displacement of the fringes using Fourier analysis

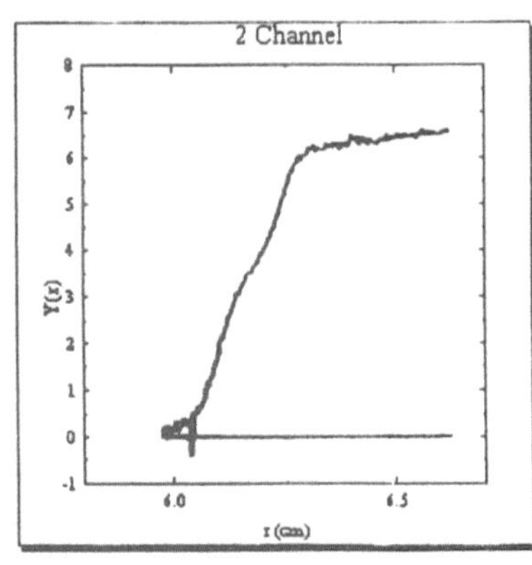

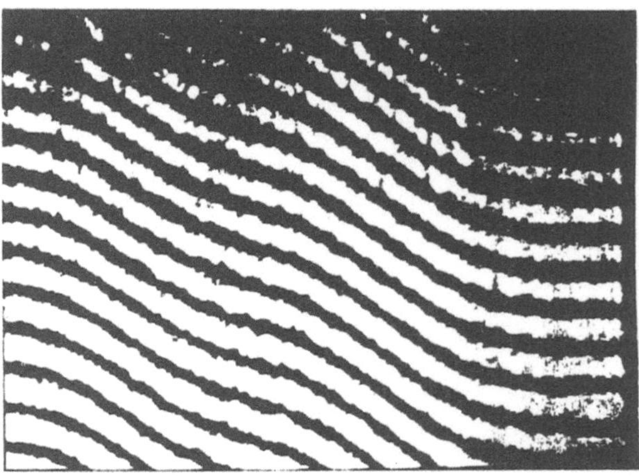

Table 1 Determination of the Extinction Coefficient of PSA in pH 7.4 PBS, 25 °C[a])

ΔJ, Fringe shift (fringes, 670 nm)	Absorbance (O.D. @ 280 nm)	Absorptivity (AAA) (L/g-cm)	Absorptivity (XLA) (L/g-cm)	Extinction Coefficient (XLA) (L/mol-cm)
253 ± 0.03	1.55 ± 0.01	1.84 ± 0.04	1.70 ± 0.01	48,359

[a]) XL-A measurements and absorptivity estimates are based on a mean value of 8 separate readings. AAA absorptivity estimates are based on a mean value of 10 separate readings. The error for each is expressed in standard deviations about the mean.

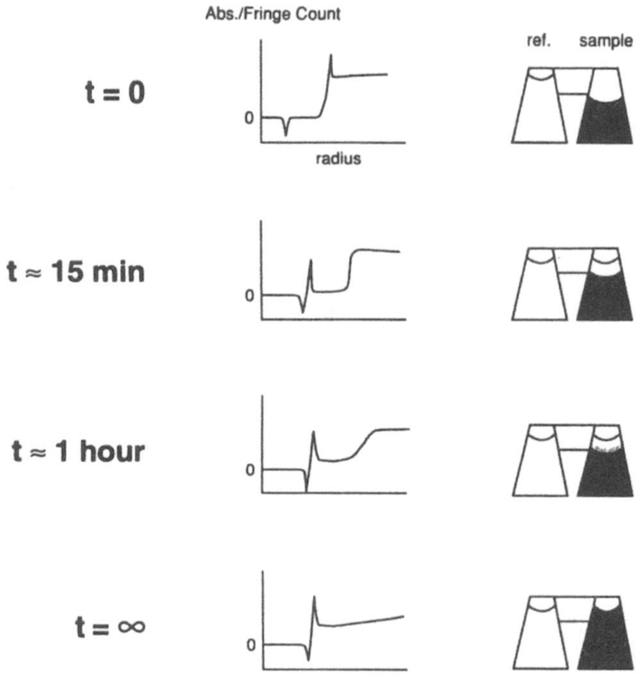

Fig. 4 The method of sedimentation boundary analysis at different time points. Boundary formation and spreading are shown in a synthetic boundary centerpiece and as a plot using either absorbance or interference optics (i.e., OD or fringe displacement)

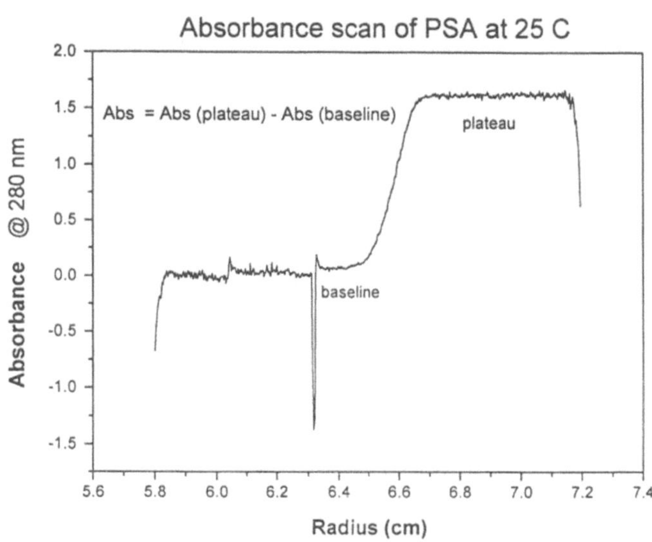

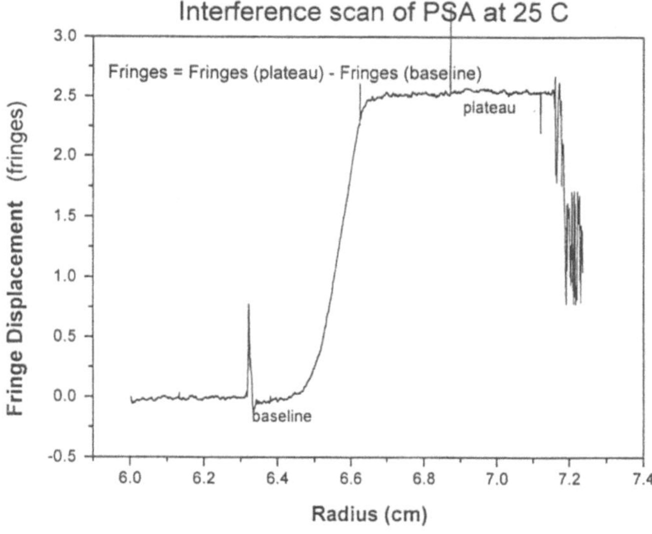

Fig. 5 Representative interference and absorbance scans of 0.77 mg/mL PSA in pH 7.4 PBS buffer. Data was collected at 4000 rpm and 25 °C

by the molecular weight, i.e., 28 430 g/mol estimated by ion spray mass spectrometry [5].

The set of sedimentation boundary analysis runs used in determining the extinction coefficient of PSA was made with a synthetic boundary centerpiece. The method utilizes centrifugal force to form a solvent boundary by layering a reference buffer onto a sample via a thin capillary connecting both sectors of a synthetic boundary centerpiece (see Fig. 4). Absorbance and fringe displacement are read as the difference of the upper plateau and the baseline. Rotor speed is selected such that the upper plateau and the baseline remain flat during data collection, thereby ensuring more precise measurements (see Fig. 5).

PSA was run at an approximate concentration of 0.77 mg/ml in pH 7.4 PBS buffer at 25 °C. The rotor was spun at 8000 rpm for 10 min in order to accelerate transfer of the PBS buffer onto the solution of PSA. The rotor speed was then adjusted to 4000 rpm for the remainder of the run. A series of sequential absorbance/interference scans were collected approximately every 5 min for a period of one hour. An average of eight absorbance/interference scans were used to estimate the OD and fringe displacement.

Results

Absorbance and fringe displacement measurements for PSA are reported in Table 1. The concentration is determined from the fringe displacement using Eq. (4). An absorptivity of 1.70 L/g-cm is determined directly from Beer's Law by using the measured absorbance and a pathlength of 1.2 cm for the cell. An absorptivity of 1.84 L/g-cm for PSA, using a concentration estimated by amino acid analysis, is included as a comparison and shows close agreement between the two methods [6].

Acknowledgement The author thanks Dr. Hans Müller, Bayer AG for the sample used in Fig. 1, Dr. Walter Mächtle, BASF for the samples used in Fig. 3, and Dr. Thomas Stamey, Stanford University School of Medicine for providing us with the PSA.

Reference

1. Perlmann GE, Longsworth LG (1948) J Am Chem Soc 70:2719–2724
2. Solar HA (ed) 1970 Handbook of Biochemistry 2nd Edition CRC Press: C-67
3. Yphantis DA (1964) Biochemistry 3: 297–317
4. Laue TM, Anderson AL, Demaine PD (1994) Progr Colloid Polym Sci 94:74–81
5. Belanger A, van Halbeek H, Graves HCB, Grandbois K, Stamey TA, Huang L, Poppe I, Labrie F (1995) Prostate (in press)
6. Stamey TA, Teplow DB, Graves HCB (1995) Prostate (in press)

Progr Colloid Polym Sci (1995) 99:167–186
© Steinkopff Verlag 1995

H. Cölfen
S.E. Harding

A study on Schlieren patterns derived with the Beckman Optima XL-A UV-absorption optics

Received: 18 April 1995
Accepted: 5 June 1995

Dr. H. Cölfen (✉)
Max-Planck-Institute for Colloid
& Interface Research
Colloid Chemistry Department
Kantstraße 55
14513 Teltow, Germany

S.E. Harding
National Centre for Macromolecular
Hydrodynamics
University of Nottingham
Sutton Bonington LE12 5RD
United Kingdom

Abstract This study describes how typical "Toepler" Schlieren patterns can be derived with the UV-absorption optics of the Beckman Optima XL-A ultracentrifuge without any modification of the optical system. Such a Schlieren effect can greatly enhance the application range of the XL-A ultracentrifuge which cannot yet be used to its possible full potential due to the present lack of a refractometric optical system and – not only for sedimentation velocity experiments – particularly the Schlieren optical system. It is shown that the Schlieren patterns detected with the UV-absorption optical system are caused by light refraction and not by light scattering or experimental artefacts. Two possible explanations for the generation of the Schlieren effect are given. Both explanations have in common that the refracted light cannot be detected by the photomultiplier system and is hence lost for detection. The intensity of the refracted light which can be registered with the photodetector of the optical system is dependent on the angle of incidence of the light, e.g., the local slope of the sedimenting boundary, and hence leads to a detection of the refractive index change with radial position: i.e.,

a Schlieren pattern. The advantages, disadvantages and differences of the XL-A Schlieren effect compared to conventional "Philpot–Svensson" Schlieren optics are discussed demonstrating that both optical systems deliver the same results for examples of sedimentation velocity experiments on several polysaccharides and bovine serum albumin. An estimation of the sensitivity of the Schlieren effect as a function of the wavelength selected is given as well as suggestions of how the Schlieren effect can be increased or suppressed. Considerations about the quantitative evaluation of XL-A Schlieren patterns show that the Schlieren effect is dynamic with respect to the concentration dependence. This creates some potential for new applications but is also a serious warning to those people who want to derive quantitative information from XL-A absorption traces as these can be superimposed by a Schlieren effect at certain wavelengths.

Key words Schlieren optics – absorption optics – analytical ultracentrifugation – sedimentation velocity – light refraction – optima XL-A ultracentrifuge

Introduction

Throughout the several decades of Analytical Ultracentrifugation, the typical Schlieren peak has not only been the pattern most people associate Analytical Ultracentrifugation with, but Schlieren optics has furthermore proved to be the most suitable optical system for sedimentation velocity experiments and some other applications (e.g., density gradient experiments, experiments with gels, etc.). This is especially valid for polysaccharides and synthetic polymers in UV-absorbing samples which have not been detectable with a normal UV-absorption optical system up to now. All classical types of Analytical Ultracentrifuges, e.g., the Beckman Model E, the MOM ultracentrifuge, the Heraeus AZ-9100 and the MSE Centriscan (the latter already equipped with a multiplexer controlled scanning Schlieren optics for the use of six place rotors) or the MSE MK II are equipped with at least this type of optical system as a standard. But the Beckman Optima XL-A (the only Analytical Ultracentrifuge available after a long time of stagnation in development) only at present has UV-absorption optics although in the near future there will be commercially available Rayleigh interference optics for it based on the prototype of Laue [1]. Although this interference optical system – without doubt, being the most accurate system for sedimentation equilibrium work – is designed as an on-line system suitable for evaluation with computers [1] and hence suitable for differentiation leading to a Schlieren pattern or for the time derivative analysis of solutions with concentrations as low as 0.02 mg/ml [2], many important applications – well established in the past – still cannot be covered by this optical system due to its present physical limitations. One example might be the detection in highly concentrated solutions where no Rayleigh interference fringes are formed anymore, but Schlieren patterns are still detectable. Another critical example might be steep concentration gradients for which the detection accuracy of a CCD-camera based Rayleigh interference optics is considerably limited. Further, in systems where a z-average molecular weight is desirable from sedimentation equilibrium, Schlieren optical records yield this directly. So there is still a demand for a Schlieren optical system for the Beckman Optima XL-A ultracentrifuge. At present, this means that many scientists need to keep their old "dinosaurs" with all their well known disadvantages, just because of the lack of an alternative. To the authors' knowledge, no real commercial progress in designing a Schlieren optical system for the Beckman Optima XL-A has been made up to now, although the Schlieren optical system is virtually identical to the Rayleigh interference optics with the addition of a knife edge or phase plate in the focal point of the collimating lens which focuses the parallel light illuminat-

ing the sample in the cell. For a good description of the function of the optical elements of the Schlieren optical system see [3]. This lack of provision of a reliable Schlieren optical system for the Optima XL-A might be an effect of the considerably decreased optical path-length due to the now significantly decreased instrument size as Schlieren optics has up to now proved to be a long-range optical system with large focal length lenses requiring long optical paths in the instruments still in use (Model E, MOM, etc.). Further improvements of the Schlieren optical system with respect to its sensitivity and accuracy [4] – which can be considerably enhanced – indicated that Schlieren optics does not lend itself easily to miniaturization because of decreased sensitivity. In contrast to the present commercial desire for apparatus miniaturization, a study on an ultrasensitive Schlieren optical system [4] clearly pointed out that a more sensitive Schlieren optical system not only requires a larger phaseplate or knife edge, but furthermore requires longer optical paths. This means that the application of lenses with smaller focal lengths and hence decreased optical paths as realized with the Rayleigh interference optics for the Optima XL-A simply cannot be transferred to a Schlieren optical system without significant loss of accuracy. This might be one of the reasons why a Schlieren optical system, although formally identical to the already established Rayleigh interference optical system, has not been launched for the Optima XL-A up to now. Furthermore, problems in the picture evaluation process of Schlieren patterns from a phase plate (detection of the zero'th order of the Schlieren line in a non-uniformly illuminated greyscale picture) has restricted their availibility for on-line picture evaluation, although some partly successful approaches have been made [5, 6]. Nevertheless, none of these approaches leads to universally applicable on-line picture evaluation because the brightness values which must be used to define the relevant picture information (e.g., the Schlieren curve) are not that different in a Schlieren grey scale picture as they are in a Rayleigh interference pattern. The high contrast of interference fringes has been exploited by Rowe et al. [7] using the Fresnel fringes in phase-plate Schlieren patterns to derive the zero'th order with increased accuracy. Such an evaluation technique is actually adapted to a knife edge based on-line Schlieren system for an MSE-MKII ultracentrifuge [8]. Another knife edge based Schlieren optical system principally suitable for on-line evaluation is the scanning Schlieren system of the old MSE Centriscan ultracentrifuge [9]. Furthermore, it was reported that the UV-absorption optics with photoelectric scanner of the Model E can be modified by introducing a knife edge as analyzer element so that it is possible to scan up to five Schlieren patterns simultaneously when using the multiplexer with the six place rotor [10]. This system should as well be principally suitable for on-line evaluations.

From the present point of view, it seems to be unlikely that first a classical phase plate or "Philpot–Svensson" Schlieren optical system (see e.g. [11]) for the Beckman Optima XL-A will be launched and second fully automatically and preferably on-line evaluation of Schlieren patterns can be maintained as is required by modern expectations. These considerations already show that new ways need to be explored if a Schlieren optical system can ever get a chance to be established as an on-line system available for computerized evaluation in the near future. Up to now, many of the benefits of Analytical Ultracentrifugation can still not be used by those scientists who definitely require refractometric optics [12], namely all scientists interested in the investigation of synthetic polymers in organic solvents, or polysaccharides or many colloidal systems with the Analytical Ultracentrifuge. But especially the latter field is steadily expanding. Hence the motivation for this study was to start some basic research on the generation of Schlieren patterns leading to a solution for all scientists working with the Beckman Optima XL-A with its present standard UV-absorption optical system. Hopefully, the results of this study will encourage others to further explore the Schlieren effect on the Beckman Optima XL-A and to carry out further studies. Perhaps the results given may even serve as basis to design a reliable Schlieren optical system for the XL-A ultracentrifuge.

Materials and Methods

A Beckman Optima XL-A (Beckman, Spinco Division, Palo Alto, California, USA) Analytical Ultracentrifuge equipped with standard UV-absorption optics as described elsewhere [13] was used to investigate the Schlieren effect. The basic experiments for this study have been carried out on an XL-A from 1991 (one of the original four prototype or "beta site" units) in Nottingham which does not have a blocking filter in the monochromator. Control experiments were performed on an XL-A from 1992 in Dr. A.J. Rowes' laboratory in Leicester, UK, and on an XL-A from 1993 in Dr. A. Seiferts laboratory (Potsdam, FRG) already equipped with the blocking filter, and even a newer XL-A from 1994 in Dr. H.G. Müllers' laboratory (Leverkusen, FRG) with blocking filter. For the controlexperiments on the newer XL-A's, the blocking filter was disabled to maintain comparable experimental conditions on all machines. All experiments were performed at 20 °C at various speeds of 12 600, 30 000, 50 000 and 60 000 rpm depending on the sample and the type of experiment. All runs have been of the sedimentation velo-

city type with the exception of one sedimentation equilibrium experiment on bovine serum albumin. The buffer was used as reference solvent in every case. Generally, a wavelength scan was first performed at 3000 rpm. Afterwards, three wavelengths have been selected (where the sample shows little or no absorption) to derive the Schlieren patterns. The only exception is the bovine serum albumin sample which has been used to demonstrate the equivalence between the conventional absorption optical traces (scanned at an absorbance wavelength) and the Schlieren patterns. To maintain short scanning times at the cost of noise, scans were performed in continuous mode with a step size of 0.005 cm and only one average to get scanning times of about 2 min/scan.

The samples and buffers used in this study were:

- bovine serum albumin 96–99%, Sigma
- bovine serum albumin partially degraded
- Chitosan, practical grade from crab shells, Sigma
- κ-carrageenan, No. C-1013, Sigma
- Na-alginate, Fisons
- Dextran 20 ($M_w \approx 20\,000$ g/mol), Pharmacia Fine Chemicals, Uppsala, Sweden
- Dextran 150 ($M_w \approx 150\,000$ g/mol), Pharmacia Fine Chemicals, Uppsala, Sweden
- Xanthan Keltrol RD, Kelco International

The buffers used were a phosphate/chloride buffer, pH 7, $I = 0.3$ for all samples except chitosan which was dissolved in an acetate/chloride buffer, pH 4.5, $I = 0.1$. The loading concentrations of the samples were 1 or 2 mg/ml with the exception of one run employing high bovine serum albumin (BSA) concentrations of 5, 10 and 20 mg/ml.

To compare the results of the XL-A Schlieren effect, a Beckman Model E Analytical Ultracentrifuge equipped with conventional Schlieren optics and working with a phase plate as analyzer element has been used applying a phase plate angle of 70° for all velocity runs. The sensitivity of the XL-A Schlieren effect, was estimated by performing an equilibrium experiment at 12 600 rpm with a 1 mg/ml BSA solution (degraded BSA) at 19 °C simultaneously on the XL-A and a Model E with calibrated Schlieren optics [14] at the phase plate angles 40–85° in 5° steps. For all Model E experiments, the same samples and experimental conditions have been applied as for the XL-A experiments. For the Model E experiment, a 2° monosector 12 mm KEL-F centerpiece was used, whereas in the XL-A a 12 mm 2.5° KEL-F double-sector center-piece was employed. Both cells were filled with 120 μl BSA solution. The slight difference in the filling heights resulting from different volumes due to the different sector angles of both centerpieces could be neglected because the filling height was small.

To investigate if the Schlieren effect is a general effect occurring on UV-absorption optical systems or if it is related to the special optical arrangement in the Optima XL-A, a control experiment with a 2 mg/ml BSA solution in phosphate/chloride buffer was performed at 50 000 rpm using a MSE Centriscan ultracentrifuge equipped with scanning absorption and scanning Schlieren optics.

The Schlieren photos of the Model E ultracentrifuge have been evaluated using a method similar to that reported by Gauglitz et al. [15]. The photos were scanned into a computer with a high resolution and then evaluated using image analysis software for electron micrographs (Zeiss AnalySIS 2.0, Soft Imaging Software GmbH, D-48153 Münster). All XL-A scans were evaluated with the Microcal Origin software (Beckman, Palo Alto, California, USA) using the data pointer to derive radial positions etc. The sedimentation coefficients have been evaluated using self-written software for all experiments (XL-A absorption, XL-A Schlieren and Model E Schlieren) [16].

Results and discussion

The Schlieren effect

If bovine serum albumin (BSA), a well characterized protein is investigated by means of a sedimentation velocity experiment scanning at the absorption maximum of 276 nm (absorption maximum, see wavelength scan Fig. 1a), typical absorption traces are derived (Fig. 1b).

If now the wavelength is varied to one where the sample does not absorb or where at least the absorption is minimal (for example 400 and 600 nm), typical Schlieren patterns are derived (Fig. 1b).

There are several characteristics of the observed Schlieren peaks which already become obvious from this first experiment.

– The Schlieren peak well represents the first derivative of the absorption profile if the shape of the peak is considered. Both traces become broader with increasing time. This implies that the peak must somehow be based on light refraction although the photomultiplier of the XL-A is only able to detect light intensities.

– For steep gradients (for example as experienced at the beginning of the experiment) where dn/dr is at its maximum, the Schlieren peak maximum is considerably shifted towards the cell bottom and does not represent the boundary position which would be expected from the normal absorption scan. This becomes obvious from the scans at 276 nm and 400 or 600 nm in Fig. 1b. The effect vanishes with time as the boundary gets broader. The

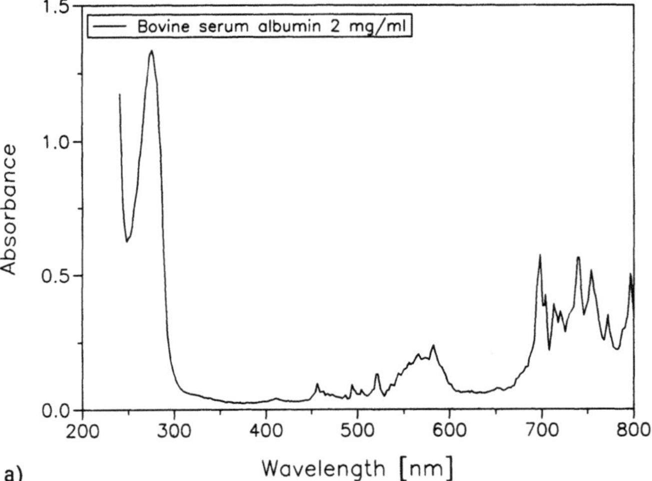

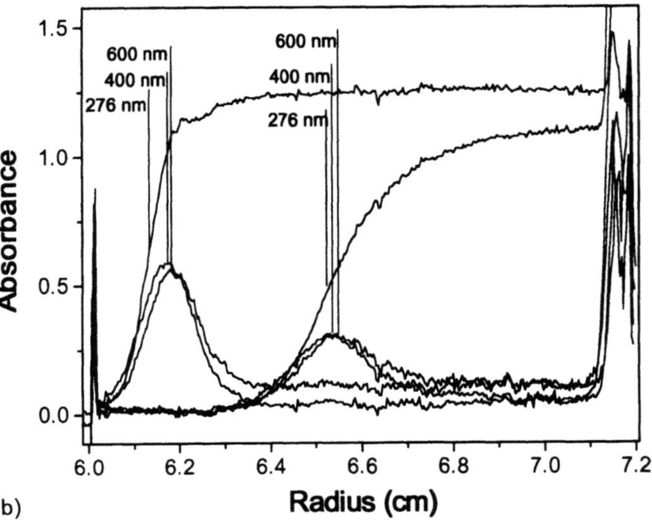

Fig. 1a) Wavelength scan for a 2 mg/ml bovine serum albumin solution in phosphate/chloride buffer; **b)** Radial scans during a sedimentation velocity experiment with a 2 mg/ml BSA solution in phosphate/chloride buffer at 50 000 rpm and 20 °C. The scans were taken at 276, 400 and 600 nm. The time interval between the individual scans at these wavelengths is 2 min. Hence there is a small but constant shift between the boundary positions scanned at different wavelengths due to the sedimentation of the protein

consequence hereof is that the very first peaks of the experiment should not be taken into account if a sedimentation coefficient is to be evaluated from such Schlieren traces, unless the boundary remains steep as is the case for hypersharp peaks (See the Xanthan example given later in this article).

The next point of interest was if the sedimentation coefficients for BSA evaluated from absorption and Schlieren traces agree (if the first Schlieren peaks for the steep concentration gradient are not taken into account for the evaluation). In the following experiment with BSA,

Schlieren and absorption traces have been derived under identical conditions just by a variation of the scanning wavelength. The corresponding traces are shown in Fig. 2.

If the sedimentation coefficients (uncorrected) are evaluated, a value of 4.69 S is derived from the absorption traces at 276 nm, whereas the corresponding value from the Schlieren traces at 400 nm is 4.60 S. These values are in good agreement, demonstrating that the Schlieren effect can be exploited to determine sedimentation coefficients.

The wavelength dependence of the Schlieren effect

To investigate the nature of the Schlieren effect an experiment was carried out to look at the wavelength dependence. By plotting the optical density of a fixed point on the Schlieren peak as a function of the wavelength, it should be possible to distinguish whether the Schlieren traces are caused by true refraction, light scattering or even experimental artefacts. If light scattering is responsible for the Schlieren effect, one would expect the registered optical density to follow the Rayleigh law $1/\lambda^4$. Light scattering effects must be considered as a possible reason because the photomultiplier of the XL-A can only register differences in light intensities and the absorption optical system of the XL-A does not have any obvious analyzer element like a phase plate or a knife edge. If true refraction would cause the Schlieren effect, the registered optical density should be proportional to $1/\lambda^2$ due to the wavelength dependence of the refractive index increment [17]. If some experimental artefacts would be the reason, none of these dependencies should be expected.

Fig. 2 Absorption and Schlieren traces for 2 mg/ml BSA in phosphate/chloride buffer at 20 °C, 50 000 rpm scanned at 276 nm (absorption) and 400 nm (Schlieren) with a scan interval of 20 min

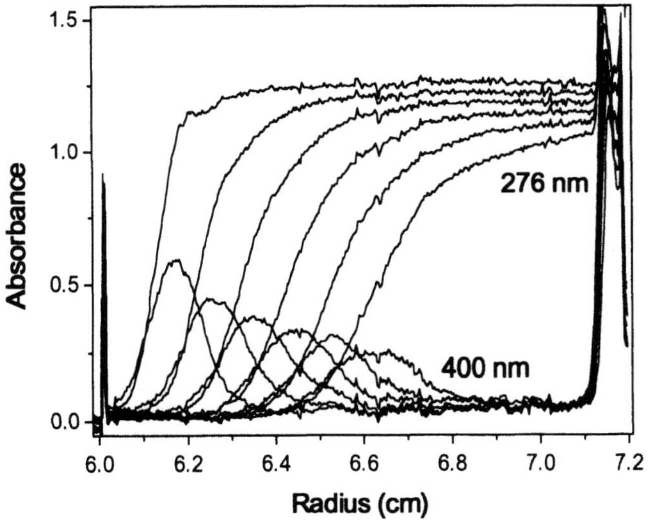

As proteins exhibit a UV-absorption, they are not particularly well suited to investigate the wavelength dependence of the registered "absorption" as it must be expected that there is a transition between normal absorption and the Schlieren effect. Therefore a sample was chosen which does not show significant UV-absorption – Xanthan, a polysaccharide. Xanthan is particularly well suited to investigate the Schlieren effect because it shows a hypersharp Schlieren peak which can still be detected at polymer concentrations lower than 0.05 mg/ml [18]. The wavelength scan of a 0.8 mg/ml Xanthan solution in phosphate/chloride buffer is shown in Fig. 3a.

It can be seen from Fig. 3a that the absorption between 220 and 740 nm is sufficiently low to investigate the wavelength dependence of the Schlieren effect. The following figure shows the sedimentation profile of Xanthan scanned at 500 nm as an example.

Before the wavelength dependence was investigated, an experiment with the same solution and under the same conditions was carried out on a Beckman Model E ultracentrifuge equipped with conventional "Philpot-Svensson" phase plate Schlieren optics to investigate, if the same peak shape is derived here. The result is shown in Fig. 3c.

It can clearly be seen that the peak shapes detected in Figs. 3b and c are identical. Nevertheless, there is an obvious difference between these plots. In Fig. 3c, the baseline is shifted to higher dn/dr values due to the increased acting pressures at the higher centrifugal forces (compression of the liquid and hence density and refractive index increase) as it is well known for such experiments. But in Fig. 3b, the baseline is almost 0. The same can be seen in Figs. 1b and 2, although in these cases the speed was rather high with 50 000 rpm.

Schlieren traces as shown in Fig. 3b have been derived for the wavelength range between 220 and 740 nm in 20 nm steps. At wavelengths higher than 740 nm, the data became so noisy that a proper evaluation of the experiments was not possible anymore. Unfortunately, the peak maximum of the hypersharp Xanthan pattern is not very well characterized in terms of its optical density. To minimize the errors in the proper determination of the absorbance value of the peak maximum, 18 peaks of successive scans have been evaluated for the wavelengths in the range between 220 and 560 nm. Furthermore, the half peak height (position of the half peak height regardless if the fine upper range of the peak can be seen or not) has been evaluated in the same manner to get slightly safer values.

If one looks at the peak shapes at different wavelengths, one observes a sudden change in the overall sedimentation diagram at wavelengths of 560 and 580 nm. All sedimentation diagrams at wavelengths up to 560 nm look like that in Fig. 3b, although it is obvious that the peaks

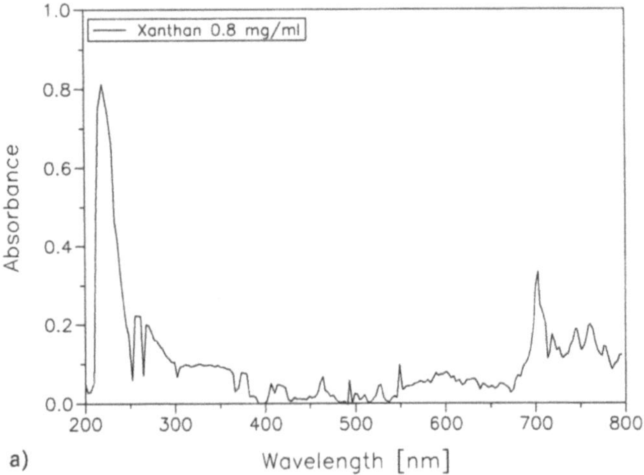

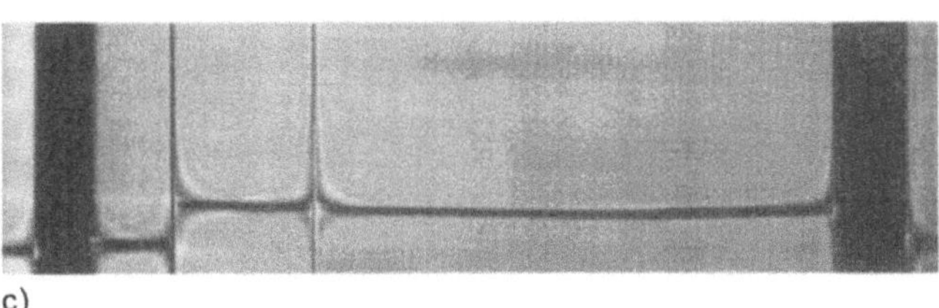

Fig. 3a) Wavelength scan for a 0.8 mg/ml Xanthan solution in phosphate/chloride buffer; **b)** Schlieren traces for 0.8 mg/ml Xanthan in phosphate/chloride buffer at 20 °C, 30 000 rpm scanned at 500 nm with a scan interval of 23 min; **c)** Schlieren photo of a 0.8 mg/ml Xanthan solution in phosphate/chloride buffer at 20 °C and 30 000 rpm using a conventional Schlieren optics with a mercury lamp and green filter ($\lambda \approx 540$ nm) on a Beckman Model E ultracentrifuge; **d)** Schlieren traces for 0.8 mg/ml Xanthan in phosphate/chloride buffer at 20 °C, 30 000 rpm scanned at 580 nm with a scan interval of 23 min

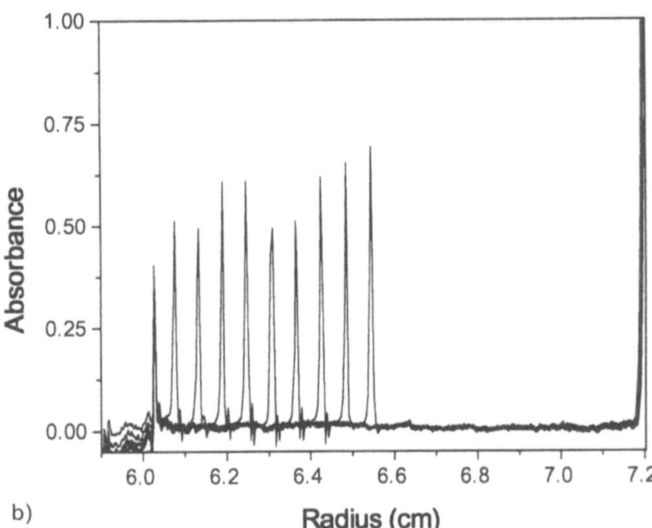

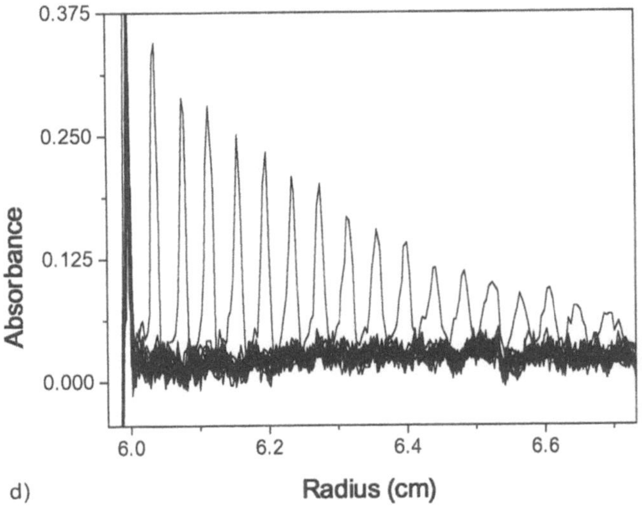

become slightly broader with increasing wavelength. This is most obvious at the base of the peaks. However, at 580 nm the peak height decreases with the progress of sedimentation as it is shown in Fig. 3d. With progression of the experiment the peaks become broader due to diffusion.

If the wavelength is further increased, the peaks also become broader whereas their height is steadily decreased. At 740 nm in this example, the data quality becomes so bad (peak maximum close to the noise amplitude) that a proper evaluation is not possible anymore.

As the peak height decreases with the progress of the experiment for scanning wavelengths between 580 and 740 nm (see Fig. 3d), only the very first peak was considered for evaluation. Therefore, it has to be expected that the results of the wavelength dependence of the Schlieren effect are of rather poor quality.

If now the wavelength dependency of the optical density of the peak maximum or half peak height is investigated, it becomes quite obvious that light scattering cannot be responsible for the generation of the Schlieren peaks. No proportionality between the optical density and $1/\lambda^4$ was found, as can be seen in Fig. 4a.

If the second possible cause of the wavelength dependence is explored – i.e., the "refraction" effect, a more or less linear dependence is found (within the rather large error intervals) indicating the dependence on λ^{-2} as shown in Fig. 4b.

This must mean that light refraction is responsible for the generation of the Schlieren peaks. Therefore the observed peaks are real Schlieren peaks representing the first derivative of the concentration gradient rather than experimental artefacts or a light scattering effect.

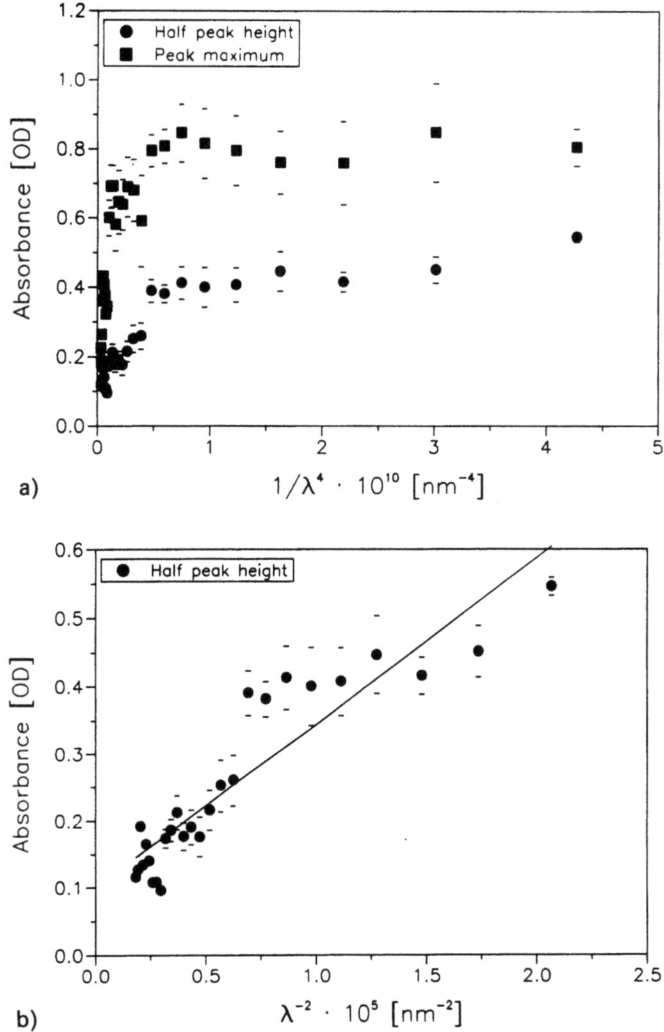

a)

b)

Fig. 4a) Dependence of the registered optical density/"absorbance" for a Schlieren peak on $1/\lambda^4$ as a functional check for light scattering as reason for the generation of the Schlieren peaks. The $-$ signs represent the upper and lower limits of the standard deviation; **b)** Dependence of the registered optical density/"absorbance" for a Schlieren peak on $1/\lambda^2$ as a functional check for light refraction as reason for the generation of the Schlieren peaks. The $-$ signs represent the upper and lower limits of the standard deviation

If light refraction causes a Schlieren pattern registered with a UV-absorption optics, the important question arises of how normal absorption traces are influenced by light refraction, e.g., can absorption and light refraction be superimposed? In such a case, no accurate absorbance measurements would be possible because the registered absorbance at a distinct position is the sum of the real absorbance and the Schlieren effect although the Schlieren effect is less pronounced in terms of optical density than the normal absorbance profile. However, especially sedimentation equilibrium experiments should be affected by

such a superposition and the question is, how can accurate quantitative absorption measurements ever be made?

To answer these questions the series of sedimentation velocity experiments carried out with Xanthan before has been repeated with BSA because a transition between absorption and Schlieren traces has to be expected for this protein example. As for Xanthan, the scanning wavelength was varied between 220 and 740 nm in 20 nm increments. Some of the results are presented in Fig. 5.

Although these scans are rather noisy, the transition between the normal absorption trace and the Schlieren peaks can be seen. At 280 nm, the profile still corresponds to a normal absorption profile, whereas at 300 nm – a wavelength where the UV absorption is considerably decreased (see also Fig. 1a) – the transition state seems to be represented. Although the contribution of the Schlieren effect is already obvious, the pattern still does not correspond to the Schlieren pattern derived for 320 nm or higher wavelengths. From this result, it seems that the transition between the two traces can be rather sharp and not too continuous if the absorption decrease between the considered wavelengths in the wavelength scan is significant (e.g., transition absorption – no absorption, see Fig. 1a). This is encouraging because such results imply the possibility of a multiple use of the XL-A UV-absorption optics as Schlieren and absorption optics without one of them suffering from the presence of the other effect (Schlieren and absorption). It might affect however, experiments on molecules with absorbing chromophores at wavelengths > 320 nm (e.g., heme proteins, NADH bound systems

Fig. 5 Wavelength dependence of the traces registered with the UV-absorption optics at 280, 300 and 320 nm. The sample was a 1 mg/ml BSA (degraded) solution in phosphate/chloride buffer at 20 °C and 50 000 rpm

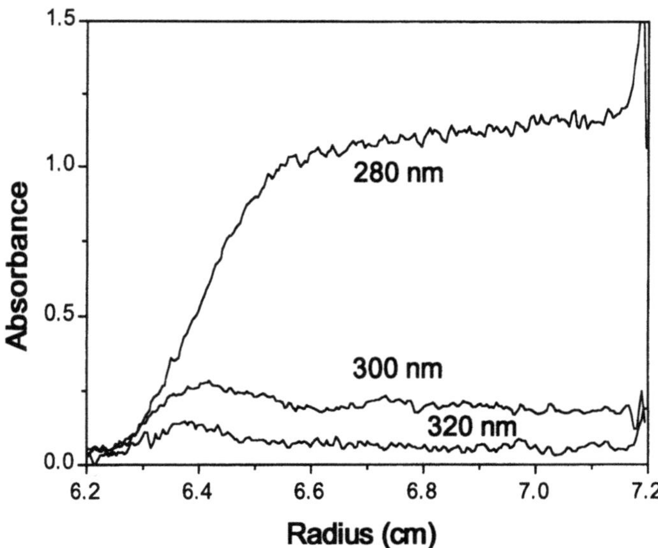

or fluorescein labelled systems). Nevertheless, systematic studies need to be performed now. At least there seems to be a chance to find a wavelength, where either the absorption is very predominant (Schlieren effects can be neglected) or the absorbance can be neglected to give Schlieren peaks purely based upon light refraction.

Schlieren peaks with polysaccharides

From the results given above, it becomes clear that it should be possible to derive Schlieren patterns for samples which do not absorb at all or at least not significantly at the wavelength selected. To verify this experimentally, a comparative study was carried out on several polysaccharides. As these samples do not show a significant absorption in the UV or visible light, they should be well suited to generate Schlieren patterns. Therefore, sedimentation velocity experiments were performed with the same polysaccharide solution under identical conditions using the XL-A with absorption optics and the Model E equipped with conventional Schlieren optics employing a mercury light source with green filter ($\lambda \approx 540$ nm). To get an idea of the sensitivity of the Schlieren effect, the phase plate angle of the Model E Schlieren optics was adjusted to 70° to maintain comparable conditions of the conventional Schlieren photos.

The subject of interest in these experiments was not only if the sedimentation coefficients in the buffer at 20 °C ($s_{20,b}$) derived from the XL-A and Model E Schlieren peaks agree well, but furthermore to investigate if different peak shapes are properly displayed on the XL-A. The Xanthan results presented above can serve as an example for a correctly displayed hypersharp peak. The other polysaccharides have been selected to enhance the range of examples to a symmetrical normal peak, broad peaks and asymmetrical peaks.

Alginate

The first example given is Na-alginate (see also [19]) which shows a fairly symmetrical peak.

If the sedimentation coefficients $s_{20,b}$ are evaluated, a value of 1.49 S is obtained from the XL-A Schlieren traces whereas a value of 1.46 S is derived from the Model E Schlieren photos.

κ-carrageenan

The next polysaccharide of interest was κ-carrageenan which shows an asymmetrical Schlieren peak.

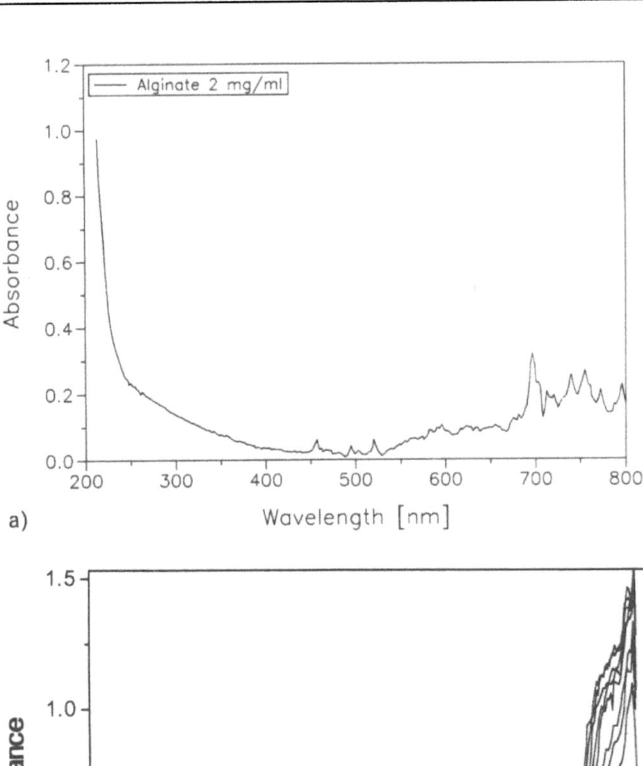

a)

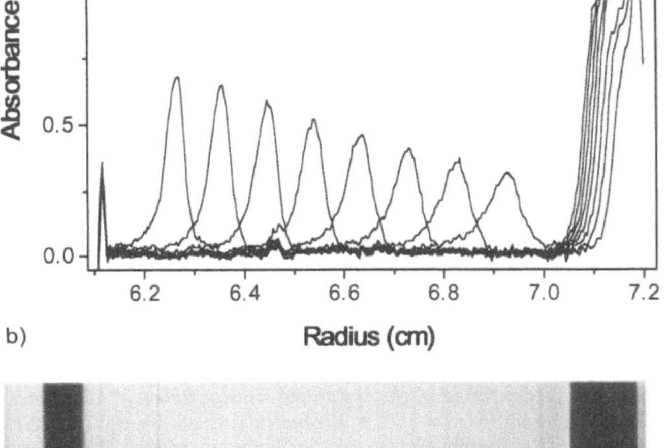

b)

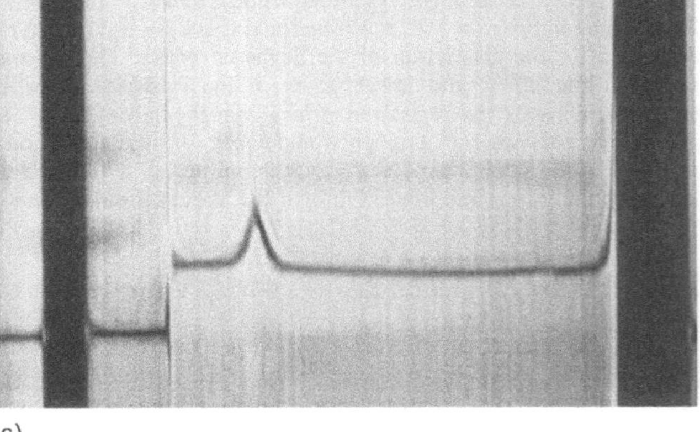
c)

Fig. 6a) Wavelength scan for a 2 mg/ml Na-alginate solution in phosphate/chloride buffer; **b)** Schlieren traces for a 2 mg/ml Na-alginate solution in phosphate/chloride buffer at 20 °C, 60 000 rpm scanned at 500 nm with a scan interval of 40 min; **c)** Schlieren photo of a 2 mg/ml Na-alginate solution in phosphate/chloride buffer at 20 °C and 60 000 rpm using the conventional Schlieren optics of the Beckman Model E ultracentrifuge

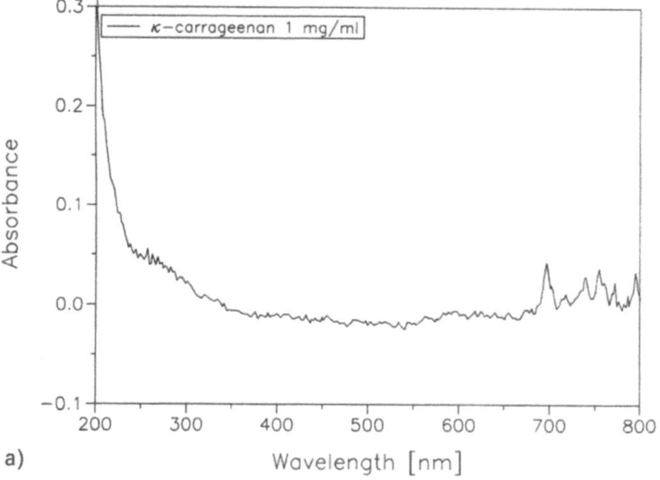

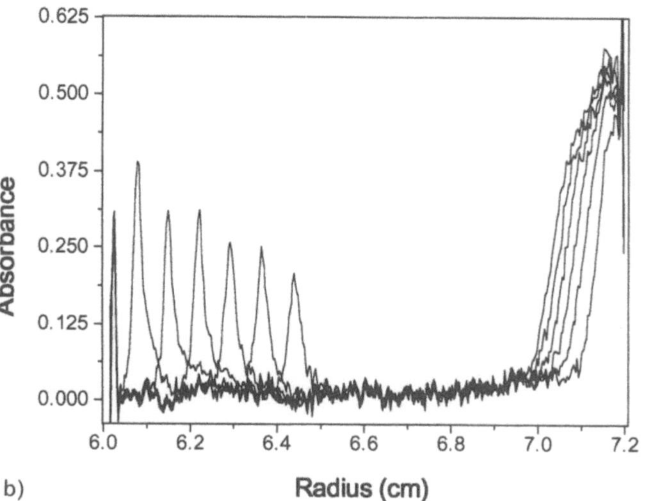

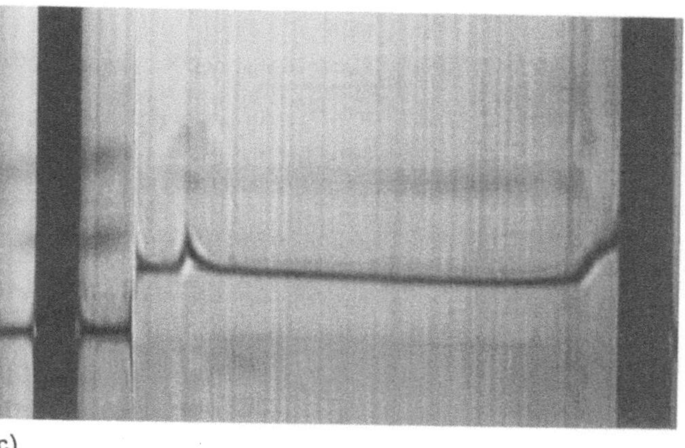

Fig. 7a) Wavelength scan for a 1 mg/ml κ-carrageenan solution in phosphate/chloride buffer; b) Schlieren traces for a 1 mg/ml κ-carrageenan solution in phosphate/chloride buffer at 20 °C, 60000 rpm scanned at 500 nm with a scan interval of 33 min; c) Schlieren photo of a 1 mg/ml κ-carrageenan solution in phosphate/chloride buffer at 20 °C and 60000 rpm using the conventional Schlieren optics of the Beckman Model E ultracentrifuge

The $s_{20,b}$ value derived from the XL-A Schlieren peaks is 1.46 S whereas from the Model E Schlieren photographs, a value of 1.48 S is derived.

Chitosan

Chitosan has been selected as an example for an asymmetrical Schlieren peak.

The $s_{20,b}$ from the XL-A Schlieren peaks is 1.36 S, that derived from the Model E Schlieren photos is 1.41 S. The apparent extra peak near the cell bottom in the Model E photo (see Fig. 8c) is caused by an experimental artefact (scratches or dirt on the window, etc.) rather than by a sedimenting species. This artefact did not move during the whole experiment.

Dextran ($M_w = 20000$ g/mol)

As an example for a polydisperse sample yielding a broad Schlieren peak, a low molar mass dextran ($M_w = 20000$ g/mol) was investigated.

For this example, it was rather difficult to trace the exact boundary position, especially in the Model E Schlieren photographs. Therefore the difference between the sedimentation coefficients from the XL-A and Model E are significant. Nevertheless, the XL-A value seems to be more reliable because the boundary position could be traced much better here. For the XL-A traces, a $s_{20,b}$ value of 2.18 S was derived whereas that from the Model E Schlieren photos was 2.83 S.

Dextran ($M_w = 150000$ g/mol)

As a last example, another dextran with $M_w = 150000$ g/mol was investigated.

The $s_{20,b}$ value for the XL-A traces is 4.18 S, whereas that for the Model E Schlieren photographs is 4.22 S.

If the sedimentation coefficients derived from the XL-A Schlieren peaks are compared with those from conventional Schlieren photos, good agreement can be seen with the one possible exception of the low molecular weight dextran sample. This means that – at least for sedimentation velocity experiments – the XL-A can be used exploiting the Schlieren effect.

If one now looks at the sensitivity of the Schlieren effect, one can compare the height of the Schlieren peaks with respect to their base of the XL-A and Model E Schlieren peaks. For every sample it becomes obvious that the XL-A is more sensitive than the conventional Schlieren optics employing a phase plate angle of 70°. The "effective" phase

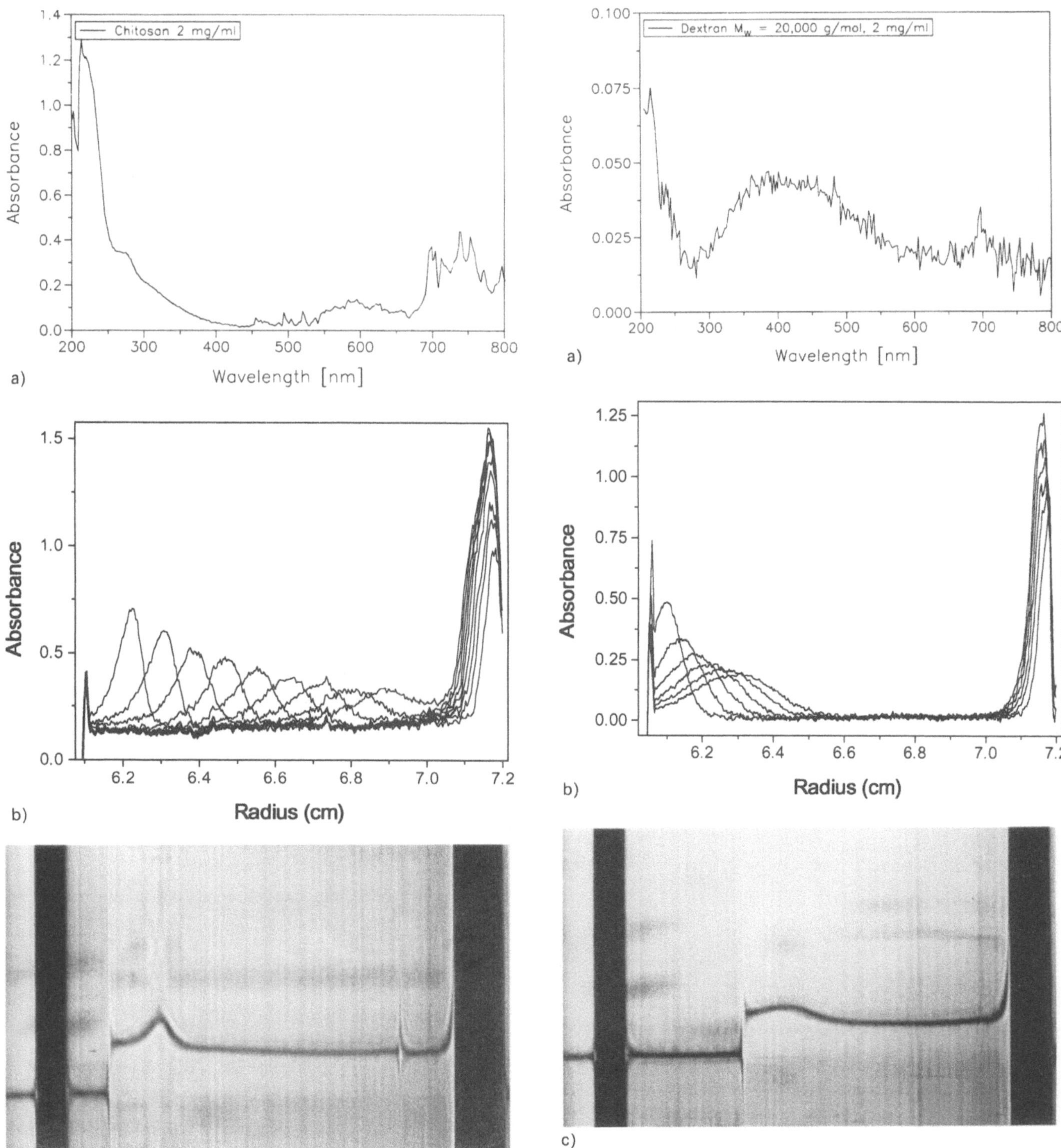

Fig. 8a) Wavelength scan for a 2 mg/ml chitosan solution in acetate/chloride buffer; **b)** Schlieren traces for a 2 mg/ml chitosan solution in acetate/chloride buffer at 20 °C, 60 000 rpm scanned at 500 nm with a scan interval of 40 min; **c)** Schlieren photo of a 2 mg/ml chitosan solution in acetate/chloride buffer at 20 °C and 60 000 rpm using the conventional Schlieren optics of the Beckman Model E ultracentrifuge

Fig. 9a) Wavelength scan for a 2 mg/ml dextran ($M_w =$ 20 000 g/mol) solution in phosphate/chloride buffer; **b)** Schlieren traces for a 2 mg/ml dextran ($M_w = 20\,000$ g/mol) solution in phosphate/chloride buffer at 20 °C, 50 000 rpm scanned at 500 nm with a scan interval of 20 min; **c)** Schlieren photo of a 2 mg/ml dextran ($M_w = 20\,000$ g/mol) solution in phosphate/chloride buffer at 20 °C and 50 000 rpm using the conventional Schlieren optics of the Beckman Model E ultracentrifuge

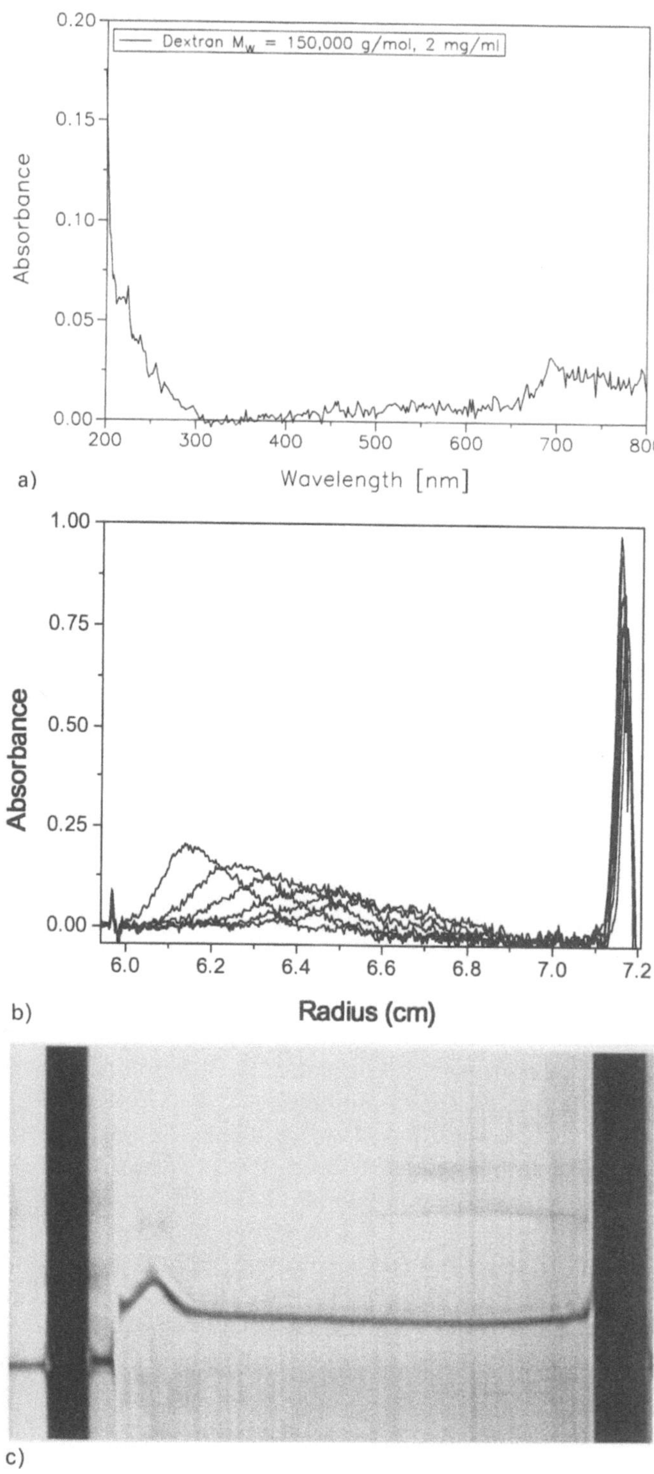

Fig. 10a) Wavelength scan for a 2 mg/ml dextran (M_w = 150 000 g/mol) solution in phosphate/chloride buffer; **b)** Schlieren traces for a 2 mg/ml dextran (M_w = 150 000 g/mol) solution in phosphate/chloride buffer at 20 °C, 50 000 rpm scanned at 500 nm with a scan interval of 20 min; **c)** Schlieren photo of a 2 mg/ml dextran (M_w = 150 000 g/mol) solution in phosphate/chloride buffer at 20 °C and 50 000 rpm using the conventional Schlieren optics of the Beckman Model E ultracentrifuge

plate angle of the XL-A can be estimated to be about 60° at 500 nm although it must be pointed out that the sensitivity for some samples seems to be even higher (κ-carrageenan and the low molecular mass dextran).

If the XL-A and Model E Schlieren traces in Figs. 6 to 10 are compared, it can be stated that in every case, the shapes of the Schlieren peaks agree. The same holds for the region of the cell bottom. Due to a higher sensitivity (compared to the Model E Schlieren optics with a phase plate angle of 70°) the XL-A displays more detail.

With the exception of chitosan, it is found again that the baseline of the XL-A Schlieren patterns is zero although a high speed has been selected which leads to a significant displacement of the baseline in the Model E Schlieren photos. This can be explained by the type of measurement in the XL-A: The absorption of the reference solvent is substracted from the sample absorption at that position leading to a zero baseline regardless of the speed.

Some considerations about the quantitivity of the Schlieren effect

It is an important question for future applications if the Schlieren effect is quantitative or not. To investigate this, an experiment was performed measuring three different concentrated BSA solutions (20, 10 and 5 mg/ml) in the same run at 20 °C and 60 000 rpm using a short scanning interval of only 3 min between two cells scanned at the same wavelength (at the cost of some noise). The short scanning interval is necessary to ensure that the peaks are essentially measured at the same time and can hence be compared. The question now is if the solute concentration can be derived by integrating the area between the baseline and the Schlieren peaks or not? This can be addressed by evaluating a term proportional to the effective phase plate angle of the XL-A Schlieren pattern for each of the samples. In general: is the term proportional to the phase plate angle for the 20 mg/ml BSA equal to that for the 10 mg/ml and the 5 mg/ml BSA solution? In such a case the XL-A Schlieren effect is quantitative. For the evaluation of this term, namely $\tan\Theta \cdot \text{const.}$, the following formula can be used [14]:

$$\tan\theta \cdot \text{const.} = \frac{c_0 \cdot \Delta n}{A_{\text{Sample}} \cdot (r_{\text{Sample}}/r_{\text{Meniscus}})^2,} \tag{1}$$

where Δn is the refractive index difference of the solute (0.00186 per 1% by wt. BSA), c_0 the initial solute concentration in % by wt., Θ the angle of inclination of the Schlieren diaphragm, A_{Sample} = area between Schlieren equilibrium gradient and baseline at a distinct phase plate angle in a specified radial range, r_{Sample} = radial position of interest in the sample gradient, r_{Meniscus} = radial position of the meniscus. The constant contains factors like the optical lever, the magnification factors of the camera and cylinder lens system and the height of the center-piece in case of a conventional Schlieren optics but cannot be determined for the XL-A in absence of these optical components. Nevertheless, all these factors must remain constant throughout the entire run as they are all optical constants.

First the area under the XL-A Schlieren peaks was determined using the Microcal Origin software (Beckman, Palo Alto, USA). A baseline has been determined in a way connecting both ends of the peak (where the slope does not change anymore) by a line. The area between this line which serves as a substitute for the baseline (it can be obtained when running a double sector cell in conventional Schlieren optics) and the Schlieren peak was determined with the integration function of the Origin software. These areas have been determined for four subsequent scans for each of the original concentrations to minimize evaluation errors. The results are given in Table 1.

It is obvious that the values presented in Table 1 are rather erroneous from what can be expected if the integration procedure with an artificial baseline is taken into account. Nevertheless, if one looks at the ratios $\tan\Theta\cdot$const. for two different BSA concentrations (with some exceptions) the effective phase plate angle term is increased by about a factor of 3 if the solute concentration is doubled or increased by factor of 9 if the solute concentration is increased by factor of 4. So, it may be possible that a term of the kind $3^{(\text{high solute concentration}/2\cdot\text{low solute concentration})}$ can be used to estimate the increase in the effective phase plate angle Θ of the XL-A with alteration of the solute concentration. What is clear without doubt from these results is that the Schlieren effect is not quantitative. This has important consequences:

– Sedimentation equilibrium experiments as well as quantitative sedimentation velocity experiments exploiting the Schlieren effect are not possible (at least at higher concentrations) because the concentration gradient causes local alterations of the effective phase plate angle in sedimentation equilibrium experiments. This effect does not disturb the evaluation of the sedimentation coefficient from the movement of the boundary and hence does not affect the use of the Schlieren effect for sedimentation velocity experiments too much. Also, the overall shape of the peak registered is not altered significantly as it could be shown in the many examples given before.

– The effective phase plate angle is increased more than linearly with increasing concentration. This is very advantageous for high concentration work (concentrations much higher than 20 mg/ml used in this study should be possible in 12 mm centerpieces) as the effective phase plate angle is increased for the higher concentrations simultaneously decreasing the sensitivity of the Schlieren effect. On the other hand, going to lower concentrations the effective phase plate angle decreases more than linearly which increases the sensitivity of the Schlieren effect. This enables the use of the Schlieren effect to investigate low concentrated polymer solutions in the 12 mm XL-A double sector cells which cannot be otherwise investigated using conventional Schlieren optics – even when employing 30 mm cells. This could be demonstrated for Xanthan by Dhami et al. [18]. Such a finding means that the observed XL-A Schlieren effect is a dynamic effect as far as the concentration dependence is concerned. An explanation for this can be due to light scattering: light scattered from neighboring radial positions enters the photomultiplier and thus more light is registered than would correspond to pure refraction. This decreases the registered optical density. As light scattering is concentration dependent, it can be understood why the Schlieren effect becomes more insensitive with higher concentration. Only at very low concentrations can it be expected that pure light refraction is registered.

To visualize the view that the Schlieren effect is dynamic, it is advantageous to look at the Schlieren patterns for the three different concentrated BSA solutions measured under identical conditions.

It can be seen that the area under the peak for the 5 mg/ml sample (0.2340 relative units) is already close to that under the 10 mg/ml sample (0.2732 relative units). The same is valid for the 10 and 20 mg/ml samples (0.3347 relative units). It can be stated that there is no obvious proportionality between the peak area and the solute concentration.

From the results above, the dynamic nature of the XL-A Schlieren effect with regard to the solute concentra-

Table 1 The term $\tan\Theta\cdot$const. according to Eq. (1) for three different concentrations of BSA solutions investigated at a wavelength of 500 nm in an Optima XL-A ultracentrifuge. The ratios of $\tan\Theta\cdot$const. for two initial BSA concentrations are given in addition

	$\tan\Theta\cdot$const. for BSA 20 mg/ml	BSA 20/ 10 mg/ml	$\tan\Theta\cdot$const. for BSA 10 mg/ml	BSA 10/ 5 mg/ml	$\tan\Theta\cdot$const. for BSA 5 mg/ml	BSA 20/ 5 mg/ml
Scan # 1	0.69598	2.9	0.23811	3.3	0.07292	9.5
Scan # 2	0.69540	3.1	0.22121	2.9	0.07632	9.1
Scan # 3	0.51323	2.3	0.21880	2.5	0.08754	5.9
Scan # 4	0.58470	2.7	0.21729	2.3	0.09535	6.1

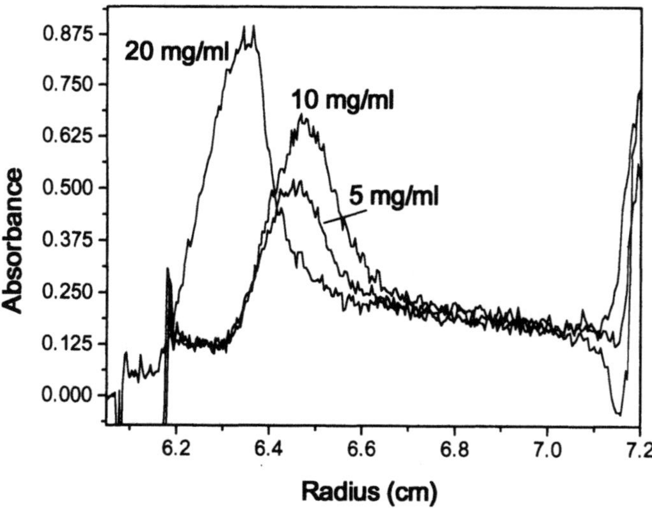

Fig. 11 Schlieren traces for three different concentrated BSA solutions scanned at 60 000 rpm, 20 °C and 500 nm. All traces correspond to the same scanning time (± 3 min). The filling height of the 20 mg/ml sample was slightly higher than that of the other samples

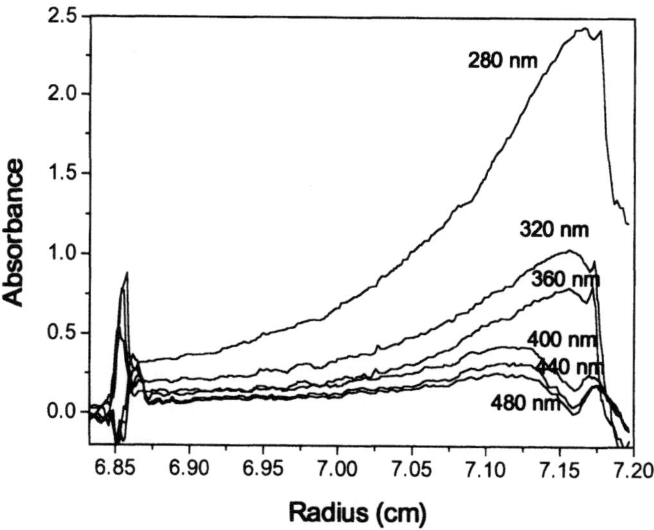

Fig. 12 Equilibrium scans for a 1 mg/ml BSA (degraded) solution at 20 °C, 12 600 rpm and 20 °C scanned at different wavelengths. The 280 nm scan corresponds to the conventional UV-absorption trace of the protein

tion can clearly be confirmed. Hence, it should not be possible to derive relevant information from sedimentation equilibrium experiments. To investigate this further, a series of sedimentation equilibrium experiments has been carried out with a 1 mg/ml BSA solution (degraded BSA) scanning at different wavelengths. From the experimental scans (see Fig. 12), it already becomes obvious that the M_z values calculated for different wavelengths should differ in contrast to those values from different phase plate angles with conventional Schlieren optics.

The difference between the scans at different wavelengths especially becomes obvious in the region of the cell bottom where the concentration is the highest. At wavelengths above 400 nm the cell bottom is not resolved anymore which implies light scattering in that region. Although the scattering should be decreased with increasing wavelength due to the $1/\lambda^4$ dependence, its influence only becomes obvious above 400 nm due to the superposi-

tion with the Schlieren effect which is decreased with increasing wavelength. Beyond 400 nm, the Schlieren effect seems to be predominant.

If now apparent $M_{z,app.}$ values are calculated from these traces (leaving the 280 nm scan out which corresponds to absorption rather than to the Schlieren effect), the results in Table 2 are obtained.

It is obvious that the XL-A values show a systematic decrease with increasing wavelength. This corresponds to the flatter equilibrium gradients at higher wavelengths (see Fig. 12), whereas the M_z-values from conventional Schlieren optics at different wavelengths are constant though having large statistical errors. The $M_{z,app.}$ at 320 nm is higher than the Model E reference. This may be explained by some contribution from UV-absorption of BSA at that wavelength. Such results should serve as a serious warning to all workers seeking accurate quantitative information from the XL-A absorption optics as the

Table 2 M_z-values calculated from conventional Schlieren optics (Model E, lower half of the table) compared to wavelength dependent apparent $M_{z,app.}$-values from the XL-A Schlieren effect (upper half). The sample was 1 mg/ml BSA (degraded) in phosphate/chloride buffer investigated at 19 °C and 12 600 rpm using $\rho = 1.0115$ g/ml and $\bar{v} = 0.732$ ml/g

XL-A	320 nm	360 nm	400 nm	440 nm	480 nm	520 nm	560 nm
$M_{z,app.}$ [g/mol]	49 300	37 500	36 600	30 200	21 900	24 100	16 400
M_z [g/mol]	41 600	34 000	37 100	39 500	40 200		
Model E	40°	45°	50°	55°	60°		

pure absorption signal might be superimposed by the Schlieren effect or the light-scattering dependent on the scanning wavelength.

Sensitivity of the Schlieren effect

To get an estimate of the sensitivity of the Schlieren effect, it is necessary to continuously vary the solute concentration from a given value to zero. This can be established best by performing a high-speed sedimentation equilibrium experiment where the meniscus concentration is zero. The experiment was performed on the XL-A as well as on a Model E with calibrated Schlieren optics. It is assumed that the concentration in the equilibrium gradient is equal at the same radial position for the Model E and the XL-A runs: A Schlieren pattern from the XL-A is examined to find the radial position r_{Sample} where the solute gradient differs first from the baseline (this corresponds to the lowest detectable refractive index difference in the Schlieren pattern from the XL-A). It is then possible to calculate a Δn value from the Model E Schlieren photograph taken at a distinct phase plate angle by measuring the area between the Schlieren peak and the baseline in the range from the meniscus to r_{Sample}. and using [14]:

$$\Delta n = \frac{A_{Sample}}{A_{Cal}} \frac{\phi x}{T c_0} \left(\frac{r_{Sample}}{r_{Meniscus}}\right)^2, \qquad (2)$$

where ϕ is the wedge angle of the Model E calibration cell (0.00906 rad), x the distance between the calibration marks of the calibration cell (0.9995 cm), T = height of the centerpiece (1.2 cm) and A_{cal} = Area of the calibration cell according to the instructions given in [14] at the same phase plate angle.

The calculated Δn values are those values which are detected by the conventional Schlieren optics at the speci-

fied phase plate angle and that radial distance where the XL-A gradient differs from the baseline first at the chosen wavelength. Hence a value greater than zero indicates that the conventional Schlieren optics is more sensitive under the conditions chosen. The XL-A has a decreased sensitivity of the specified Δn value compared to the conventional Schlieren optics under the given conditions. The results of corresponding systematic experiments are given in Table 3. The X indicates that there is no significant difference between the XL-A and Model E anymore which gives the effective phase plate angle of the XL-A at the specified wavelength. A zero value indicates that the XL-A is more sensitive than the Model E under the given conditions. As control and estimation of the error of the results in Table 3, the Δn value at the cell bottom is given for the conventional Schlieren optics which should be a constant value for all phase plate angles.

First, it can be seen from Table 3 that the Δn value at the cell bottom is roughly constant though having significant error especially at higher phase plate angles. This reflects the sources of error for the area measurements under the Schlieren profiles: at low phase plate angle, the area is quite large but the gradient is rather broad so that it is difficult to trace exact positions. On the other hand, at higher phase plate angles with a sharper gradient, the area to integrate becomes considerably smaller. The latter effect seems to be predominant so that the values for the higher phase plate angles in Table 3 should be treated as more erroneous. Furthermore, the errors in the Δn values at the different wavelengths must be considered as not too small due to the very small areas to be measured between the Schlieren gradient and the baseline. This is expressed for example in the non-systematic values for 400–480 nm in Table 3.

However, the results in Table 3 give a realistic estimation of the effective phase plate angle of the XL-A

Table 3 Δn values calculated from area measurements between the Schlieren curve and the baseline via Eq. (2). The Δn values in the table have been multiplied by 10^5. The dn/dr value at the meniscus has been taken as the baseline value. If an X is indicated, the effective phase plate angle of the XL-A Schlieren patterns roughly equals the corresponding phase plate angle from conventional Schlieren optics. As a control the Δn value at the cell bottom is given for the different phase plate angles

	Δn at 320 nm	Δn at 360 nm	Δn at 400 nm	Δn at 440 nm	Δn at 480 nm	Δn at 520 nm	Δn at 560 nm	Δn at 600 nm	Δn cell bottom
40°	4.42	6.99	10.75	10.74	10.73	15.74	18.35	68.00	865.60
45°	3.76	4.03	5.93	5.93	5.93	8.95	8.96	47.27	845.23
50°	2.88 ×	3.21 ×	7.09	7.08	7.08	9.07	8.76	44.36	847.03
55°	0	0	1.94 ×	1.94 ×	1.94 ×	6.63	9.37	48.92	827.61
60°	0	0	0	0	0	4.19	7.00	47.35	891.38
65°	0	0	0	0	0	1.74 ×	5.81	60.21	860.91
70°	0	0	0	0	0	0	5.95	57.00	849.99
75°	0	0	0	0	0	0	5.17 ×	59.95	946.92
80°	0	0	0	0	0	0	0	72.76	946.77
85°	0	0	0	0	0	0	0	35.67 ×	745.17

Schlieren patterns at different wavelengths in agreement with the first estimation given for sedimentation velocity experiments with the polysaccharides above. As it can be expected from the wavelength dependence of the refractive index increment dn/dc [17]

$$\frac{dn}{dc} = \left(\frac{dn}{dc}\right)_0 (0.94 + 2\cdot 10^6 \cdot \lambda^{-2}) , \qquad (3)$$

with $(dn/dc)_0$ = refractive index increment at 578 nm, the sensitivity of a refractive index based optics for concentration detection is generally decreased with increasing wavelength which has also been pointed out for the Rayleigh interferometer [20].

This view is clearly supported by the results in Table 3. A systematic increase in Δn is found with a significant jump between 560 and 600 nm. It can be concluded that the sensitivity of the XL-A Schlieren effect is decreased drastically in this range. The effective phase plate angle of the XL-A consequently shows the same trend as the Δn values. It is increased with increasing wavelength. Overall the results in Table 3 clearly indicate that the XL-A Schlieren effect is quite sensitive when working with effective phase plate angles of only 50° at an appropriately low wavelength where the sample does not absorb. As can be seen from all previous XL-A scans, the Schlieren pattern is much better defined (sharp gradient although it might be a bit noisy) as is the case with Schlieren photos at phase plate angles of 50°.

It is also useful to compare the distances where the gradient in the Schlieren curve deviates from the baseline first for the XL-A and the Model E Schlieren traces as an independent check for the effective phase plate angle. The higher the radial position of the first deviation, the more insensitive is the optical system. The results are given in Table 4.

Qualitatively the results in Table 3 are confirmed by those independently derived in Table 4. The effective phase plate angle of the XL-A at a specified wavelength can be found by looking at which phase plate angle of the conventional Schlieren optics nearly the same radial position of the first deviation from the baseline is observed in the patterns from Model E and XL-A. Furthermore, it can be stated that the conventional Schlieren optical system with phase plate angles of 40° is not much more sensitive than

the XL-A Schlieren effect at 320 nm. The first deviation from the baseline for the XL-A at 320 nm is at 6.876 cm that of the Model E with 40° phase plate angle 6.864 cm. The difference is 0.012 cm which corresponds to a phase plate angle change of about 10°, in agreement with the finding of an effective XL-A phase plate angle of 50° at 320 nm in Table 3. Nevertheless, the uncertainty in the definition of the radial position of the first deviation from the baseline becomes obvious as the differences for a given phase plate angle increment are not systematic in every case.

The nature of the Schlieren effect

The first question to answer in this context is if the Schlieren effect is a general effect (e.g., observable with every UV-absorption optics under appropriate conditions) or an effect generated by the special optical design of the XL-A UV-absorption optics. Therefore, a velocity experiment has been performed on a MSE Centriscan ultracentrifuge with a 2 mg/ml BSA solution using the UV-absorption optics, but with the Schlieren filter (green light) instead of the UV-filter in front of the lightsource – but no knife edge of course. No Schlieren traces could be derived. This agrees with the experience of other groups who work with a Model E equipped with a photoelectric scanner and have never observed Schlieren peaks when varying the wavelength to one where the sample does not absorb [21, 22]. Hence an effect like the light reflection with refraction which is described by Fresnels' laws of reflection cannot be responsible for the Schlieren effect as we initially thought before control experiments were performed.

So, if the reason for the Schlieren effect is the particular design of the XL-A absorption optical system, the question is raised of what component might serve as analyzer element as the UV-absorption optics does not contain a knife edge or phase plate [13]. Furthermore the photomultiplier as detector is only capable to detect light intensities but no light refraction. To generate a Schlieren peak detectable by a scanning absorption optical system, the light must be deviated completely out of the optical system depending on the particular amount of the light refraction. This

Table 4 Comparison of the radial position where the gradient differs first from the baseline for XL-A Schlieren traces (XL-A) at different wavelengths and conventional Schlieren patterns (Model E) at different phase plate angles

λ [nm]	320	360	400	440	480	520	560	600		
XL-A [cm]	6.876	6.882	6.902	6.902	6.902	6.919	6.923	7.047		
Model E [cm]	6.864	6.869	6.876	6.902	6.905	6.915	6.915	6.928	7.019	7.084
Phase plate	40°	45°	50°	55°	60°	65°	70°	75°	80°	85°

would then cause an apparent local absorption leading to a Schlieren peak when scanning radially.

The analyzer element must be between the cell and the photomultiplier to generate a Schlieren effect. The only optical component in such a position is the slit-lens assembly consisting of a mask, a camera lens assembly and the 25 μm slit above the photomultiplier [13]. Hence there are two possibilities for analyzer elements:

– The analyzer element is the mask above the camera lens assembly serving to limit the angular light diversion from the sample to $\pm 4°$ regardless of the radial position to limit the disturbing effect of light scattering.

– The analyzer element is the 25 μm slit directly above the photomultiplier tube

To distinguish between these two possibilities, the appropriate experiment would be to perform a run without the mask and see if Schlieren patterns are generated or not. However, this experiment cannot be done by a normal XL-A user without endangering a serious misalignment of the scanning optics of the XL-A. Thus, the only possibility to consider the nature of the Schlieren effect here is to give arguments for or against each of the two possibilities.

To establish whether the Schlieren effect can be observed on every XL-A, control experiments have been performed on the XL-A of Dr. Rowe, University of Leicester, Dr. Müller, Bayer AG and that of Dr. Seifert, Universität Potsdam. On all three machines, no Schlieren effect was observed with usual polymer concentrations. Only at high polymer concentrations was it possible to derive Schlieren patterns on the machine of Dr. Rowe [8]. It might be that this is possible on the two other machines as well but this has not been explored in detail. Such difference between the four machines is suprising as the optical components of all four XL-A's should be essentially the same. The presence of the blocking filter on the newer machines cannot account for the Schlieren effect because the filter has been disabled to maintain comparable conditions. The fact that the Schlieren effect cannot be observed on all machines implies that something must be essentially different between the machine in Nottingham and the three other ones. This is an agrument against the mask being the analyzer element as this component is the same on all machines.

To find the difference between the machines, the height of the slit-lens assembly from the bottom plate to the top of the mask has been determined. For the Nottingham machine, it was found to be 13.15 mm, for Dr. Müllers machine 14.25 mm, for Dr. Seiferts machine 14.20 mm and for Dr. Rowes machine 14–14.5 mm. Obviously, the Nottingham machine differs in the height of the slit-lens assembly. This is an argument for the slit above the photomultiplier being the analyzer element. The light entering the cell from the monochromator must be parallel to the meniscus [13]. As the sedimenting sample is homogeneous at constant radius, the light refraction caused by the solute concentration gradient can only occur in the plane of the axis of rotation and the vector of the centrifugal force (the parallel light band illuminating the sample is considered to be infinitesimaly small to visualize better what is occuring). As the radial scanning follows the vector of the centrifugal force, all refracted light beams should be detected unless they are refracted so much that they are beyond the radial limit of the scanning system. But this is most unlikely due to the small optical pathlength between cell and slit-lens assembly.

However, it is clear that light must be lost completely in the optical system to generate a Schlieren peak. Otherwise, one should expect a negative Schlieren peak with the same area as that of the refracted light (see Fig. 13).

If the light beam b) is refracted, its intensity is lost at the position a) where the unrefracted light beam would have been picked up initially. This causes a loss of light with respect to the reference solvent and hence an absorption is detected at position a). If now the radial scanning moves to position b) a negative absorption would be detected (intensity increase with respect to the reference solvent) to the same order of magnitude as the absorption at position a) because not only the unrefracted light beam

Fig. 13 Schematical drawing of the light detection with the radially scanning absorption optics of only one refracted light beam (b) with all other parallel light beams being unrefracted

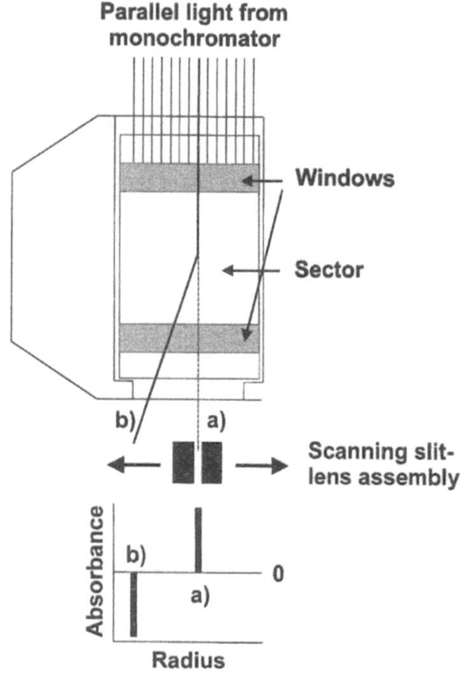

at position b) is detected but furthermore the refracted light beam. Such a pattern has to be expected if the mask is the analyzer element and the camera lenses are able to focus all light onto the slit because the mask scans radially and therefore picks up the refracted light as well.

An explanation of how the light can be lost completely for detection is given in Fig. 14. The slit-lens assembly has been designed to be adjustable in height to allow refocusing of the camera lenses to compensate for variations in the refractive index of the glass at different wavelengths [13]. This is necessary only very rarely for the normal use of the XL-A scanning in the UV. Nevertheless, this adjustment points out that already a major wavelength variation might be sufficient to cause a defocusing of the camera lenses at that particular wavelength. If the very narrow slit above the photomultiplier (25 μm) is taken into account it can be imagined that a defocusing of the camera lenses may have drastic effects. This defocusing can be achieved by altering the height of the slit-lens assembly as shown schematically in Fig. 14.

For a properly focused camera lens assembly, all light passing the mask is focused onto the slit and is therefore detected with the photomultiplier (Fig. 14a). If now the height of the slit-lens assembly is decreased (or increased), the unrefracted light beam still reaches the slit. But the refracted light beam is no longer able to pass the slit (Fig. 14b). Even if the light had been reflected (see dashed line in magnification in Fig. 14b), it would still not pass the slit. Hence the refracted light in this example would be completely lost for the system. This is an extreme example to visualize the effect. Now situations can be considered where the beam is only partially able to pass the slit and is hence detected with decreased intensity (Fig. 14c). This situation is then in analogy to a conventional Schlieren optics working with a knife edge.

The suggested reason for the Schlieren effect to be due to a defocused slit-lens assembly is the only way for a complete loss of light, meaning that the situation indicated at position b) in Fig. 13 cannot occur. This would mean that the mask cannot be the analyzer element due to the following argumentation: If the collimation of light by the camera lenses is so good that all light entering the mask can pass the 25 μm slit, the situation outlined in Fig. 13 must occur. If some light beams cannot pass the mask at a certain radial position, an absorption must be registered by the photomultiplier. Nevertheless the refracted light would be captured at another radial position. So it seems more than likely that the light must be lost between the mask and the 25 μm slit. And for this situation, the explanation with a defocusing of the camera lenses holds (Defocusing is understood in the general sense that the camera lenses cannot properly focus the light onto the slit anymore either due to the slit being out of focus or having such an angle of incidence for the light that it cannot be focussed anymore). This explanation is supported by the experimental finding that the height of an XL-A capable to generate Schlieren patterns is 13.15 mm, whereas it is 14.25 resp. 14.20 mm or 14.0 to 14.5 mm on the machines which cannot (the latter seem to have a properly aligned slit-lens assembly). The next experiments to be carried out would be to alter the height of the slit-lens assembly to see if the Schlieren effect can be generated or suppressed. But these experiments cannot be carried out on an XL-A in daily routine use. Another useful experiment also not possible to perform on XL-A's in steady use would be to vary the aperture of the mask and then see if the Schlieren effect is altered. If the mask would be the analyzer element, it should be possible to increase the sensitivity of the Schlieren effect by decreasing the width of the aperture. But if all light illuminating the sample is parallel and perpendicular to the vector of the centrifugal force the slit must be the analyzer element.

One condition where the mask could serve as analyzer element would be if light is deviated for more than the 4° (to limit disturbing light scattering) in the plane perpendicular to the radial scanning of the slit-lens assembly. In such a case, the light would be lost completely to the system as well. But if it is now considered that the sample in an ultracentrifuge is homogeneous at a certain radial position (unless convection phenomena occur which disturb the sample homogeneity at constant radius near the cell walls), no refraction caused by changes in the solute

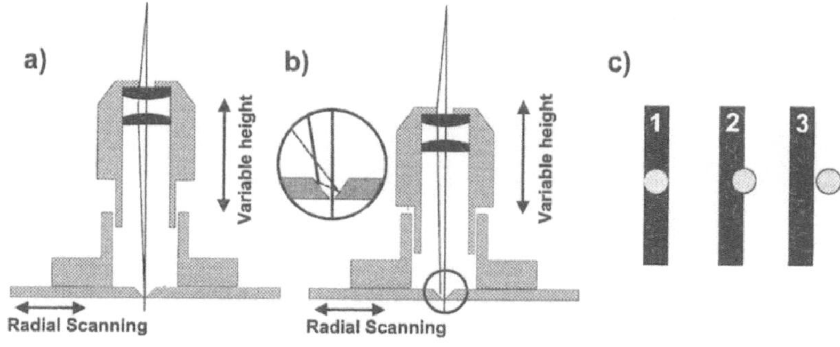

Fig. 14 Slit lens assembly of the Optima XL-A with the camera lens assembly in focus a) and off focus b). An undeviated light beam is presented as well as a deviated one. c) represents how the 25 μm slit (black) can serve as analyzer element with (1) unrefracted light, (2) slightly refracted light and (3) refracted light completely lost for detection

concentration can occur at this radial position. Even if the concentration dependence of light scattering would somehow cause a registered radial optical density profile, the wavelength dependence of the optical density must be related to $1/\lambda^4$. This is not the case (see Fig. 4a).

However, we can now think of potential alterations to the geometry of the XL-A ray optics which make it also possible for the mask to serve as analyzer element. This possibility is due to the feature that the monochromator can be twisted [23]. Normally, the monochromator is positioned in a way that the parallel light illuminating the sample is also parallel to the meniscus. For simplicity let us consider the band of parallel light to be narrow. If now the monochromator is twisted, the situation outlined in Fig. 15 may occur.

If the tilted light from the monochromator is refracted it may now occur that this refracted light can no longer be picked up by the slit-lens assembly and is hence completely lost for detection. This situation is indicated in the lower half of Fig. 15. In such a case, the generation of a Schlieren pattern can be expected. However, this extreme case seems to be very unlikely because the monochromator is more or less fixed in its mount by an additional pin. Therefore it seems to be much more likely that it is only possible to "point" the monochromator towards the cell bottom or the axis of rotation. In such a case the band of light would still coincide with the scanning area of the slit-lens assembly (black area in Fig. 15), but the illuminating light would simply not be parallel to the meniscus anymore. For this it

Fig. 15 Illumination of the sample in an ultracentrifuge cell if the monochromator is twisted (Upper figure). The lower figure shows the corresponding situation for the radially scanning slit-lens assembly. The black area shows the area where the slit-lens assembly would be able to pick up light

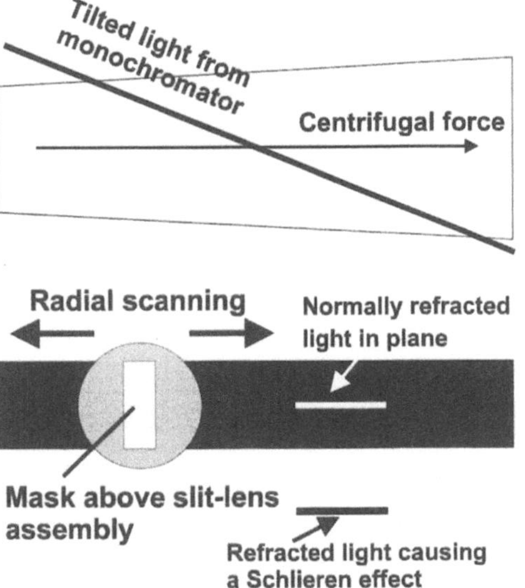

can be expected that the mask can no longer serve as an analyzer element. However, by means of such "pointing" of the monochromator, it should be possible to alter the sensitivity of the Schlieren effect. "Pointing" towards the cell bottom should increase the Schlieren effect whereas that towards the axis of rotation should decrease or even supress it.

At the moment, it cannot be distinguished unequivocally if the slit or the mask of the slit-lens assembly is the analyzer element although the generation of the Schlieren pattern seems clear. But the bulk of evidence appears to be in favor of the view that the 25 μm slit is responsible for the loss of light for detection rather than the mask unless extreme conditions are applied.

Conclusions

It could be demonstrated that the Schlieren effect observed with the Optima XL-A absorption optics is based on light refraction. The shape of Schlieren peaks as well as the sedimentation coefficients calculated from the boundary rate of movement agree with those from conventional Schlieren optics. As the generation of the Schlieren effect requires scanning at a wavelength where the sample does not absorb, the Schlieren effect is capable of extending the range of applications for the XL-A considerably (investigation of polysaccharides, synthetic polymers in UV-absorbing solvents, etc.). However the present effect suffers from not being quantitative for higher concentrations. Instead it is dynamic, i.e., sensitive at low concentrations and insensitive at higher ones. The reason for this is that light scattering is particularly significant in the high concentration range. But for highly dilute solutions, the Schlieren effect should be so predominant that it can be treated as quantitative. However, this limits the concentration range for all investigations requiring quantitative information (sedimentation equilibrium, etc.). If the disturbing light scattering could be suppressed, the Schlieren effect should be quantitative for all concentrations.

For sedimentation velocity work and particularly the determination of the sedimentation coefficient from the rate of movement of the boundary, the present dynamic nature of the Schlieren effect is an advantage rather than a limitation. It could be demonstrated that low concentrations are accessible with the XL-A Schlieren effect in 12 mm cells, which cannot be investigated in a conventional Schlieren optical system when using 30 mm cells [18]. On the other hand, concentrations as high as 20 mg/ml (and this concentration can probably still significantly be increased) can still be investigated without the need of changing the 12 mm centerpieces.

Progr Colloid Polym Sci (1995) 99:167–186
© Steinkopff Verlag 1995

The dynamic nature of the Schlieren effect opens up the potential for completely new applications. One may be the investigation of mixtures (interacting or not) of a highly concentrated non UV-absorbing species with a small amount of a UV-absorbing one. Every component can now be observed with sufficient accuracy just by varying the wavelength of the monochromator between UV and visible. Such mixtures could not be investigated by the combination of UV-absorption and Rayleigh interference optics because of the concentration limitations of the latter. Of course, the Schlieren effect can also be applied to mixtures with lower concentrations than that outlined before as well as to the broad range of possible species combinations in a mixture, regardless of whether the components absorb UV or not.

Another advantage of the XL-A Schlieren effect compared to conventional optics is that the baseline is almost 0 due to the difference measurement with the reference solvent. Hence, pressure effects do not shift the baseline when increasing the speed which happens with conventional Schlieren optics so that double sector cells must be used if quantitative information is desired. Another advantage of the XL-A Schlieren effect is that the gradients are well defined (though they might be noisy) in contrast to the broad Schlieren curves at low phase plate angles.

The sensitivity of the Schlieren effect can be altered with wavelength. The highest sensitivity can be achieved for the lowest wavelength where the sample does not absorb anymore. The effective phase plate angle under such conditions was found to be $50°$ for the example of BSA. The fact that the Schlieren effect can be generated over a broad range of wavelengths enables other new applications with colored samples which can be investigated at a wavelength other than that of their absorption in the visible range.

As the transition range between normal absorption and the Schlieren effect was found to be small (for the example of BSA after only about 20 nm), the application of normal absorption and the Schlieren effect is possible without losing too much of the accessible wavelength range where no superposition of absorption and the Schlieren effect occurs and hence relevant information can be derived. Nevertheless, this requires that the absorption is decreased suddenly with increasing wavelength. If that is not the case, broad transition ranges can be expected. At least for such samples this study serves as a serious warning to those who seek to derive quantitative information from XL-A absorption scans. This is also valid for all other samples if it is scanned at a wavelength where absorption and Schlieren effect are superimposed.

A disadvantage of the XL-A Schlieren effect is that for steep concentration gradients, the radial position of the peak is slightly displaced towards the cell bottom compared to the normal absorption trace due to the light refraction. As this is not the case for broader peaks, the first peaks of the experiment cannot be considered for the evaluation of the sedimentation coefficient if the peak considerably gets broader with time. If that requirement is fulfilled, the sedimentation coefficients agree well with those derived from conventional Schlieren optics without this limitation.

It is of advantage that the Schlieren effect can be exploited for an on-line evaluation as the data aquisition system of the on-line absorption optics is used. The ASCII data file format ensures an easy import into evaluation software and the numerous difficulties of the image analysis of conventional Schlieren patterns are avoided.

Up to now, the most important question about the nature of the Schlieren effect could still not be answered unequivocally. Although all observations made in this study support that a defocusing of the camera lenses of the slit-lens assembly is responsible for the generation of Schlieren patterns and hence the 25 μm slit above the photomultiplier is the analyzer element, the mask above the assembly could in principle be the analyzer element as well. The experiments necessary to distinguish between these two possibilities cannot be carried out by a user who needs his XL-A for daily routine work, but should not require too much effort for the commercial manufacturer (Beckman Instruments) to perform on one of their test units. The benefit of these experiments should be great for numerous users who still wait for the Schlieren optics to be launched on the XL-A. It should be possible to design a scanning Schlieren optics for the XL-A without significant alterations of the existing UV-absorption optics on the basis of this study, regardless if the mask or the slit above the photomultiplier is the analyzer element. If the mask should be found to be the analyzer element, variable width of the aperture of the mask should provide a variable sensitivity of the Schlieren effect additional to the variation of sensitivity with wavelength alteration. On the other hand, if the height of the slit-lens assembly is responsible for the generation of the Schlieren effect (by defocusing the light off the 25 μm slit), an automatic alteration of the slit-lens assembly height should provide a sensitivity alteration of the Schlieren effect or even its supression (If the light is properly focused onto the slit). Whatever the answer to the question about the precise generation of the Schlieren effect, it should enable inclusion of a commercial dynamic Schlieren optics on the XL-A. This should satisfy the needs of all users who must investigate non-UV absorbing samples on the XL-A, but who do not need to derive quantitative information from the Schlieren gradients (sedimentation equilibrium). But in principle, a quantitative Schlieren optical system should also be possible on the basis of the observations in this paper.

Acknowledgements The authors wish to thank Prof. Dr. Thomas M. Laue, University of New Hampshire, Dr. A.J. Rowe, University of Leicester, and Dr. H.G. Müller, Bayer AG Leverkusen for the helpful discussions about the nature of the Schlieren effect and useful suggestions as well as for partially performing control experiments. We also thank Dr. A. Seifert, Deutsches Institut für Ernährungsforschung Rehbrücke for the permission to use his XL-A. Prof. Dr. J. Behlke, MDC Berlin, Prof. Dr. M.D. Lechner, University of Osnabrück, Dr. W. Mächtle, BASF AG and Dr. R. Gauglitz, Free University Berlin are acknowledged for their offers to perform control experiments on their ultracentrifuges. We also thank Mr. Pete Husbands for technical assistance and Mr. Rajesh Dhami for discussions and supplementary experiments [18].

References

1. Laue TM, Anderson AL, Demaine PD (1994) Progr Colloid Polym Sci 94:74–81
2. Stafford WF III (1992) In: Harding SE, Rowe AJ, Horton JC (Eds) Analytical Ultracentrifugation in Biochemistry and Polymer Chemistry, Royal Chemical Society, Cambridge: 359–393
3. Beckman Instruments, Spinco Division (1964) Model E Analytical Ultracentrifuge Instruction Manual E-IM-3, Palo Alto, California
4. Cölfen H, Borchard W (1994) In Bonner RF, Cohn GE, Laue TM, Priezzhev AV (Eds) Biochemical Diagnostic Instrumentation, SPIE Proceeding, Vol 2136, SPIE, Bellingham, Washington: 307–314
5. Mächtle W, Klodwig U (1989) Colloid Polym Sci 267:1117
6. Cölfen H (1994) Bestimmung thermodynamischer und elastischer Eigenschaften von Gelen mit Hilfe von Sedimentationsgleichgewichten in einer Analytischen Ultrazentrifuge am Beispiel des Systems Gelatine/Wasser, Verlag Köster, Berlin
7. Rowe AJ, Wynne-Jones S, Thomas DG, Harding SE (1992) In: Harding SE, Rowe AJ, Horton JC (Eds) Analytical Ultracentrifugation in Biochemsitry and Polymer Chemistry, Royal Chemical Society, Cambridge: 49–62
8. Rowe AJ (1995) personal communication
9. MSE Technical Publication No. 73, Crawley, UK
10. Reisner AH, Rowe J (1971) Anal Biochem 41:1–15
11. Lloyd PH (1974) Optical Methods in Ultracentrifugation, Electrophoresis and Diffusion, Oxford University Press, Oxford, United Kingdom
12. Mächtle W (1991) Progr Colloid Polym Sci 86:111–118
13. Giebeler R (1992) In: Harding SE, Rowe AJ, Horton JC (Eds) Analytical Ultracentrifugation in Biochemistry and Polymer Chemistry, Royal Chemical Society, Cambridge: 16–25
14. Beckman Instruments, Spinco Division (1963) Calibration Cell for the Model E Analytical Ultracentrifuge, Technical Bulletin E-TB-003C, Palo Alto, California
15. Gauglitz R, Miertschink M, Marx G (1989) Colloid Polym Sci 267:1108
16. Cölfen H (1994) unpublished
17. Perlman GE, Longsworth LG (1948) J Am Chem Soc 70:2719
18. Dhami R, Cölfen H, Harding SE (1995) this volume
19. Horton JC, Harding SE, Mitchell JR, Morton Holmes DF (1991) Food Hydrocolloids 5:125–127
20. Laue TM (1992) In: Harding SE, Rowe AJ, Horton JC (Eds) Analytical Ultracentrifugation in Biochemistry and Polymer Chemistry, Royal Chemical Society, Cambridge: 63–89
21. Beyer P; Group of M.D. Lechner, University Osnabrück (1995) Private communication
22. Gauglitz R; Free University Berlin (1995) Private communication
23. Laue TM; University of New Hampshire (1995) Private communication

Progr Colloid Polym Sci (1995) 99:187–192
© Steinkopff Verlag 1995

R. Dhami
H. Cölfen
S.E. Harding

A comparative "Schlieren" study of the sedimentation behaviour of three polysaccharides using the Beckman Optima XL-A and Model E analytical ultracentrifuges

Received: 4 April 1995
Accepted: 16 May 1995

Abstract Recently (H. Cölfen and S.E. Harding, *Progr Colloid Polym Sci* 99: 167–186 (1995), it was discovered that sedimentation velocity experiments using the Beckman Optima XL-A ultracentrifuge, scanning at wavelengths where the macro-molecules did not absorb, yielded typical Schlieren-like traces. Evaluation of the apparent sedimentation coefficient of these traces at a finite concentration for a number of polymers gave results which agreed to within ± 0.05 S with results from experiments performed simultaneously on the Beckman Model E equipped with a 'classical' Svensson–Philpott phase-plate Schlieren optical system.

The aim of this present study was to investigate the concentration dependence of the sedimentation coefficient of three polysaccharides (xanthan, locust bean gum and arabinoxylan) using the Beckman Optima XL-A equipped with absorption optics and Model E with Schlieren optics to compare the values for the (infinite dilution) sedimentation coefficient, $s^o_{20,w}$ and the sedimentation concentration dependence regression coefficient, k_s. The values obtained by both methods for the three biopolymers compare very well (to within approximately 10% for $s^o_{20,w}$ and 20% for k_s). The Beckman Optima XL-A enabled concentrations as low as 0.041 mg/ml (in the case of xanthan) to be measured and gave more precise data than the Model E.

This work provides further evidence that the Beckman Optima XL-A may be used as a potential Schlieren optical system based analytical ultracentrifuge, hence broadening its range of applications to include non-absorbing macromolecular systems such as polysaccharides and synthetic polymers.

Key words Optima XL-A – Schlieren – polysaccharides – sedimentation coefficient

Dr. R. Dhami · S.E. Harding (✉)
National Centre for Macromolecular Hydrodynamics
The University of Nottingham
Sutton Bonington Campus
Loughborough LE12 5RD,
United Kingdom

H. Cölfen
Max-Planck-Institute for Colloid and Interface Research
Colloid Chemistry Department
Kantstraße 55
14513 Teltow, Germany

Introduction

In the last few years there has been a renewed interest in the technique of analytical ultracentrifugation as a means of characterising macromolecular systems. The introduction of the Optima XL-A by Beckman Instruments, Inc. (Palo Alto, California, USA) played an undeniable role in this renewal. The Optima XL-A provides a scanning absorption optical system with on-line data capture and analysis facilities. Therein lies the limitation of this ultracentrifuge. Unless the system of interest possesses a chromophore the Optima XL-A is experimentally out of bounds. Consequently, this instrument has had little impact on the study of synthetic polymers, polysaccharides and other non-chromophoric macromolecules which do not absorb in the visible and 'useable' ultraviolet region.

A recent study [1] has shown that 'Schlieren-like' traces occur in the XL-A during sedimentation velocity experiments, at wavelengths where the solution is effectively transparent, i.e. where the absorption is effectively zero. This is particularly surprising when one considers that this instrument does not contain a knife-edge or phase plate as analysing element. An explanation for this potentially useful phenomenon has been offered in terms of the mask above the slit lens assembly essentially acting as a knife-edge. It was demonstrated that the apparent sedimentation coefficient of a number of different polysaccharides determined from these 'Schlieren' boundaries at a single concentration in the Optima XL-A agreed with s values obtained in the Beckman Model E to within ± 0.05 S; this gave a strong indication that the images observed were genuine 'knife-edge' or Toepler [2, 3] Schlieren boundaries.

This present study aims to extend these observations by a comparative investigation of the concentration dependence of the sedimentation coefficient of three polysaccharides:- on xanthan, arabinoxylan and locust bean gum, using the Optima XL-A and Beckman Model E analytical ultracentrifuges.

Materials and solution preparation

The commercial food grade xanthan, used in this study was obtained from Kelco International (London, U.K). The arabinoxylan and locust bean gum preparations were kindly supplied by A. Ebringerova (Slovak Academy of Sciences, Bratislava) and Prof. E.R. Morris (Silsoe College, Cranfield University, U.K.), respectively.

All polymer solutions were prepared by dissolving a known amount of solute in deionised distilled water, at 4°C with gentle stirring, to give the desired concentration. The xanthan and xylan solutions were dialysed exhaustively against an $I = 0.3$, pH 6.8 phosphate chloride buffer of the following composition, NaCl 14.615 g, KH_2PO_4 1.561 g, and $Na_2HPO_4 \cdot 12H_2O$ 4.595 g in 1 litre [4].

Experimental

Beckman Optima XL-A

0.3 ml of polymer solution were loaded into the right-hand channel of a 12 mm double sector cell, with 0.35 ml of buffer injected into the left-hand channel. Prior to high speed velocity experiments, low speed (3000 rpm) wavelength scans were performed to determine the appropriate scanning wavelength(s). Wavelengths were chosen at which absorption was negligible (Fig. 1a).

Operating speeds of 30 000 rpm at a scanning interval of approximately 10 min were used for xanthan and locust bean gum samples, while a rotor speed of 60 000 rpm and scanning interval of 15 min were employed for arabinoxylan solutions. All experiments were performed at 20 °C. By overlaying successive scans with the Beckman Microcal ORIGIN software, the data was analysed using the QUICKBASIC algorithm, XLA-VEL [5] to find the sedimentation coefficients.

Beckman Model E

A Beckman Model E analytical ultracentrifuge equipped with a resistance-temperature indicator control unit, an automatic photographic system and the novel LED light source [6] was used for sedimentation velocity experiments. For concentrations < 1.0 mg/ml, 0.6 ml of polymer solution were loaded into a 30 mm path-length single sector ultracentrifuge cell. For concentrations above 1 mg/ml, 0.35 ml of polymer solution were injected into a 12 mm path-length cell. Rotor speeds were as for the XL-A experiments. Photographs were taken at intervals of 10–30 mins and analysed by enlargement directly onto a graphics tablet interfaced to an IBM-PC. Sedimentation coefficients were determined using the QUICKBASIC algorithm MOD-EVEL [5].

Results and discussion

Figure 1b demonstrates the Toepler–Schlieren images from the XL-A for locust bean gum scanned at the 'transparent' (see Fig. 1a) wavelength of 523 nm. Although a little noisy, this could have been improved by taking the scan at finer radial intervals or taking more averages. Despite the noise the classical diffusion broadening of the boundaries is clearly evident.

The hyper-sharp sedimenting boundaries characteristic of the much more highly concentration dependent xanthan [7] can be seen at all the concentrations studied. Overlayed successive scans for 0.041, 0.3, and 0.6 mg/ml xanthan solutions are presented in Figs. 2a, b, and c respectively.

The difference in the shape of the Toepler–Schlieren boundaries are reflected in the concentration dependence of the sedimentation coefficients. Values of $s_{20,w}$ corrected to standard conditions [8] were plotted as a function of concentration corrected for moisture content and radial

Fig. 1 Sedimentation velocity experiment on locust bean gum. Loading concentration $c = 0.7$ mg/ml, temp $= 20.0\,°C$. (a) wavelength scan (200–600 nm) and (b) successive radial scans at $\lambda = 523$ nm, rotor speed 30000 rpm, scanning interval 16 min. Direction of sedimentation is from left to right

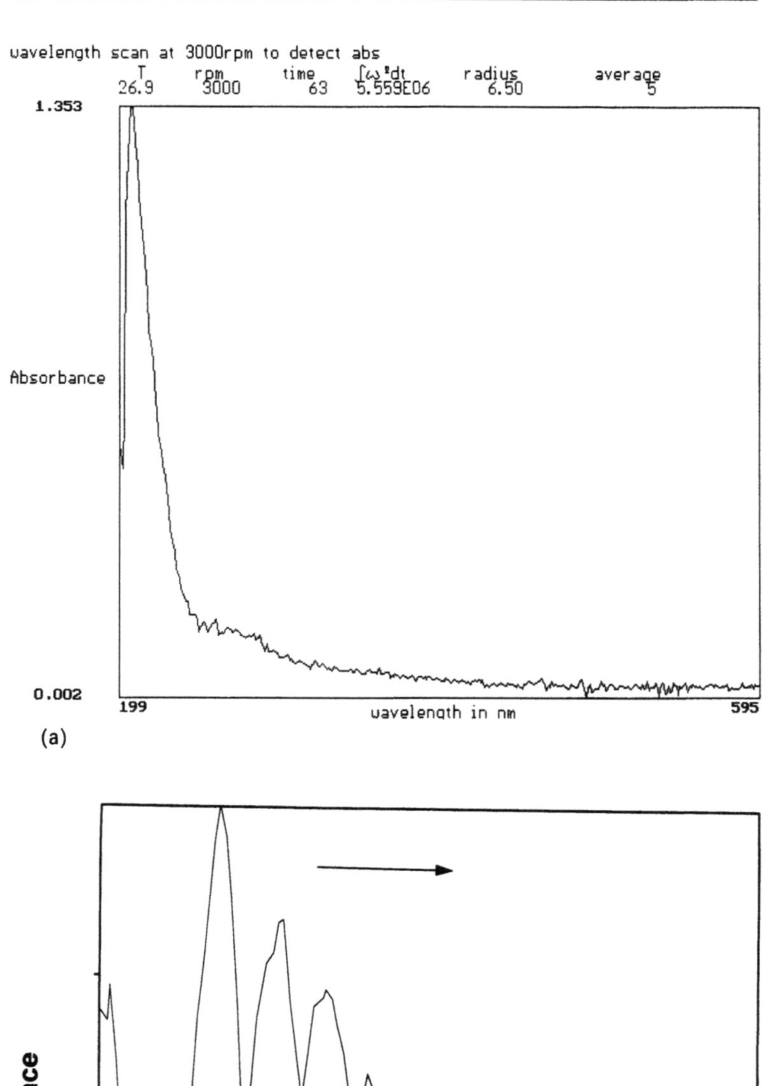

(a)

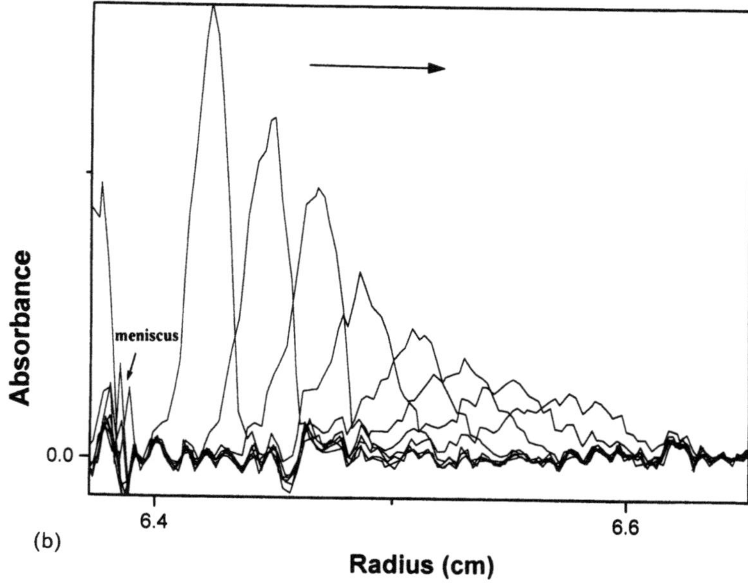

(b)

dilution (Figs. 3–5). The $s^{o}_{20,w}$ and the k_s, sedimentation concentration dependence regression coefficient, were determined from the intercept and the gradient, respectively in each case. The results are summarised in Table 1. It should be stressed that these data are for comparative purposes only: a simple linear regression to non-reciprocal data is unlikely to be the optimum form of data analysis.

Besides the obvious differences in the behaviour of xanthan and the other saccharides, the most significant features of these result is the agreement between the Toepler–Schlieren XL-A boundary shapes, sedimentation coefficients and concentration dependences with the conventional Philpot–Svensson phase-plate results from the Model E. In the case of locust bean gum the hydrodynamic parameters, when one takes into account standard errors, are essentially identical. While no real conclusions can be drawn for arabinoxylan, since only three data points are available from the Model E, it is clear that the sedimentation coefficients at the concentrations measured are comparable.

R. Dhami et al.
Optima XL-A studies on polysaccharides

Fig. 2 Sedimentation diagrams of xanthan solutions (a) $c = 0.041$, (b) $c = 0.3$, and (c) $c = 0.6$ mg/ml. Note the classical 'hypersharp' boundaries. Rotor speed = 30 000 rpm. Temp = 20.0 °C. Scanning interval = 8 min

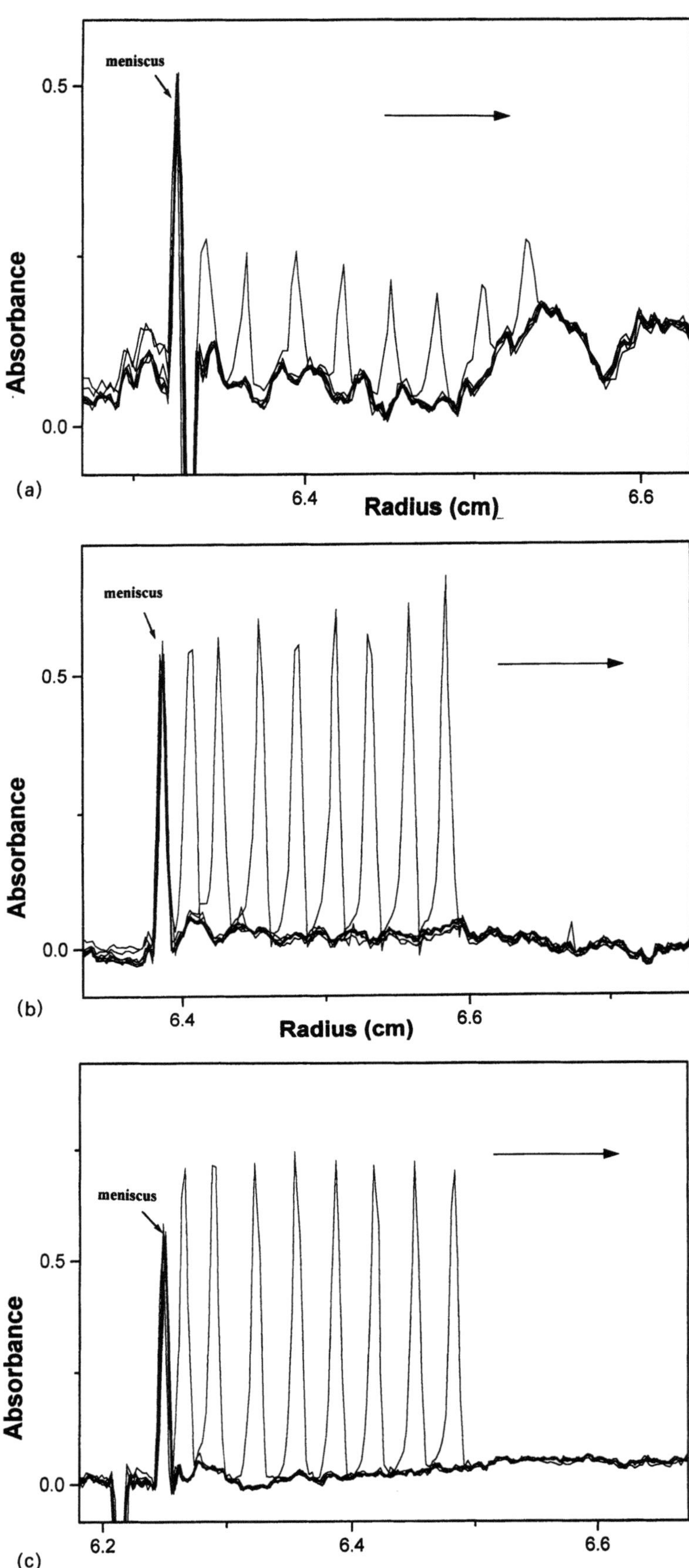

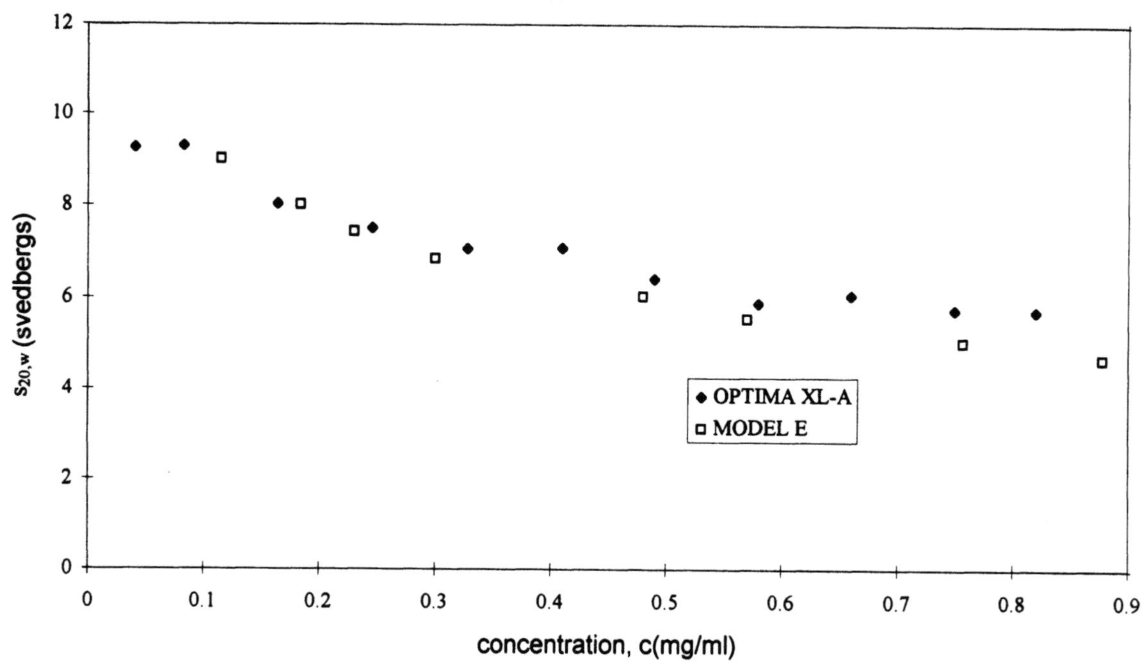

Fig. 3 A plot of sedimentation coefficient, $s_{20,w}$ versus concentration, c for locust bean gum obtained using the Optima XL-A and the Model E

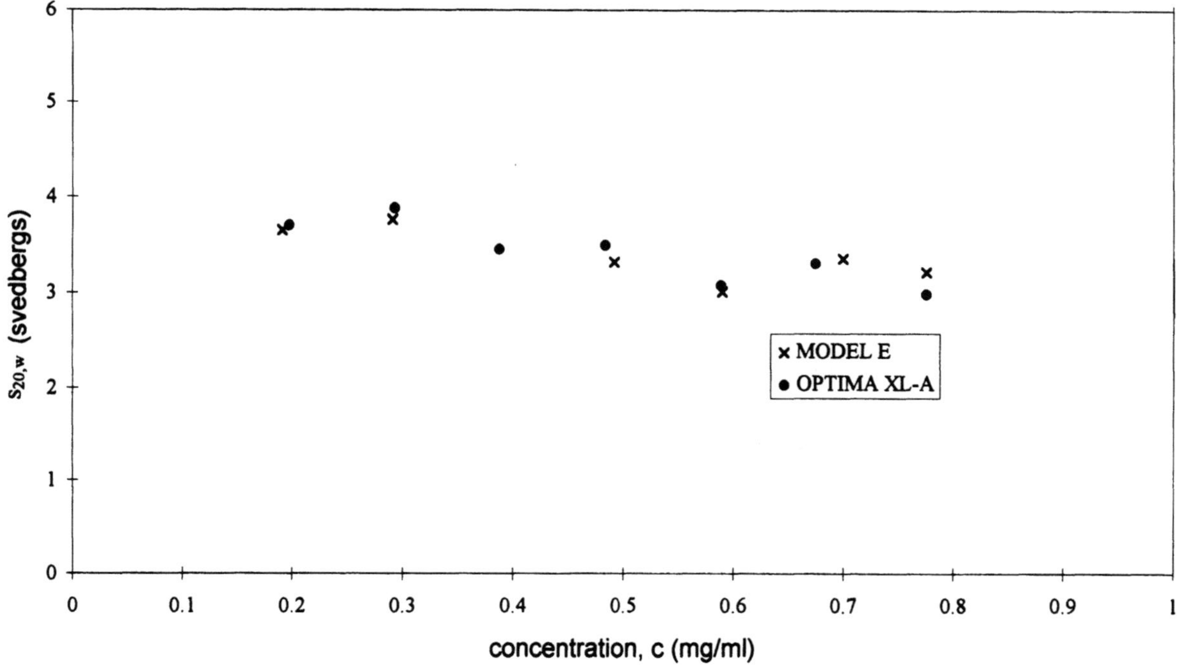

Fig. 4 A plot of sedimentation coefficient, $s_{20,w}$ versus concentration, c for xanthan obtained using the Optima XL-A and the Model E

For xanthan, simple linear extrapolation (see note of caution above) of both data sets also yields similar results: k_s and $s^{\circ}_{20,w}$ obtained on the Optima XL-A being equal to 600($\pm$64) ml/g and 9.38($\pm$0.21)S *cf.* Model E 590($\pm$68) ml/g and 8.87($\pm$0.31)S. However, closer scrutiny of the Model E results demonstrated that there is a strong concentration dependence at the low concentration region <0.6 mg/ml. Extrapolation to infinite dilution over this portion of the data shifts the k_s and $s^{\circ}_{20,w}$ to higher values and thus accounts for the difference in the

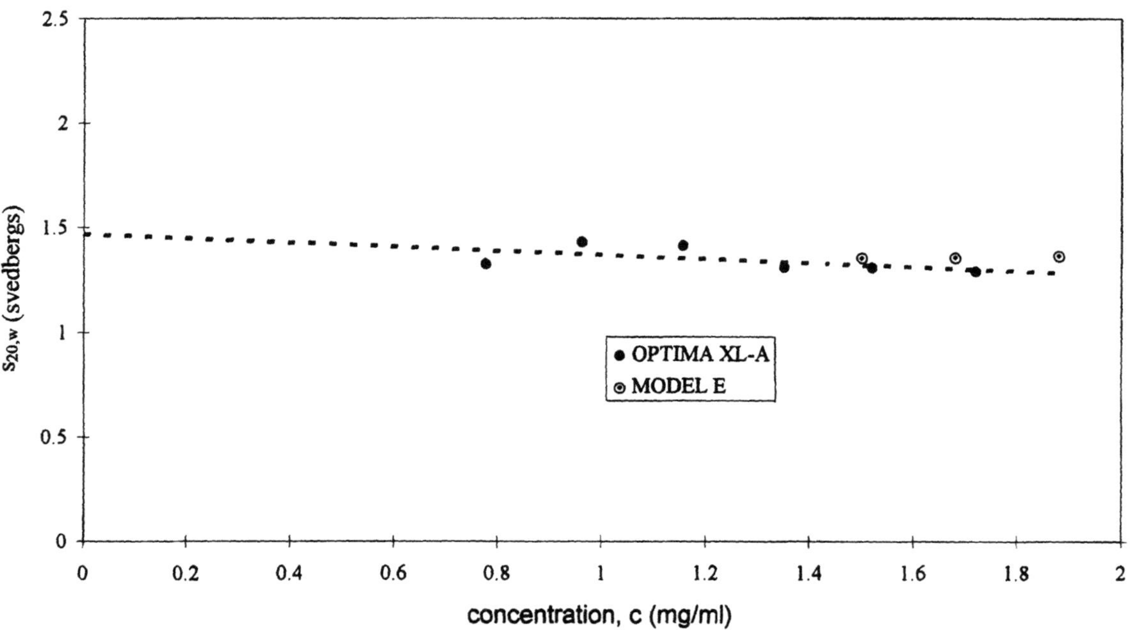

Fig. 5 A plot of sedimentation coefficient, $s_{20,w}$ versus concentration, c for arabinoxylan obtained using the Optima XL-A and the Model E

Table 1 Comparison of the hydrodynamic parameters for locust bean gum, xanthan and arabinoxylan determined using the Toepler–Schlieren-like traces of the Optima XL-A and the conventional Philpot–Svensson phase-plate Schlieren optical system of the Model E

Polymer	k_s (ml/g)		$s^\circ_{20,w}$ (Svedbergs)	
	Model E	XL-A	Model E	XL-A
Xylan	–	–	–	1.5 ($\pm$ 0.1)
Xanthan	1100 ($\pm$ 110)	660 ($\pm$ 70)	10.2 ($\pm$ 0.3)	9.4 ($\pm$ 0.2)
Locust-bean gum	250 ($\pm$ 100)	340 ($\pm$ 80)	3.9 ($\pm$ 0.2)	4.1 ($\pm$ 0.2)

for comparative purposes only. The simple linear extrapolations from which these data are calculated are not necessarily the optimum ones

results quoted in Table 1. It would appear that the sensitivity of the Optima XL-A is greater than that of the Model E, especially when one considers that only 12 mm pathlength cells are compatible with the Optima XL-A, as compared to the 30 mm cells which may be used in the Model E. While the Model E can detect xanthan concentrations of approximately 0.1 mg/ml, the Optima XL-A is capable of tracing the sedimentation of a xanthan solution as low as 0.041 mg/ml. At such a low concentration the $s_{20,w}$ is comparable to $s^\circ_{20,w}$ even for the highly non-ideal xanthan. The concept of measuring $s^\circ_{20,w}$ by a single experiment is a particularly attractive one to the physical chemist. A more detailed study on the sensitivity of the XL-A Schlieren optics is given in [1].

Figures 3–5 demonstrate that the Optima XL-A data is more precise than that obtained using the Model E. This may be attributed to the poorer temperature control of the RTIC unit of the Beckman Model E resulting in fluctuations in the solvent viscosity, and hence the resulting sedimentation coefficient.

This work provides further evidence of the use of the Optima XL-A as a Schlieren optical system analytical ultracentrifuge. The implication of this study is that the Optima XL-A may be applied to systems which previously had been inappropriate, due to their lack of absorbing chemical groups, such as polysaccharides and synthetic polymers.

References

1. Cölfen H, Harding SE (1995) Progr Colloid Polym Sci 99:167–186
2. Toepler A (1866) Annin Phys Chem ([V] 7) 127, 556
3. Lloyd PH (1974) Optical methods in Ultracentrifugation, Electrophoresis and Diffusion, Oxford Univ Press, Oxford
4. Green AA (1933) J Am Chem Soc 55, 2331–2336
5. Cölfen H, unpublished
6. Cölfen H, Husbands P, Harding SE (1995) Progr Colloid Polym Sci 99:193–198
7. Dhami R, Harding SE, Mitchell JR, Jones T, Hughes T, To KM (1995) Carbohydrate Polymers 24, (in press)
8. Tanford C (1961) Physical Properties of Macromolecules, Chapter 6, Wiley, NY

the fact that the type of modulable stroboscope flashlight described in [1] as the light source for the Schlieren optical system and as a light source for the UV-absorption optics of a Heraeus AZ 9100 ultracentrifuge [7], is currently used in the new Beckman Optima XL-A for the UV-absorption optics. In the 4 years the Optima XL-A has been used in our laboratory, the light source has never had any failure. Current work is now being carried out to apply high power modulable laser diodes emitting in the visible as light sources for the Schlieren optical system [8].

Although all the alternatives described above have proved to be successful, they still have the disadvantage of being expensive and partly requiring modifications of the Schlieren optical system of the ultracentrifuge. Especially the latter might be a major reason for not commonly applying these light sources because a biochemist or any other user of an Analytical Ultracentrifuge wants to use the centrifuge as measuring device rather than spending too much effort to modify it. Therefore, the motivation of this study was to give cheap and easy to mount alternatives to the standard mercury lamp which do not require any alteration of the existing optical system and do not require major changes in the optical alignment procedure. The modifications have been carried out using a Beckman Model E, still the most widespread ultracentrifuge in the world. Although the Beckman Optima XL-A is available now with UV-absorption optics and in the near future with Rayleigh interference optics, the Schlieren optical system will still not be redundant for a large variety of questions to be answered applying Analytical Ultracentrifugation. A promising approach has been made to introduce a Schlieren optical system for the Beckman Optima XL-A, generating a Schlieren pattern with the already existing UV-absorption optics of the ultracentrifuge [9]. Nevertheless, it is not yet known, if this so called Schlieren effect is able to completely replace the conventional Schlieren optical system working with a knife edge or phase plate. Therefore the suggested LED light source is not only an improvement for the present machines but might furthermore play a role as modulable light source for a future Schlieren optics for the XL-A.

Results

The light emitting diode (LED)

An ultrabright LED (Toshiba) with a diameter of 10 mm and a light intensity of 26 Candela 9000 fold the intensity of a normal LED and an opening angle of 10° is used. First experiments concerning the general suitability of this LED as a modulable light source for the Model E ultracentrifuge have already been reported [4]. Although it is

preferred to use green, white or yellow light to get a better contrast of the Schlieren patterns as well as a better dn/dc ratio, no alternatives to the red light of the ultrabright LED are available at present. This might change within the near future as manufacturers like for example Hewlett Packard already have announced the launch of ultrabright amber LED's with 14 Candela and opening angles of 6° [10]. It is likely that other brighter LED's will follow. Nevertheless, it has been shown that even red light gives reasonable picture quality [4]. Figure 1 shows all parts, which are necessary, to modify the existing light source of the Beckman Model E ultracentrifuge to an LED light source.

The housing location spigot is reduced to 8 mm length to allow the connection to the LED. The resistor of 180 Ω is used to maintain the optimum light intensity of the LED after connecting it to a 12 V power supply. The LED then works with a voltage of 2 V and a current of 45 mA. The only modifications of the lamp housing necessary before the assembly of the light source is to drill a hole $\varnothing$ 11 mm into the center of the already existing hole. Furthermore, a hole with 6 mm diameter is drilled into the side of the housing 16 mm from the base, through to the verticle hole and thus allowing the connection of the LED. Figure 2 shows how to assemble the light source.

The tube 4 has to be cut to a length so that the LED 2 ends up at the top of it. The tube and the LED are inserted into the 11 mm diameter hole at the bottom of the lamp housing. After the cylindrical lens 3 with $f = 10$ mm (Spindler and Hoyer, Göttingen, FRG) is inserted through the large aperture from the side of the lamp housing, the tube with the LED is pushed to the end automatically

Fig. 1 Parts required for the modification of the original Beckman Model E light source into an LED light source. 1 = Original housing of the mercury lamp, 2 = resistor 180 Ω, 3 = ultrabright LED, 4 = cylindrical lens, 5 = tube

Progr Colloid Polym Sci (1995) 99:193–198
© Steinkopff Verlag 1995

H. Cölfen
P. Husbands
S.E. Harding

Alternative light sources for the Schlieren optical system of analytical ultracentrifuges

Recieved: 8 March 1995
Accepted: 23 May 1995

Dr. H. Cölfen (✉)
Max-Planck-Institute for Colloid- and
Interface Research
Colloid Chemistry Department
Kantstraße 55
14513 Teltow, Germany

P. Husbands · S.E. Harding
National Centre for Macromolecular
Hydrodynamics
Department of Applied Biochemistry
and Food Science
University of Nottingham
Sutton Bonington LE12 5RD
United Kingdom

Abstract This paper presents alternative light sources to the rather unreliable high pressure mercury lamps, commonly used for the Schlieren optical system of Analytical Ultracentrifuges. The alternatives are an ultra-bright light emitting diode (LED) and a halogen lamp, which are both commonly available and very cheap. The non-uniform light emitted by these light sources is either transformed to uniform diffuse light with a ground glass or only the uniform parts of the light beam are used. This maintains a sufficient picture quality with constant light intensities for the lifetime of the light source which is much higher than that of a mercury lamp. For the example of a Beckman Model E ultracentrifuge, it is demonstrated that only minor alterations to the original lamp housing are necessary to transform the mercury light source into a halogen or LED one. The modified light source can be used in the same way as the original one and original alignment procedure of the Schlieren optical system can still be applied. Comparative Schlieren patterns are presented to demonstrate the picture quality achieved with each light source. The advantages and disadvantages of both light sources are discussed.

Key words Analytical ultracentrifugation – Schlieren optical system – LED – halogen lamp – mercury lamp

Introduction

For several decades the Schlieren optical system has been used exhausively as the principle detection system for an analytical ultracentrifuge, mainly for sedimentation velocity experiments. The typical Schlieren peak of a sedimenting boundary might be the most well known optical pattern derived by ultracentrifuge detection optics and is very suitable for the definition of the boundary. But one of the major disadvantages with using this optical system so far has been that a high pressure mercury lamp is the light source. This lamp emitts intense heat and hence requires water cooling, has a rather limited lifetime of only about some months or even only weeks, has a decreasing intensity with increasing lifetime, requires frequent cleaning caused by dirt in the cooling water, works with dangerous high voltage, releases harmful mercury after breakage and is rather expensive. Therefore, it is not suprising that scientists have tried to replace this light source for the Schlieren optical system by more reliable alternatives. Several successful attempts have been reported in the literature applying modulable stroboscope flashlights [1] or modulable He-Ne lasers [2–5]. Both types of light sources have proved to have lifetimes of at least several years and constant intensities although a modification of the Schlieren optical system is necessary if lasers are used [4, 6]. The reliability of these light sources is expressed by

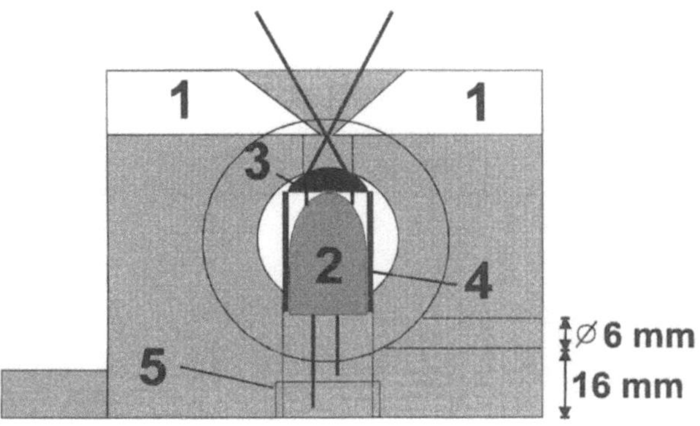

Fig. 2 Schematic side view of the modified LED light source. 1 = original jaws, 2 = ultrabright LED, 3 = cylindrical lens $f = 10$ mm, 4 = tube, inner diameter 10 mm, 5 = location spigot reduced to 8 mm in length. Figure not true to scale

Fig. 3 Parts required for the modification of the original Beckman Model E light source into a halogen light source. 1 = Original housing of the mercury lamp, 2 = halogen lamp, 3 = ground glass

aligning the cylindrical lens to the right position. The LED light source is then ready to use.

The LED light source has been mounted into a Beckman Model E ultracentrifuge and operated without any failure for already now 7 months with daily extensive sedimentation velocity studies. As an LED is a fast modulable light source, it was tried to modulate the light source directly with the TTL signal from a multiplexer. For this purpose, the resistor has been removed and the TTL-signal served as modulated power supply. In this study, we used the Zcontrol 1.0 multiplexer, described in [6]. To produce the trigger impulse for the multiplexer, the electronic circuit described in [4, 5] was applied. A six place rotor with appropriately modified code ring was used for the tests. It was possible to pulse the LED in a way that a cell in a defined rotor hole could be observed. Hence, the LED light source can be used as a modulable light source for multiplace rotors. The light intensity registered on the screen was comparable to that which is derived running a two place rotor with one cell + counterbalance cell.

The halogen lamp

As a second and even less costly alternative to the mercury lamp, a commonly available conventional 150 W halogen lamp has been selected. Like the mercury lamp, this light source dissipates heat but to a much smaller extent. Therefore water cooling is not necessary which is the prerequisite of the application of the mercury lamp. As with the LED light source, only a few parts are required which are presented in Fig. 3.

The original lamp housing 1 has to be modified in a way that an undercut has to be machined out for the ground glass. The latter has to be machined to the appro-

priate size and can then be readily inserted. This design can also be used instead of the cylindrical lens in Fig. 2 for the LED light source. After a hole of 12 mm diameter has been drilled into the light source body to open up the water jacket, the halogen lamp is inserted from the side. Because it fits exactly, it is automatically aligned. To insulate the light source connections from the metal lamp housing, two pieces of Nylon 66 with a 10 mm diameter hole were machined to insert into both ends of the light source as shown in Fig. 4b. Figure 4a shows the schematic drawing of the modified light source, Fig. 4b shows the assembled one.

To cool the light source, an 18 W fan is placed 10 cm away from the light source. The fan blows steadily as long as the light source is switched on. During continuous operation, the light source does not become hotter than 55 °C, a temperature at which it is still possible to make alterations for alignment purposes (usually at the jaws when wearing thin protective gloves). The halogen light source was tested for 100 h of continuous operation without any damage. We have used it without failure for 5 months already for sedimentation velocity experiments.

Discussion

Both light sources presented in this article have a lower light intensity than the original mercury light source. The difference is not too significant for the halogen lamp but for the LED, the intensity of the light source itself can be significantly lower. The reason that the LED still can be applied as light source is the rather small light emission angle of the light of only 10°. Most of the emitted light can

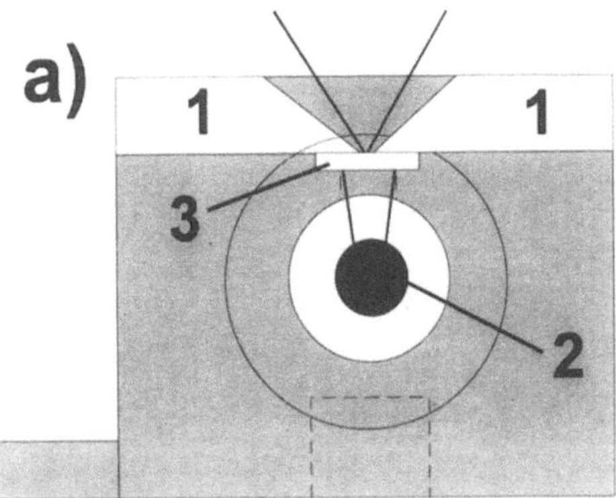

Fig. 4 a) Schematic side view of the modified halogen light source. 1 = original jaws, 2 = halogen lamp, 3 = ground glass; b) Photo of the assembled halogen light source

Table 1 Comparison of the capabilities of three light sources for the Schlieren optical system of a Beckman model E ultracentrifuge

	Slit width [mm]	Lowest phase plate angle for sharp pattern	Exposure time [s]	Development time [min]
Original mercury light source	0.08	40°	30	3
LED-light source	0.85	65°	40	5
Halogen light source	0.25	65°	30	4

be used for the detection optics whereas in the case of the halogen and mercury lamps, most intensity cannot be used leading to the intense heat dissipation.

Both the LED and the halogen light sources do not have a uniform light intensity which makes it necessary to either make the light diffuse with a ground glass (in such case, the light scattering ground glass serves as light source), or only select those parts with uniform light intensity out of the beam. Both decreases the light intensity further. In the case of the LED, the loss of light intensity is so significant that the photographic system is not able to register the image with reasonable exposure times unless a ground glass is used which only slightly diffuses the emitted light. To increase the light intensity of the LED light source, the light has been focussed onto the slit between the jaws by means of the 10 mm cylindrical lens (see Fig. 2). But even by these means, the light intensity registered on the screen was not sufficient to allow the jaws to be closed enough to maintain a sufficient picture quality (even at very high phase plate angles). Therefore a compro-

mise had to be found. The jaws of the light source were opened as much as possible to maintain a reasonably sharp Schlieren line down to a phase plate angle of 65°. This limits the picture quality leading to broadened menisci and less sharp Schlieren lines. A further requirement was an exposure time of not more than 40 s. This is the longest available exposure time on the automatic photographic system of the Beckman Model E.

This compromise for the LED was not necessary to the same extent for the halogen light source due to the much higher emitted light intensity. The differences in the setup for the three light sources are summarized in Table 1. From this table, it becomes obvious that the jaws for both the halogen and the LED light source cannot be closed as much as is possible for the mercury lamp. Depending on the gap, this leads to a limited performance of the optical system leading in turn to a broadened Schlieren curve for the bigger gaps.

To compare the qualities of the Schlieren patterns for the three light sources, an 8 mg/ml bovine serum albumin solution in water was centrifuged at 60 000 rev./min and 20 °C in a 12 mm single sector cell. A phase plate angle of 70° was used. To maintain a clean Schlieren pattern for the mercury light source, a new lamp has been installed. The green filter has been used for the mercury lamp only. The film used was Kodak technical pan, the developer Kodak HC110 and the fixer Ilford Hypam. The Schlieren pictures derived are presented in Fig. 5.

It can be seen that the picture quality is sufficient for each light source, although the meniscus is broadened considerably for the LED and halogen light source due to the more opened slit. Nevertheless, the contrast of the picture for the halogen lamp was by far the best. The contrast for the mercury and LED light source pictures was comparable although the light intensity of the mercury lamp is so much higher than that of the LED. Fresnel interference fringes can be observed with all light sources but their number is the largest with the mercury lamp.

From the point of heat dissipation, both light sources suggested here are far superior than the mercury lamp. The

Fig. 5 Schlieren patterns of
a sedimenting bovine serum
albumin solution in water
centrifuged at 20 °C and 60 000
rev./min derived with
a) Schlieren optical system
with a mercury lamp as light
source, exposure time 30 s,
b) Schlieren optical system
with an LED-light source,
exposure time 40 s and c) with
a halogen lamp as light source,
exposure time 30 s. A different
BSA preparation was used here.
The phase plate angle was 70°
for all examples

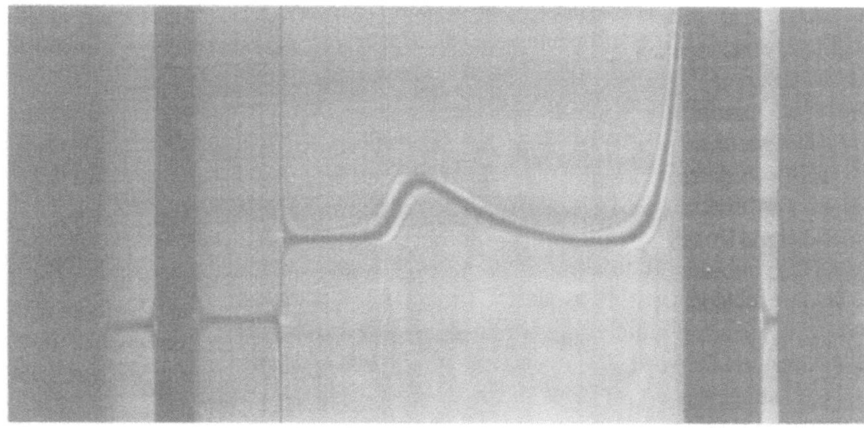

a)

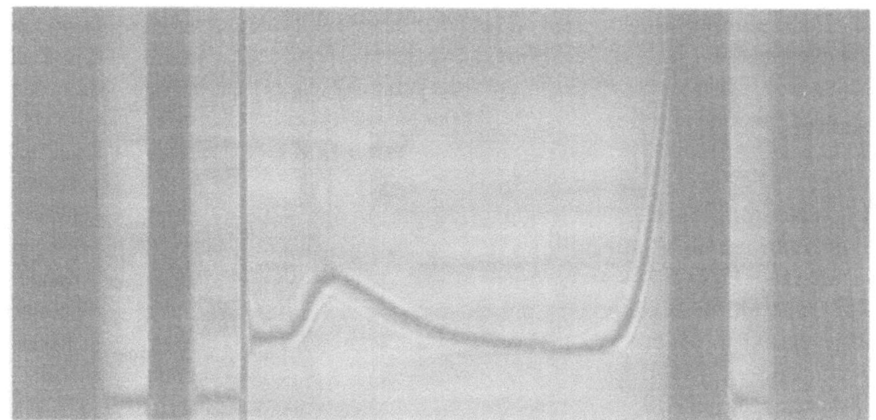

b)

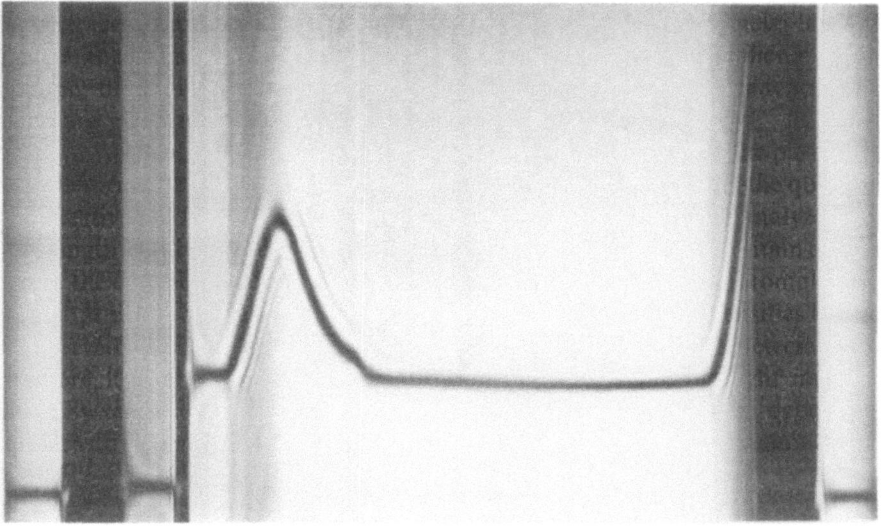

c)

LED light source does not dissipate any heat and can be
operated in continuous operation for years. Both, the
halogen and the mercury light source have to be switched
off after an experiment in order to enhance their life
expectancy. This effect is much more drastic for the mercury lamp. Hence, from this point of view as well as from
the life expectancy, the price and availibility, both suggested light sources can advantageously be used in place of

the mercury lamp. Nevertheless, the LED light source, although being of the lowest heat dissipation, the lowest power consumption and operation voltage at a low price and which furthermore has modulation capabilities cannot yet completely substitute the mercury lamp due to the low light intensity. The potential benefit of an LED for the Schlieren optical system is however clearly indicated: With the rapidly growing optoelectronic market, more powerful LED's should be available in the near future. Until then, it seems to be most sensible to substitute the mercury lamp by a halogen lamp which delivers comparable results and regarding picture contrast, even better pictures than the mercury lamp. The advantages of the halogen light source are still obvious: No cooling water, reduced power consumption and operation voltage, far less heat dissipation, constant light intensity throughout the lifetime and the price, which is comparable to that of the LED.

For all light sources, the original optical alignment procedure can still be applied. The only difference is caused for the LED light source by its low light intensity. In order to place the phase plate into the focal point of the upper chamber lens, the line image can no longer be used to find the focal point as the light intensity is too low. Instead, a trial and error procedure has to be applied to find the focal point by moving the phase plate on the optical bench to a point, where the Schlieren line no longer moves if the phaseplate is rotated: this point is relatively easy to find.

Acknowledgements We want to thank Prof. Dr. W. Borchard from the Gerhard-Mercator University of Duisburg, Germany for leaving one of his Zcontrol 1.0 multiplexers at our disposition during this study. The help of Mr. Mike Chapman with the electronics is acknowledged.

References

1. Mächtle W, Klodwig U (1976) Makromol Chem 177:1607
2. Holtus G, Borchard W (1989) Colloid Polym Sci 267:1133
3. Holtus G (1990) Dissertation, Duisburg
4. Cölfen H (1994) Bestimmung thermodynamischer und elastischer Eigenschaften von Gelen mit Hilfe von Sedimentationsgleichgwichten in einer Analytischen Ultrazentrifuge am Beispiel des Systems Gelatine/Wasser, Verlag Köster, Berlin
5. Cölfen H, Borchard W (1994) Progr Colloid Polym Sci 94:90–101
6. Cölfen H, Borchard W (1994) Anal Biochem 219:321–334
7. Kuhnert R, Rödel E, Stegemann H, Wastl G (1973) CZ-Chem Tech 2:441
8. Hinsken H, Borchard W (1995) Personal communication
9. Cölfen H, Harding SE (1995) Progr Colloid Polym Sci 99:167–186
10. Clarkin M, Medenwald W (1994) Interview in Elektronik Informationen Nr. 5, 102–103

Progr Colloid Polym Sci (1995) 99:199–208
© Steinkopff Verlag 1995

R. Gauglitz

The determination of liquid/liquid interfacial mass transfer
A new application for the analytical ultracentrifuge

Received: 4 April 1995
Accepted: 17 May 1995

Dr. R. Gauglitz (✉)
FU-Berlin
Institut für Anorganische und
Analytische Chemie
FG. Radiochemie
Fabeckstr. 34/36
14195 Berlin, Germany

Abstract In recent years we have developed a new method for the determination of liquid/liquid interfacial mass transfer in solvent extraction systems by use of the Beckman Model E Analytical Ultracentrifuge (AUC). We improved both optical systems of the AUC by use of modern technical equipment like A/D converter, PC data handling, video recording and digitizing. Software has been developed with respect to the evaluation of kinetic data and a diffusion controlled solvent extraction model was applied for the determination of transfer coefficents.

Key words Liquid/liquid extraction – interfacial mass transfer – kinetics of extraction – analytical ultracentrifuge – liquid/liquid interface

Introduction

Solvent extraction has become an important unit operation in technical and analytical separation processes [1]. Large extraction plants on a technical scale were first built in the early 1940s for uranium purification and spent fuel reprocessing within the framework of the Manhattan Project. Since then a rapid development of chemical extractant design [2] and technical contacting equipment has taken place [3]. This development continues and there is still a growing interest for new extraction processes, equipment and extractants. Most of the published papers in the field of liquid/liquid extraction a deal with thermodynamic studies of partition coefficients or partition ratios and with chemical investigations into the synthesis of new selective extracting agents. Only a few papers have covered kinetic investigations. But compared to the rapid development of contact equipment there is a growing need for kinetic information [4, 5]. Modern contact equipment like pulsed sieve plate columns, centrifugal extractors or hollow fiber modules a work in a nonequilibrium state and the time of phase contact is a dominant separation parameter. Therefore all infromation about interfacial phenomena like in-terfacial transfer kinetics, turbulence Marangoni mass transfer instabilities, release of heat, liquid double-layer properties, surface tension effects, interfacial resistance and others are important for the separation quality, for the loading capacity of the extractant, for the total phase flow and for the cost of the whole separation process. All of these facts urge for more kinetic investigations [6].

Up to now several analytical methods for the determination of liquid/liquid extraction kinetics have been developed and tested, but only two general procedures have been established world wide, the stirring cell technique and the single drop column technique [4, 5]. In the stirring cell technique both liquid phases, the aqueous phase and the organic phase with the soluted extractant, are contacted in a special designed cell and stirred in opposite directions with different speeds according to their Reynold numbers. After the system has stabilized the extractable component, i.e., a metal salt, is injected in one phase and a discontinous sampling with a following quantitative chemical analysis of the metal content of the other phase at different contact times is carried out. This experiment will be repeated with different stirring speeds in order to produce a different liquid/liquid double layer thickness. The more the stirring speed is raised, the more the interfacial resistance drops. A following analysis of experiments with different stirring speeds yields in facts about liquid/liquid transfer kinetics and it can be distinguished between a diffusion controlled or a reaction controlled solvent extraction process [7–14].

In the single drop column technique organic or aqueous drops are produced at a nozzle, a capillary or a special drop injection unit. Organic drops are produced at the bottom of the aqueous phase, which is placed in a column. After their formation they rise through the column and material transport can take place. At the end of the column they coalesce and form a separate liquid phase. The experiments are carried out with different drop production rates in order to eliminate the side effects of drop formation with a special mathematical data evaluation procedure. The drops are sampled and the metal content will be detected by quantitative chemical analysis. Experiments are carried out with different column lengths in order to realize different contact times. From the analysis of the experimental data liquid/liquid transfer kinetic phenomena can be evaluated and a distinction between a diffusion controlled or a reaction controlled solvent extraction process can be made [15–21].

Within both procedures the mass transfer takes place in a different physical surrounding with diverse parameters. Both techniques have several uncertainties with respect to the applicability of the results for different contact equipment. But nevertheless the information about the mass transfer phenomena are very important and useful for the extraction plant design.

In addition to these two techniques, we developed a method wherein two unstirred phases are contacted and the interfacial mass transfer is analyzed by use of two optic systems [22]. We work with a two-phase system being formed under particular conditions by use of the centrifugal force and we investigate the interfacial kinetic without any external influence of physical properties of the solvent extraction system, like changing the double layer thickness with rotation speed or changing drop parameters by using capillaries with different sizes for the drop formation.

Analytical technique

The investigation of extraction kinetic can be carried out with the Beckmann Model E Analytical Ultracentrifuge [23]. We use the ultracentrifuge only as an instrument for overlaying two liquids without producing turbulence at the interface during the overlay process. In addition to that the Beckmann Model E Analytical Ultracentrifuge is equipped with two optical systems the UV/VIS optic [24] and the Schlieren optical system [25]. Both systems are able to detect and record the concentration changes in the aqueous and organic phase at different times of phase contact.

The core of our experimental technique is the double sector capillary-type synthetic boundary centerpiece. This filled-epon centerpiece has a height of 1.2 cm and a radius of 1.1 cm. It is made from an epoxy resin with a filler consisting of powdered aluminum; it embodies two compartments and it has two capillaries that connect the two sectors. The compartments are trapezoidal with an angle of 2.5°. They are 1.4 cm high with a 0.3 cm bottom, a 0.2 cm top and a total volume of 0.42 cm^3. Both compartments have a small hole in the top and can be filled with a syringe (Fig. 1).

This centerpiece is placed in a special cell housing sealed with windows and gaskets and the housing is closed with a screw ring tightend by use of a torque wrench. After the cell is tightened properly both compartments of the centerpiece can be filled through the filling holes. One sector is filled with 0.12 ml aqueous phase (i.e., uranyl nitrate in nitric acid) and the other sector is filled with 0.36 ml organic phase (i.e., tri-n-butylphosphat in dodecane). The density of the organic phase has to be below 1 g/cm^3 and independent of the individual metal uptake. During the whole extraction experiment the organic phase remains lighter than the aqueous phase. After filling the sectors the filling holes are closed with gaskets and filling hole screws. The sealed cell is aligned precisely in the cell hole of a titanium rotor and the rotor is connected with the coupling stem to the drive of the centrifuge.

After the cell assembly the rotor is accelerated inside the evacuated steel chamber of the Analytical Centrifuge and the phase overlay takes place at 5000 rpm. From the beginning of acceleration the centrifugal force lasts on both of the liquid columns in the two compartments of the cell. With increasing rotation speed the centrifugal force on the higher organic column grows and the centrifugal force, lasting on the organic phase, becomes larger than the capillary force, which holds the phase back in the sector. At this point the organic liquid will be pressed slowly through the capillary from the organic sector to the aqueous sector. The overlaying starts and the solvent extraction process begins (Fig. 2). Because of the centrifugal force the surface of the aqueous column is hard and the overflowing organic extractant does not cause a visible production of turbulence at the interface. But the applied centrifugal force at 5000–10 000 rpm is not high enough to influence the motion of metal ions or small organic metal complexes with respect to their diffusionary motion. Sedimentation of metal ions within the cell dimensions needs nearly 80 000 rpm for several hours [26, 27].

The extraction process is monitored with the two optical systems of the Analytical Ultracentrifuge, the UV/VIS optic and the Schilieren optical system.

Fig. 1 Double sector capillary-type synthetic boundary centerpiece and light path through rotor and UV/VIS optics

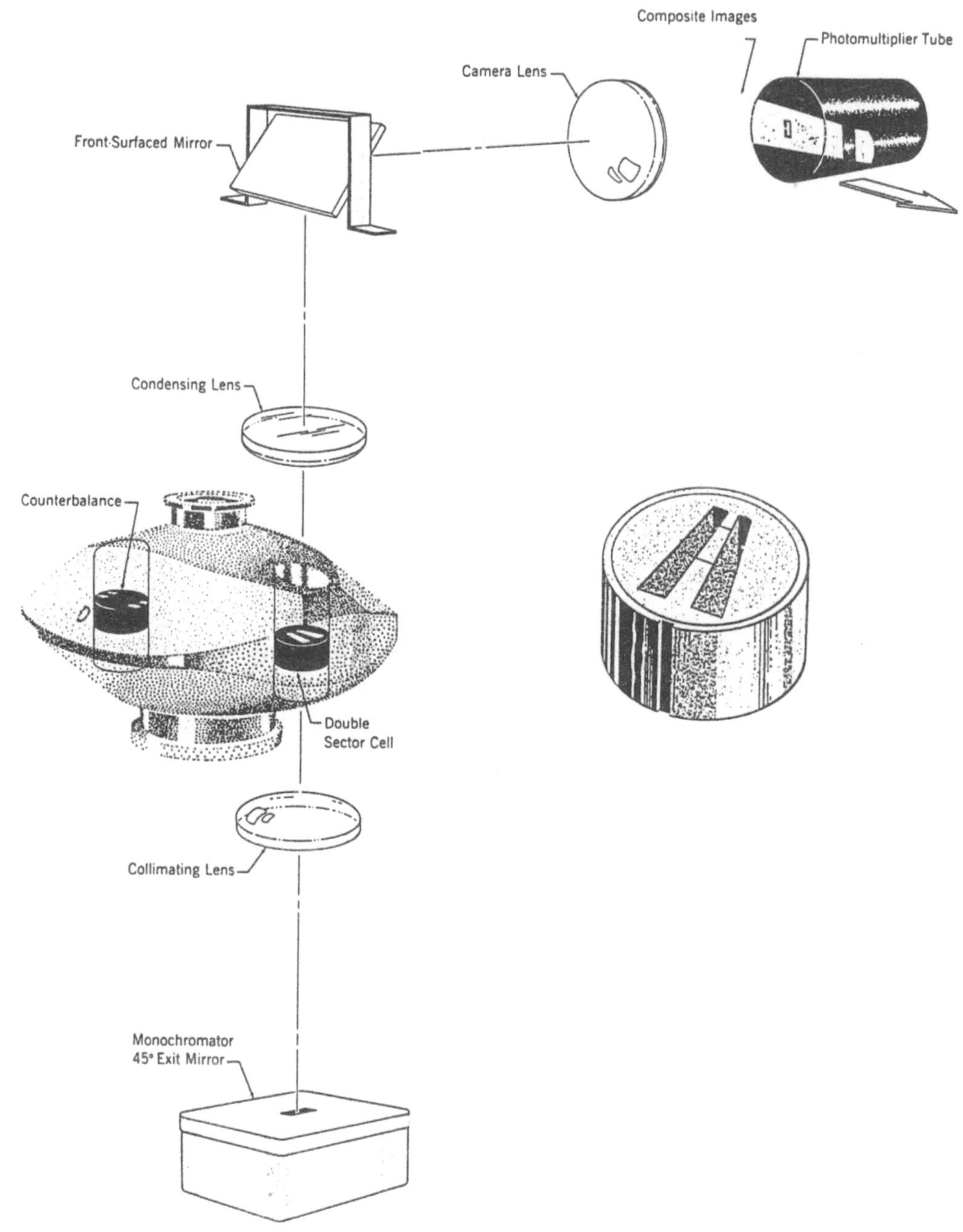

The UV/VIS optic AUC method

If the extracted complex shows an absorption in the UV/VIS region between 190 nm and 1300 nm the concentration in both phases can be monitored by light absorption. A lightbeam produced by a 100 W mercury-arc lamp is focused by the condensing mirror in the lamp house and directed to the diagonal slit entrance mirror. The entrance mirror deflects the light through the entrance slit and into the monochromator to the collimating mirror. The desired wavelength of light emerging from the exit slit passes to the 45° exit mirror, which reflects it up into the optical system. The accelerated cell and the counter balance crosses the lightbeam at every circulation and the two compartments act like a test and a reference solution. The light of the chosen wavelength is absorbed over the whole column and an absorption image of the total two-phase system is produced. This composite image of the cell sectors and the counter balance reference holes is projected through the optical system to a plane at the end of the optical tube. The magnified composite image can be scanned and monitored

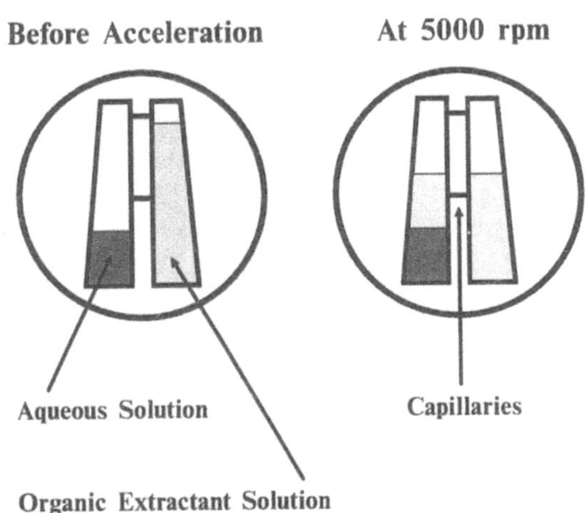

Fig. 2 Supervision on a double sector capillary-type synthetic boundary centerpiece before acceleration and at 5000 rpm

with a photomultiplier tube (PMT) which is mounted on a movable carriage. The light-sensitive end of the photomultiplier tube is covered by a mask that contains a very narrow adjustable slit (0–0.022 cm wide). A positioning system, part of the UV/VIS scanner, allows to detect the absorption at different x-positions of the column. The image of the test column and the reference column lasts as long as the accelerated cell needs to path the light beam. At 5000 rpm the cell needs 0.012 s for one surrounding, one sector lasts 84 μs in the light beam and the delay between

the pictures of both sectors is about 20 μs (Fig. 1). The photomultiplier tube receives the short bursts of light, produced by the sectors crossing the lightbeam, and converts the light pulses into current pulses that can be electronically compared [24].

Using the UV/VIS optic we get absorption scans of the extraction system. These scans were printed with a differential recorder and simultaneously transferred to a computer using an A/D-converter (Fig. 3). Absorption scans can be taken every 3 min depending on the speed the carriage moves the photomultiplier slit across the cell image. The available speed is 3.02 cm/min and the time the positioning system needs to scan the picture and rearrange the PMT to the starting position is about two minutes. A complete extraction run needs 10 to 20 scans according to the transfer rate in the system. The first 10 scans are taken one after another within the first 30 min and the other ones later on variable times. The absorption scan shows the concentration change from the bottom of the cell to the top of the cell. From the bottom to the liquid/liquid interface the absorption profile shows a decreasing metal concentration in the aqueous phase and from the liquid/liquid interface to the liquid/air interface a decreasing metal concentration in the organic extractant phase. At both interfaces a strong absorption is produced by light scattering and in the air region the system shows no absorption at all. The absorption profiles do change within time of phase contact and the metal is transferred to the organic phase according to its distribution coefficient [24].

Fig. 3 UV/VIS scan of a solvent extraction experiment

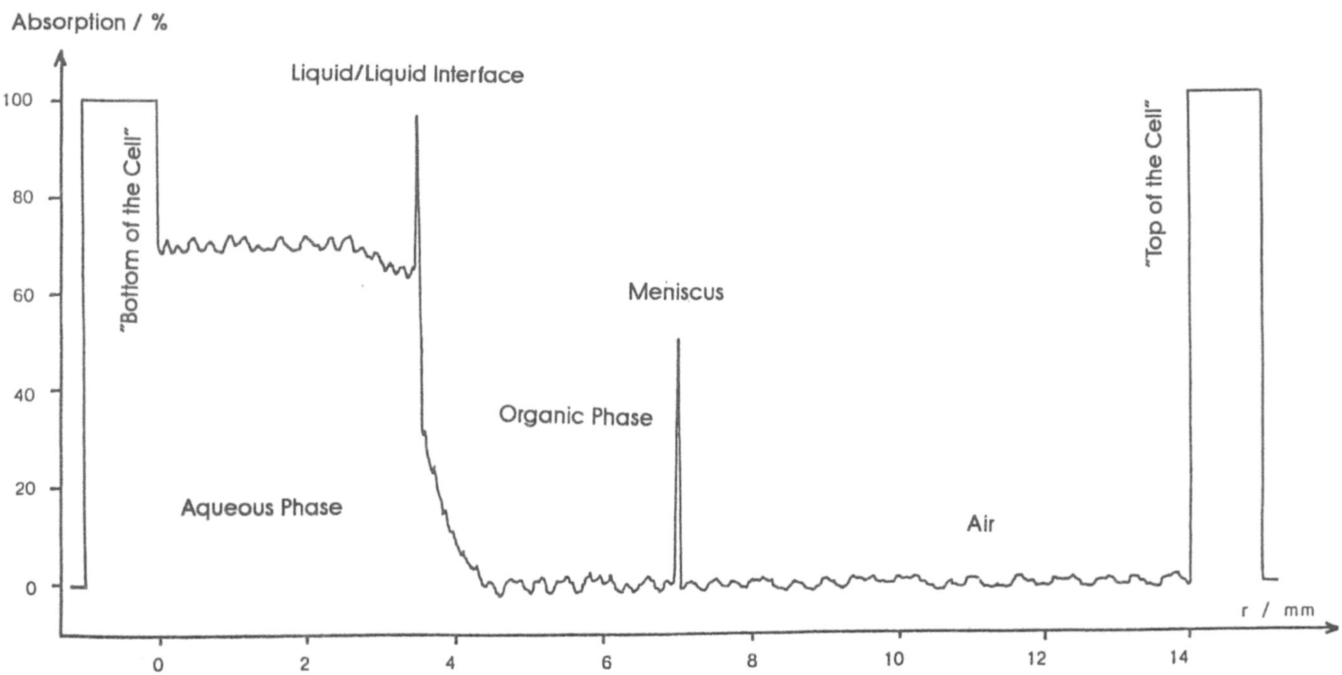

In addition to the AUC extraction experiments absorption spectra of the systems under investigation are taken with a Perkin Elmer 2000 UV/VIS spectrometer in order to prove the linearity of light absorption and the reliability of Lamber Beer's law. These spectra are taken with aqueous metal solutions and organic extractant reference solutions according to the situation in the centrifuge. Furthermore the distribution ratios of the system must be detected and the organic extractant must be equilibrated with the acid concentration of the aqueous metal solution. With respect to the following mathematical modeling it is necessary to determine the apparent diffusion coefficients of the metal ions in the aqueous and organic solution. This can be done with separate diffusion runs using the Analytical Ultracentrifuge. The determined diffusion and extraction data are detected with the same analytical technique

and therefore they include the same systematic errors. This is positive for the following mathematical modeling and leads to good comparable absorption profiles [26].

The extraction profiles are integrated from the cell bottom to the liquid/liquid interface and from the liquid/liquid interface to the liquid/air interface. These integrations can be done with a special developed data evaluation software and, as a result, we get the total change of the metal content in the aqueous and the organic phase from the first phase contact up to the time of scan recording. Later this total change of metal content is normalized to an interfacial area of 1 cm² and the integral mass flux $I_t/mol/cm^2$ is calculated. The integration of the extraction scans being taken at different times gives I_t as a function of the time of contact (Fig. 4) and the differentiation of this curve gives the flux density $J_t/mol/cm^2 s$. The flux density

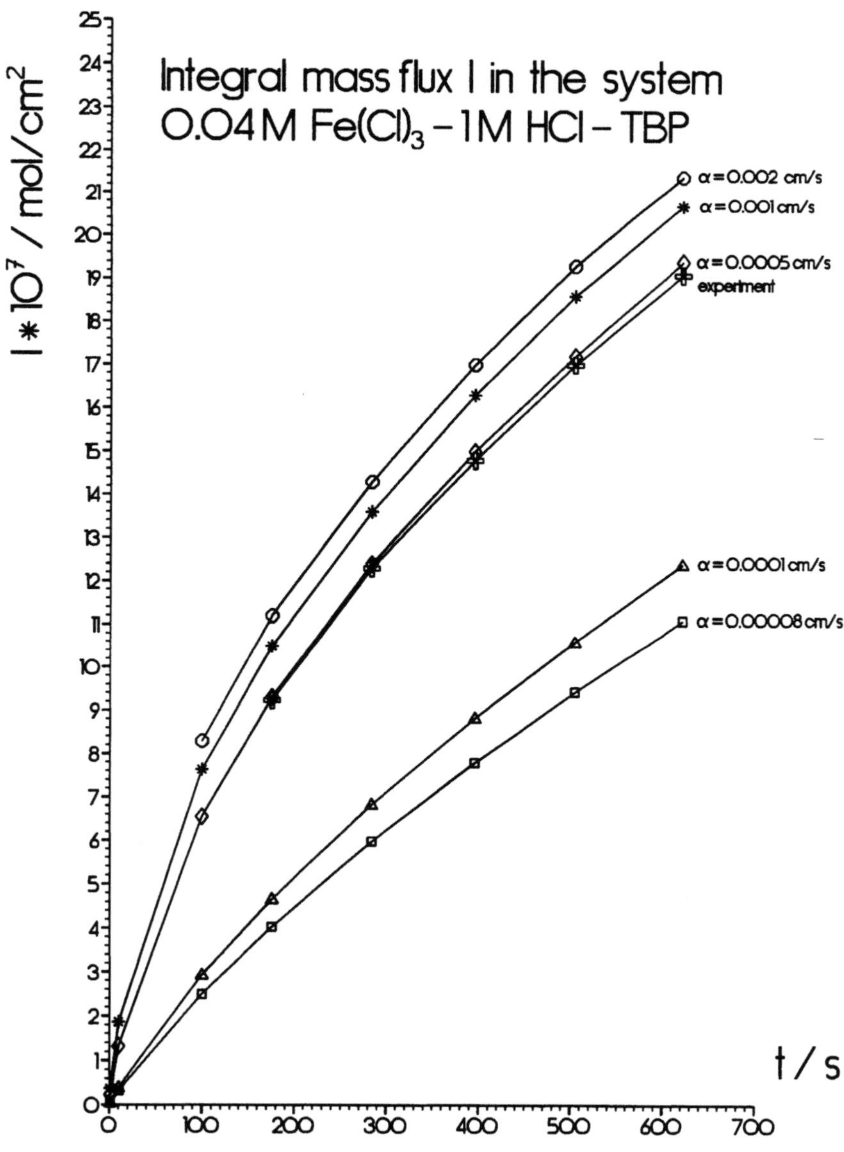

Fig. 4 Calculated values and experimental reults for integral mass transfer I_t as a function of the time of contact in the system FeCl₃-HCI-TBP

decreases from the beginning of the extraction process with the progress of the metal uptake into the organic phase. The flux density has its maximum just at the beginning of the extraction but it cannot be detected at this time with our method. Nevertheless, it is possible to extrapolate the flux density $J_{t\to 0}$ at the first phase contact from the curve of the flux density as a function of the time of contact [24].

$J_{t\to 0}$ is a very important parameter for the description of the transfer kinetic and the behavior of the liquid/liquid interface. The first order kinetic equation for a three-dimensional reaction system is known to be $-dc/dt = k*c$ with the reaction constant k/s^{-1} and the reaction rate per volume element $dc/dt/mol/cm^3$ s. The interfacial reaction is a two-dimensional reaction and therefore dc/dt must be written with the physical unit mol/cm^2 s. Instead of dc/dt we can take our detected $J_{t\to 0}$ and get the equation $-J_{t\to 0} = k*c$ for a first order interfacial reaction. Both parameters $J_{t\to 0}$ and c are known from the experiment and the reaction constant k can easily be calculated. With this method kinetic data can be evaluated directly from the experimental concentration profiles.

Another data evaluation method for the interpretation of the absorption profiles follows from mathematical modeling of the extraction process. A comparison of experimental results with calculated concentration profiles yielded from a diffusion controlled extraction model can also give much information about the kinetics of extraction.

The given solvent extraction system consists of two liquid columns with the same height limited on both sides (Fig. 5). The heavier phase is extended into the negative direction and the lighter one into the positive direction. Point 0 is the location of the liquid/liquid interface. At the beginning of the extraction the metal ions are present in the heavier aqueous phase at concentration c_0. Their diffu-

interface Eqs. (1) and (2), the resulting partial differential equations of second order Eqs. (3) and (4) can be solved by Laplace transformation. Back transformation obtained from extension into a series gives the Eqs. (5) and (6) wherein the η values are the poles of Eq. (7). Equation (7) cannot be solved directly, but the poles can be obtained with an iteration procedure done by a computer program. Using the η value the function $C(x, t)$ can be calculated for both phases and a combination with the magnification factor of the UV/VIS scans gives comparable concentration profiles. The general mass transfer model is described by Scott, Tung and Drickamer [28, 29] and the Laplace transformation is described in detail by Gauglitz [22, 30].

Figure 6a shows the results obtained from a calculation carried out for five different time intervals after an overlay. All five concentration profiles were calculated for transfer velocity of 10^{-3} cm/s. Figure 6b shows four concentration profiles at a definite time calculated with four different transfer velocities. One can clearly recognize that the run of the curve is significantly changed and that the total material transport definitely increases with rising transfer velocity. The integration of these profiles gives the integral mass flux I_t for different transfer velocities at different time intervals after an overlay.

$$J_{aq.interface} = J_{interface} = J_{org.interface} \tag{1}$$

$$-D_1 \cdot \left(\frac{\partial c}{\partial x}\right)_{-x=0,t} = \alpha \cdot (c_{-x=0,t} - m \cdot c_{+x=0,t})$$

$$= -D_2 \cdot \left(\frac{\partial c}{\partial x}\right)_{+x=0,t} \tag{2}$$

$$\left(\frac{\partial c}{\partial t}\right)_x = D_1 \cdot \left(\frac{\partial^2 c}{\partial x^2}\right)_t \tag{3}$$

$$\left(\frac{\partial c}{\partial t}\right)_x = D_2 \cdot \left(\frac{\partial^2 c}{\partial x^2}\right)_t \tag{4}$$

For the aqueous phase with $a < x < 0$

$$c(x, t) = \frac{c_0 m}{(m+1)} + \sum_{n=1}^{\infty} \frac{2\alpha c_0 (\sin k\eta_n a)(\cos \eta_n(x+a))e^{-D_1\eta_n^2 t}}{\eta_n[\alpha a(mk^2+1)(A) - \alpha ka(m+1)(B) + D_1\{(A) + \eta_n a(k(C) + (E))\}]} \tag{5}$$

For the organic phase with $0 < x < 0$

$$c(x, t) = \frac{c_0 m}{(m+1)} - \sum_{n=1}^{\infty} \frac{2\alpha c_0 k(\sin \eta_n a)(\cos k\eta_n(x-a))e^{-D_1\eta_n^2 t}}{\eta_n[\alpha a(mk^2+1)(A) - \alpha ka(m+1)(B) + D_1\{(A) + \eta_n a(k(C) + (E))\}]} \tag{6}$$

sion coefficient is D_1 in the aqueous phase and D_2 in the lighter organic phase. The transfer coefficient $\alpha/cm/s$ is the interfacial transfer velocity of the metal ions. In both phases Fick's law shall be valid and the diffusion coefficients shall be independent from the metal concentration. Taking the boundary conditions into consideration, given by continuity of the material flux passing the liquid/liquid

$$A = (\sin k\eta_n a)(\sin \eta_n a) \qquad B = (\cos k\eta_n a)(\cos \eta_n a)$$

$$C = (\cos k\eta_n a)(\sin \eta_n a) \qquad E = (\sin k\eta_n a)(\cos \eta_n a)$$

$$k = \sqrt{D_1/D_2} \qquad m = c_{aq.}/c_{org.}$$

$$k\eta\sqrt{D_1 D_2}(\sin k\eta a)(\sin \eta a) - \alpha mk(\cos k\eta a)(\sin \eta a)$$

$$- \alpha(\sin k\eta a)(\cos \eta a) = 0. \tag{7}$$

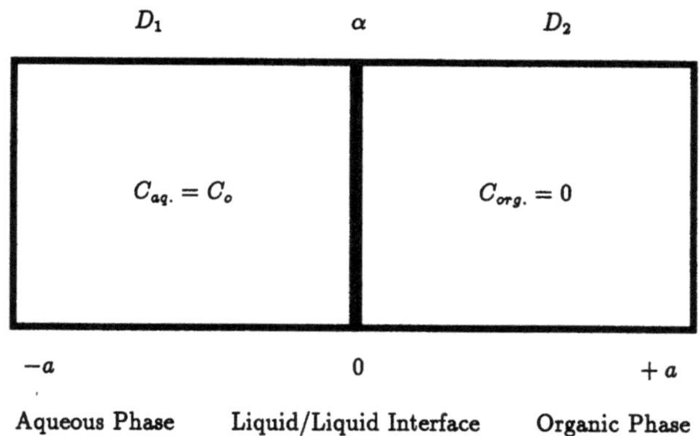

C_o Initial concentration in the aqueous phase
D_1 Diffusion coefficient in the aqueous phase
D_2 Diffusion coefficient in the organic phase
α Interfacial transfer coefficient cm/s
a Length of the phases cm

Fig. 5 Model for the solvent extraction in the Analytical Ultracentrifuge

In order to detect the transfer velocity of an extraction system, I_t can be plotted as a function of the time of contact and the resulting curves can be compared with experimental data (Fig. 4). In addition to that, a direct comparison of calculated profiles with experimental absorption scans can also result in the transfer velocity.

The UV/VIS scan evaluation of an extraction experiment can be done directly or in combination with a diffusion controlled extraction model using several different mathematical procedures. The transfer velocity, the flux density and kinetic constants can be detected and a decision between a diffusion-controlled or a reaction-controlled interfacial transfer can be done.

The Schlieren optical system AUC method

If the extracted complex shows no absorption in the UV/VIS range the concentration changes in both phases can be monitored with a second optical system detecting the refractive index gradient dn_D/dx. By use of the Schlieren optical system the change of the refractive index gradient can be observed and registered during the run of the experiment at any location within the two compartments. The refractive index n_D of a solution depends on the concentration of the soluted compound or metal, and its change is proportional to the change of concentration in the system, shown by equation $\partial n_D/\partial x = \partial n_D/\partial c * \partial c/\partial x$. The refractive index gradient curve can be seen directly with the human eye [25].

Fig. 6 Calculated results for the concentration profiles in the aqueous phase at five different times after the overlaying and at the same time with four different interfacial transfer coefficients

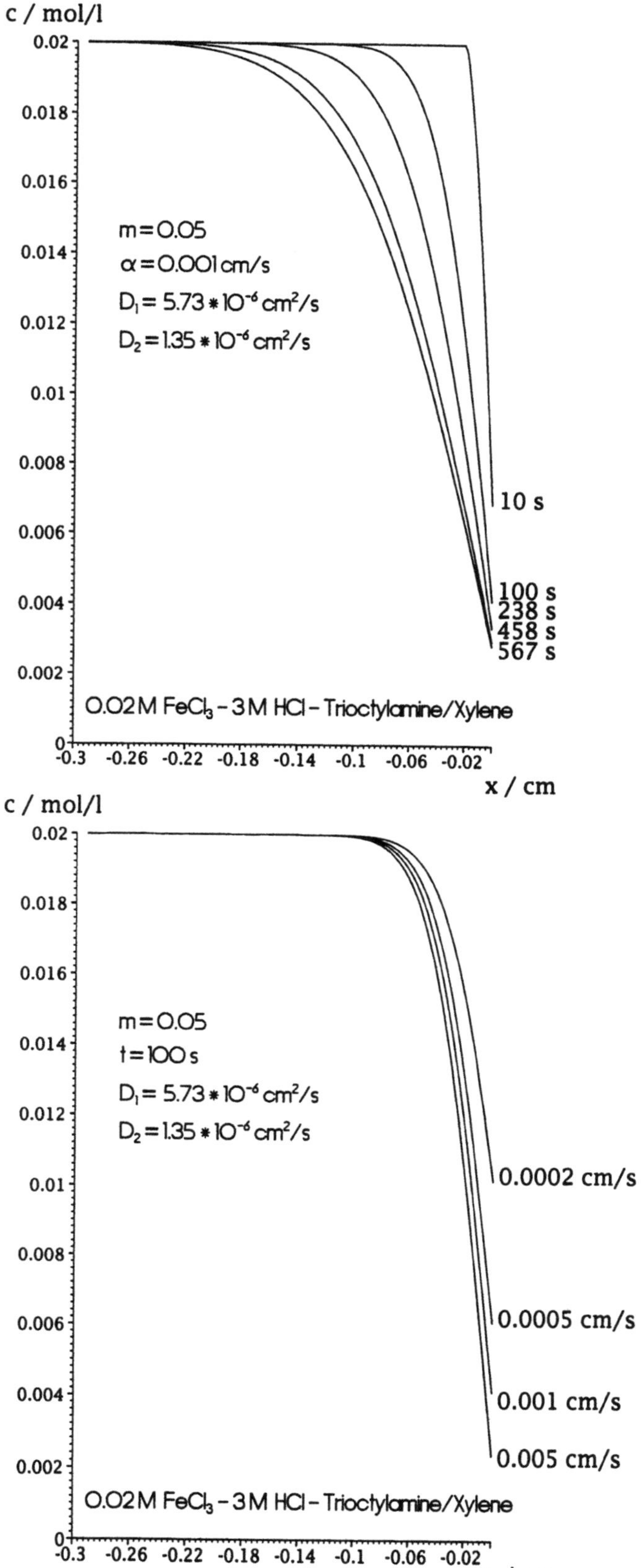

In the Schileren system, light from a 1000 W water-cooled mercury-arc lamp passes through a slit, which is oriented perpendicularly to the direction of the centrifugal field. The beam is collimated by a collimating lens located at the base of the rotor chamber. The light then passes through the cell to the condensing lens, which condenses the light and produces an image of the lightsource slit at its focal plane. If the cell is empty or filled completely with liquid of a uniform refractive index and centrifuged, a single "undeviated" slit source image will appear at the focal plane of the condensing lens. But if a refractive index gradient is present in the cell, the light rays passing through the region of the gradient will be deviated in the direction of increasing refractive index. At the focal plane of the condensing lens containing the undeviated and deviated source images the phaseplate of the Schlieren analyzer is placed and able to interfere with the light. If the slit of the Schlieren analyzer is fixed with an angle between 0° and 90°, the undeviated light passes through and forms a thin horizontal line. But if deviated lightpasses through the slit, some horizontal and vertical displacement depending on the degree of deviation and the angle of the analyzer element takes place and a schlieren curve or peak, representing the refractive index gradient in the cell is formed. A camera lens and a cylindrical lens then focus the Schlieren images onto the photographic plate or view shutter [23].

We improved the Schlieren optical system of the Analytical Ultracentrifuge by use of a videosystem, containing a video camera, a timer, a high-end video recorder, a video digitizer and a video printer, which allows the whole extraction experiment to be registered continuously. For this procedure an electric timer projects the actual time of contact in the Schlieren image, which permits single photos to be evaluated step by step after the end of the experiment. The videofilm of the experiment can be cut into single pictures by use of an ordinary video recorder, followed by digitization. The image data obtained from this procedure are transfered to a computer with respect to a following interactive data evaluation.

Figure 7 shows a Schlieren photo obtained from an extraction experiment. The boundary between the two liquids and between the lighter liquid and the air in the cell can easily be seen. The Schlieren image of the cell is framed by two black lines resulting from the counter balance. The distance of these lines is 1.40 cm in reality. In both phases the black curve of the refractive index gradient can be recognized. Additional calibration experiments done by diffusion runs allow the refractive index gradient to be referred to the concentration gradient in both of the phases. This makes the refractive index gradient curve represent the concentration gradient dc/dx at different "x"-locations in the cell compartments. From additional calculations the total material transport and the referring

Fig. 7 Schlieren picture of a solvent extraction experiment

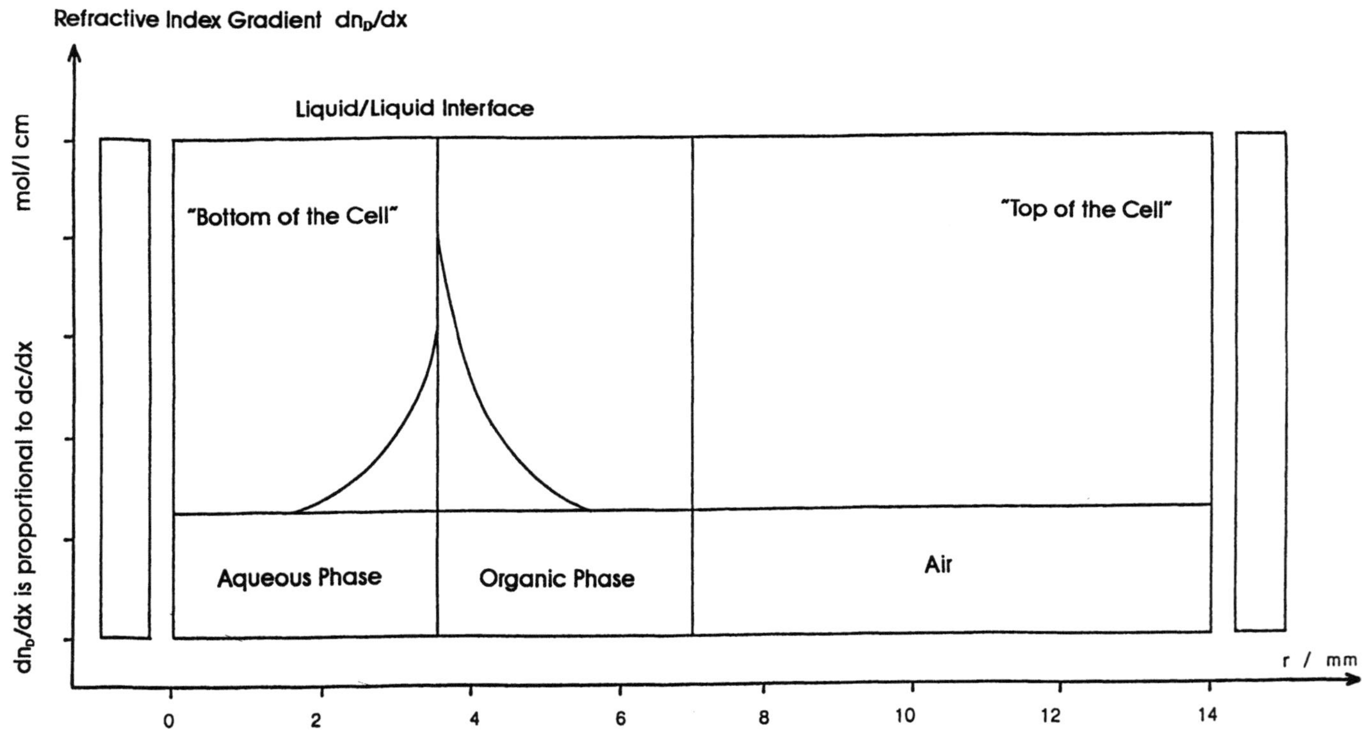

Fig. 8 Calculated results for the concentration gradient profiles in the aqueous phase at five different times after the overlaying and at the same time with four different interfacial transfer coefficients

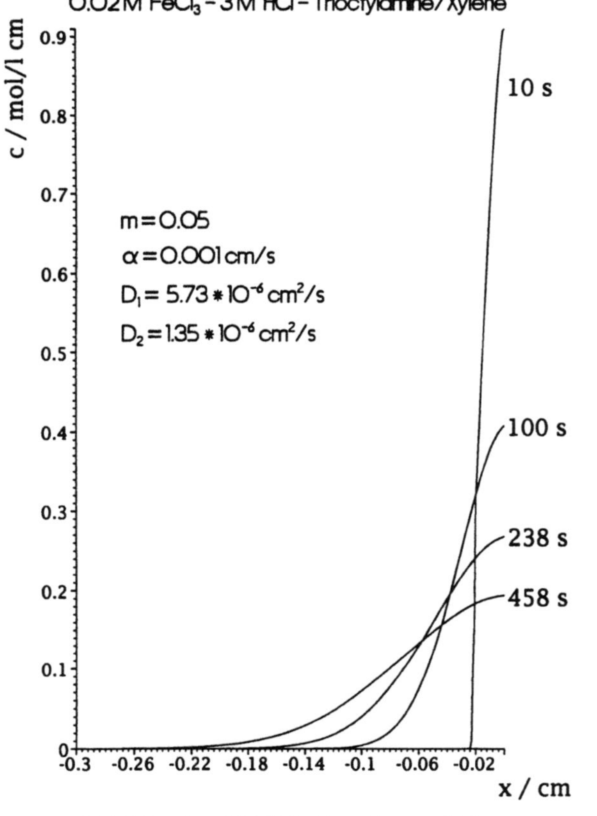

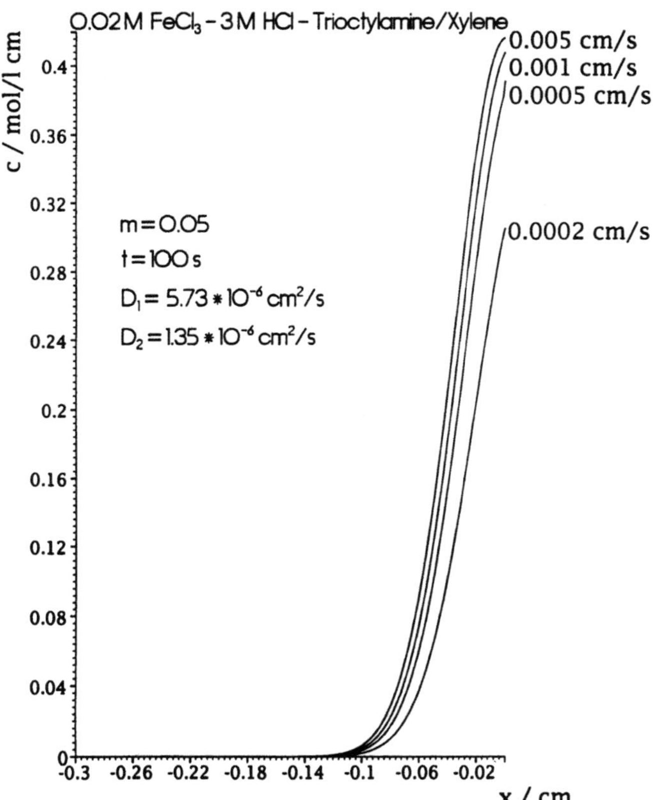

flux density can be determined. But the most important information evaluated from the Schileren pictures is the value of concentration gradient at both sides of the liquid/liquid interface.

The video recording is started before acceleration and therefore the overlaying can be detected, all phenomena taking place can be seen and every second of the extraction is documented with 30 pictures, which can be plotted with a video printer. This technique makes it possible to determine rapid liquid/liquid transfer phenomena within the first seconds of contact, i.e., interfacial reaction kinetics, release of heat due to the enthaply of extraction and production of turbulence at the interface. The flux density J_t can be calculated using the diffusion coefficient of the metal ion in the observed phase applying the equation for the diffusionary flux $J = - D * dc/dx$.

Again the diffusion controlled extraction model can be applied for the data interpretation. Therefore the derivation of the calculated concentration profiles must be combined with the individual magnifications of the Schlieren optical system. The referring concentration gradient profiles shows a drastic change with increasing liquid/liquid interfacial transfer velocity. Comparing the refractive index gradient curves obtained from the video film with those from theoretical calculations it is also possible to determine the liquid/liquid interfacial transfer velocity (Fig. 8).

Conclusion

A new application for the Analytical Ultracentrifuge has been developed. The overlaying of two immiscible liquids can be done with the common double sector capillary-type synthetic boundary centerpiece, and the investigation of liquid/liquid extraction kinetics is possible by using both of the optical systems belonging to the AUC, the UV/VIS optic and the Schlieren optical system. Several methods for the evaluation of the UV/VIS absorption scans and the refractive index gradient images has been developed. We improved the UV/VIS optic by use of a A/D converter in connection with a PC and the Schlieren optical system by use of a video system, containing a video camera, a timer, a high-end video recorder, a video digitizer and a video printer. This system allows the whole extraction experiment to be registered continuously. This method permits the extraction for unstirred liquid/liquid systems to be investigated without any change occuring at the referring boundary and the material transport takes place without any external interference at all. Several computer programs for extraction data evaluation has been developed and a model for a diffusion controlled extraction process has been applied. The new method allows the detection of transfer coefficients, flux densities, kinetic constants, concentration gradients at both sides of the liquid/liquid interface and interfacial phenomena like turbulence and release of heat.

References

1. Proceedings of the International Solvent Extraction Conference ISEC 1993 (York, England) Int Solv Extr Conf Proc, Vol I, II, III
2. Marcus Y, Kertes AS (1969) Ion Exchange and Solvent Extraction of Metal Complexes, John Wiley & Sons – New York
3. Blaß E, Goldmann G (1985) Chem Ing Tech 57:565–581
4. Hanna GJ, Noble RD (1985) Chem Rev 85:583–598
5. Danesi PR (1980) CRC-Critical Reviews in Analytical Chemistry
6. Interfacial Kinetics in Solution (1984) Faraday Discussions of the Chemical Society No. 77
7. Lewis JB (1954) Chem Eng Sci 3:248–259
8. Lewis JB (1954) Chem Eng Sci 3:260–278
9. Lewis JB (1958) Chem Eng Sci 8:295–308
10. Hahn HT (1957) J Am Chem Soc 79:4625–4629
11. Nitsch W, Roth K (1978) Colloid Polym Sci 256:1182–1190
12. Nitsch W (1979) Ber Bunsenges Phys Chem 83:1171–1177
13. Nitsch W, van Schoor A (1980) Atomenergie Kerntechnik 35:95–99
14. Nitsch W (1984) Faraday Discuss Chem Soc 77:85–96
15. Sherwood TK, Evans JE (1939) Ind Eng Chem 31:1144–1150
16. Licht W, Conway JB (1950) Ind Eng Chem 42:1151–1157
17. Coulson JM, Skinner SJ (1952) Chem. Eng. Sci. 5:197–211
18. Nitsch W (1965) Dechema Monogr 55:143–153
19. Nitsch W (1965) Z Elektrochem 69:884–893
20. Nitsch W (1966) Chem Ing Tech 38:525–535
21. Baumgärtner F, Finsterwalder L (1970) J Phys Chem 74:108–112
22. Gauglitz R (1989) Dissertation, Freie Universität Berlin
23. Beckman-Instruction-Manual E-IM-3 (1966) The Model E Analytical Ultacentrifuge Published by Spinco Division, Beckman Instruments, Inc, Stanford Industrial Park, Palo Alto, California
24. Photoelectric Scanning System for the Model E Analytical Ultacentrifuge (1968) Published by Spinco Division, Beckman Instruments, Inc, Stanford Industrial Park, Palo Alto, California (1968)
25. Wiener O (1893) Ann Phys 49:105–149 (Leipzig)
26. Elias HG (1961) Ultrazentrifugen-Methoden, Beckman Instruments GmbH, München
27. Fujita H (1975) Foundations of Ultracentrifugal Analysis Chemical Analysis Vol 42, John Wiley & Sons
28. Scott EJ, Tung LH, Drickamer HG (1951) J Chem Phys 19:1075–1078
29. Scott EJ, Tung LH, Drickamer HG (1952) J Chem Phys 20:6–12
30. Doetsch G (1967) Anleitung zum praktischen Gebrauch der Laplace-Transformation und der Z-Transformation, R Oldenburg Verlag-München-Wien

Progr Colloid Polym Sci (1995) 99:209
© Steinkopff Verlag 1995

AUTHOR INDEX

MIX
Papier aus verantwortungsvollen Quellen
Paper from responsible sources
FSC® C105338

If you have any concerns about our products,
you can contact us on
ProductSafety@springernature.com

In case Publisher is established outside the EU,
the EU authorized representative is:
Springer Nature Customer Service Center GmbH
Europaplatz 3, 69115 Heidelberg, Germany

Printed by Libri Plureos GmbH
in Hamburg, Germany